HYBRID MICROCIRCUIT TECHNOLOGY HANDBOOK

MATERIALS SCIENCE AND PROCESS TECHNOLOGY SERIES

Editors

Rointan F. Bunshah, University of California, Los Angeles *(Materials Science and Technology)*

Gary E. McGuire, Microelectronics Center of North Carolina *(Electronic Materials and Processing)*

DEPOSITION TECHNOLOGIES FOR FILMS AND COATINGS; Developments and Applications: by *Rointan F. Bunshah et al*

CHEMICAL VAPOR DEPOSITION FOR MICROELECTRONICS; Principles, Technology, and Applications: by *Arthur Sherman*

SEMICONDUCTOR MATERIALS AND PROCESS TECHNOLOGY HANDBOOK; For Very Large Scale Integration (VLSI) and Ultra Large Scale Integration (ULSI): edited by *Gary E. McGuire*

SOL-GEL TECHNOLOGY FOR THIN FILMS, FIBERS, PREFORMS, ELECTRONICS, AND SPECIALTY SHAPES: edited by *Lisa C. Klein*

HYBRID MICROCIRCUIT TECHNOLOGY HANDBOOK; Materials, Processes, Design, Testing and Production: by *James J. Licari* and *Leonard R. Enlow*

HANDBOOK OF THIN FILM DEPOSITION PROCESSES AND TECHNIQUES; Principles, Methods, Equipment and Applications: edited by *Klaus K. Schuegraf*

DIFFUSION PHENOMENA IN THIN FILMS AND MICROELECTRONIC MATERIALS: edited by *Devendra Gupta* and *Paul S. Ho*

IONIZED-CLUSTER BEAM DEPOSITION AND EPITAXY: by *Toshinori Takagi*

SHOCK WAVES FOR INDUSTRIAL APPLICATIONS: edited by *Lawrence E. Murr*

Related Titles

ADHESIVES TECHNOLOGY HANDBOOK: by *Arthur H. Landrock*

HANDBOOK OF THERMOSET PLASTICS: edited by *Sidney H. Goodman*

HANDBOOK OF CONTAMINATION CONTROL IN MICROELECTRONICS; Principles, Applications and Technology: edited by *Donald L. Tolliver*

HYBRID MICROCIRCUIT TECHNOLOGY HANDBOOK

Materials, Processes, Design, Testing and Production

by

James J. Licari

Interconnect Systems Engineering

Leonard R. Enlow

Hybrid Circuit Engineering

Rockwell International
Anaheim, California

np **NOYES PUBLICATIONS**
Park Ridge, New Jersey, U.S.A.

Library of Congress Catalog Card Number: 87-34701
ISBN: 0-8155-1152-3
Printed in the United States of America by
Noyes Publications
Mill Road, Park Ridge, New Jersey 07656

10 9 8 7 6 5 4 3 2 1

Library of Congress Cataloging-in-Publication Data

Licari James J. 1930–
    Hybrid microcircuit technology handbook : materials, processes,
design, testing, and production / by James J. Licari, Leonard R.
Enlow.
        p. cm.
    Bibliography: p.
    Includes index.
    ISBN 0-8155-1152-3
    1. Hybrid integrated circuits--Design and construction--Handbooks,
manuals, etc. I Enlow, Leonard R. II. Title. III. Title: Hybrid
microcircuit technology handbook.
TK7874.L45 1988
621.381'73--dc19   87-34701
                                                  CIP

Foreword

I have had the opportunity to write supporting statements for many publications over the past years, but after reviewing the manuscript for this book, I can honestly say that this book is one of the best treatments I have ever read of this important and timely subject. As the authors so well detail, hybrid microcircuit technology has reached a stage of strong maturity after some very difficult periods in the maturing process. Yet, the many facets of hybrids, along with the critical dependency of these various facets on one another, make for a difficult balanced presentation. The authors of this book have done an outstanding job of balancing their presentations of materials, processes, design, testing, and the rest. This will allow the reader to readily understand all of the facets of hybrid microcircuit technology as well as their interdependence. Further, the coverage in this book includes extremely important topic areas which are often neglected as part of the total technology. The authors are to be commended for their presentations on some of these areas, such as parts selection for active devices; passive devices and packages; special handling and ESD concerns; environmental and electrical testing; reliability and failure analysis considerations; and last, but not least, documentation and specifications. Regarding the all-important documentation in hybrid microcircuit technology, excellent coverage is given to both MIL-M-38510 and to MIL-STD-883, as well as MIL-STD-1772 certification requirements for hybrid microcircuit facilities. The incorporation of all of these topics, balanced so well with the usual materials, process, and design topics, will provide the reader with a total perspective not usually found for this subject area. I commend the authors for an outstanding balanced presentation and recommend this book as a "must" for the desks of all of those involved in any aspect of hybrid microcircuit technology.

Baltimore, Maryland
July, 1988

Charles A. Harper
Materials Engineering and Electronic Packaging Technologies
Westinghouse Electronics Corporation
Baltimore, Maryland

Technology Seminars, Inc.
Lutherville, Maryland

Preface

The hybrid microcircuit is, in essence, an electronic packaging and interconnection approach that assures low weight, small volume, and high density. Hybrid circuits are thus used for the most demanding applications including manned spacecraft, heart pacemakers, communications, and navigational systems. Designing and fabricating hybrid circuits require numerous and diverse skills and technologies. A thorough understanding of materials and processes, for example, is essential in the design and production of high yield, low cost, and reliable hybrid circuits. Too often hybrids have come under attack either because they should not have been used for a particular application or because an engineer, unfamiliar with the technology trade-offs, selected a design, material, or process that was marginal or incompatible. The successful hybrid manufacturer must have an in-depth understanding of the appropriate applications for hybrids and a thorough knowledge of the design guidelines and trade-offs of the numerous materials and processes that can be used. Because of the diversity of skills and sciences involved, the successful company must employ engineers and chemists with a wide variety of expertise. Indeed, there is a challenge in hybrids for almost every scientific discipline. For example:

- Electrical engineers: design and prepare circuit layouts, generate drawings, and define the electrical test parameters;
- Metallurgists: define the eutectic attachments, sealing methods, metal packages, and conductor materials;
- Organic chemists: specify adhesives, coatings, photoresists, and cleaning solvents;
- Electrochemists: responsible for platings, corrosion studies, and etching;
- Physicists: establish semiconductor and integrated circuits reliability and conduct failure analyses;

- Ceramic engineers: responsible for evaluating and selecting substrates and substrate fabrication processes, packages, and dielectrics;

- Mechanical engineers: handle tooling, wire bonding, furnace firing and screening of thick films;

- Analytical chemists: analyze sources of contamination, purity of plating baths, purity of solvents and etchant compositions.

This book integrates for the first time the basic technologies and guidelines for the design, fabrication, assembly, and testing of hybrid circuits. It presents those segments of the various technologies that are critical to the success of the circuit. The electronics field, in general, and hybrid circuits in particular, have grown so rapidly during the past twenty years and have experienced so many rapid changes that it has been difficult to assemble a book on this subject. Now that most of the processes and materials have reached a degree of maturity and are used in production by many firms, a book of this nature is appropriate. In organizing the subject matter, we were quickly faced with a dilemma. We wished to arrange the subjects in the sequence in which a hybrid circuit would be designed and built, but the steps were often interdependent. As an example, hybrid circuits begin with the design; so, logically, the design guidelines should come first. But the design guidelines could not be discussed effectively without a basic understanding of materials and processes. We therefore decided on a compromise—discussing each of the process steps and parts selection in as logical a sequence as possible, then presenting the guidelines, and concluding with chapters on documentation and failure analysis.

It is hoped that this book will be of value to those already working in the field of hybrid circuits as well as those about to enter it. This book should also benefit the thousands of peripheral companies who purchase and use hybrids or who supply materials and services to hybrid manufacturers, by providing them with a better understanding of the product they are dealing with.

The authors wish to express their indebtedness to the many individuals and companies that provided technical information and reviewed sections of the manuscript. In particular, we wish to acknowledge S. Martin (Ceramic Systems, Division of General Ceramics), D. Sommerville (Engelhard Corp.), S. Caruso (NASA/MSFC), R. Dietz (Johnson-Matthey), K. Reynolds (EMCA), L. Svach (Rockwell Collins Division), and from Rockwell International's Interconnect Systems Engineering Group: J.W. Slemmons, J. Gaglani, D. Swanson, R. Bassett, H. Goldfarb, and J. Wolfe. Wayne Kooker and Lisa Licari are to be commended for their graphics support and Sharon Ewing for editing some sections.

Anaheim, California   James J. Licari
July 1988   Leonard R. Enlow

NOTICE

x

Contents

1. INTRODUCTION 1
   Classification of Materials for Microelectronics 1
      Conductors 1
      Insulators. 3
      Semiconductors 6
   Classification of Processes 8
   Definition and Characteristics of Hybrid Circuits. 9
      Types and Characteristics 9
      Comparison With Printed Wiring Boards. 13
      Comparison With Monolithic Integrated Circuits 14
   The Hybrid Circuit Market and Applications. 15
      Hybrid Circuit Market 15
      Applications. 17
         Commercial Applications .17
         Military/Space Applications. 18
         Power Applications 20
   References 24

2. SUBSTRATES 25
   Functions 25
   Surface Characteristics 26
      Surface Roughness/Smoothness 26
      Camber 28
      Granularity 29
   Alumina Substrates 29
      Grades of Alumina. 29
      Alumina Substrates for Thick Films 31
      Alumina Substrates for Thin Films 31
      Co-Fired Ceramic Tape Substrates 35

Beryllia Substrates 35
Aluminum Nitride and Silicon Carbide Substrates 39
Ceramic Substrate Manufacture 40
Enameled Metal Substrates 41
Quality Assurance and Test Methods 42
References 43

3. THIN-FILM PROCESSES 44
   Deposition Processes 44
      Vapor Deposition 44
      Direct Current (DC) Sputtering 46
      Radio-Frequency (RF) Sputtering 49
      Reactive Sputtering 49
      Comparison of Evaporation and Sputtering Processes 51
   Thin Film Resistor Processes 52
      Thin Film Resistors 52
      The Nichrome Process 53
      Characteristics of Nichrome Resistors 57
      The Tantalum Nitride Process 58
      Characteristics of Tantalum Nitride Resistors 59
      Cermet Thin Film Resistors 62
   Photoresist Materials and Processes 63
      Chemistry of Negative Photoresists 64
      Chemistry of Positive Photoresists 69
      Processing 71
   Etching Materials and Processes 73
      Chemical Etching of Gold Films 73
      Chemical Etching of Nickel and Nickel-Chromium Films 74
      Dry Etching 74
   Thin Film Microbridge Crossover Circuits 76
   References 78

4. THICK-FILM PROCESSES 79
   Fabrication Processes 79
      Screen-Printing 79
      Drying 83
      Firing 83
      Multilayer Thick-Film Process 85
      Multilayer Co-Fired Ceramic Tape Process 89
      Low-Temperature Co-Fired Tape Process 94
   Paste Materials 95
      Types and Compositions 97
      Conductor Pastes 97
         Functions of Thick-Film Conductors 97
         Adhesion Mechanisms 99
         Adhesion Tests 100
         Metal Migration 102
         Soldering and Solder Attachment 106

Resistor Pastes 107
Dielectric Pastes 111
   Insulation Resistance 111
   Dielectric Constant 111
   Dielectric Breakdown Voltage 112
   Pinholes/Porosity 112
   Via Resolution 114
Thick-Film Capacitors 114
Non-Noble-Metal Thick Films 116
Processing of Copper Thick Films. 117
Characteristics of Copper Thick Film Conductors 119
Processing of Nitrogen-Fired Dielectrics. 124
Processing of Nitrogen-Fired Resistors. 125
Polymer Thick Films 126
PTF Conductors 126
PTF Resistors. 128
PTF Dielectrics 129
References 130

5. RESISTOR TRIMMING 132
Laser Trimming. 133
Abrasive Trimming 142
Resistor Probing/Measurement Techniques. 142
   Probe Cards 142
   Two-Point Probing. 143
   Four-Point Probing 144
   Digital Voltmeters (DVM). 145
Types of Resistor Trims 145
   Plunge-Cut. 145
   Double-Plunge-Cut. 145
   L-Cut 145
   Scan-Cut 145
   Serpentine-Cut 145
   Digital-Cut. 147
Special Requirements. 147
References 148

6. PARTS SELECTION 149
General Considerations. 149
Packages 150
   Package Types 150
      Metal Packages 153
      Ceramic Packages 157
      Epoxy-Sealed Packages. 157
      Plastic Encapsulated Packages 157
   Package Testing. 158
Active Devices 159
   Passivation. 159

Metallization 159
Transistors 160
Diodes 160
Linear Integrated Circuits 160
Digital Integrated Circuits 162
Passive Devices 164
Capacitors 164
Resistors 166
Inductors 170
Round Inductor 172
Square Inductor 172
Chip Inductors 172
Wire-Wound Inductors 172
Procurement 173
References 173

7. ASSEMBLY PROCESSES 174
Introduction 174
Die and Substrate Attachment 175
Types and Functions 175
Adhesive Attachment 176
Epoxy Adhesives 176
Adhesive Forms 178
Polyimide Adhesives 179
Adhesive Bond Strength 180
Electrical Conductivity 181
Weight Loss 183
Outgassing of Adhesives 184
Corrosivity 185
Ionic Contents 185
Metallurgical Attachment 186
Silver-Glass Adhesives 188
Wire Bonding 191
Processes 191
Thermocompression Bonding 191
Ultrasonic Bonding 195
Thermosonic Bonding 197
Microgap Bonding 199
Beam-Lead Bonding 201
Unique Bonding Processes 202
Automated Bonding 203
Tape Automated Bonding 203
Automatic Wire Bonding 208
Quality and Reliability 208
Cleaning 213
Contaminants and Their Sources 213
Solvents 213
Hydrophilic Solvents 214

Hydrophobic Solvents 214
Cleaning Processes. 218
Manual (Hand) Cleaning. 218
Batch Cleaning 218
Plasma Cleaning 222
Particle Immobilizing Coatings. 225
Parylene Coatings 225
Solvent-Soluble Coatings. 231
Particle Getters. 231
Vacuum-Baking and Sealing. 232
Vacuum-Baking. 232
Sealing 235
Soldering. 235
Welding. 238
Glass Sealing. 243
Epoxy Sealing. 244
Hermeticity and Leak Testing 244
References. 246

8. TESTING 249
Electrical Testing. 249
Electrical Testing of Die. 249
Initial Hybrid Yield (F_o). 250
First Rework Yield (F_1) 250
Electrical Testing of Hybrids 251
Visual Inspection. 256
Non-Destructive Screen Tests. 257
Thermal/Mechanical Tests. 257
Burn-In. 258
Air Ambient Burn-In 260
Liquid Burn-In 261
Air vs Liquid Burn-In. 261
Particle-Impact-Noise Detection (PIND) Testing 263
Infrared (IR) Imaging. 265
IR Microscopy 266
IR Thermography 266
Acoustic Microscopy 268
Destructive Screen Tests. 269
Destructive Physical Analysis (DPA) 269
Moisture and Gas Analysis of Package Ambients 270
References. 272

9. HANDLING AND CLEAN ROOMS 274
Handling of Hybrid Circuits and Components. 274
Cleanliness of Tools. 274
Storage 274
Clean Rooms 274
Electrostatic Discharge. 280

Development of Charge 280
Device Susceptibility 282
Static Damage. 283
   Dielectric Failure. 283
   Interconnection Metal Failure 283
   PN Junction Failures 283
Coping with ESD. 283
   Personnel Awareness 283
   Personnel Grounding 285
   Storing or Transporting of Hybrids. 285
   Humidity. 286
   Labels. 286
   Static Dissipative Work Bench Tops 286
   Conductive Floor Mats. 287
References. 287

10. DESIGN GUIDELINES 288
Hybrid Microcircuit Design Transmittal. 288
System Requirements Affecting Hybrid Circuit Design. 292
Partitioning 293
Input/Output Leads. 293
Component Density. 294
Power Dissipation 294
Mechanical Interface/Packaging Requirements 294
Material and Process Selection 294
Quality Assurance Provisions. 296
Quality Engineering/Quality Assurance Requirements 296
Screen Tests. 296
Preferred Parts List 296
Hybrid Design Process 297
Design and Layout. 297
Computer-Aided Design (CAD) 298
Artwork 299
   Photo-Plotter 299
   Rubylith 299
Design Review 300
Engineering-Model Design Verification 302
Modification and Redesign 303
Substrate Parasitics 303
Capacitance Parasitics. 303
Conclusions on Interelectrode Capacitance. 307
Capacitance Computer Program 307
   Program Description 307
Inductive Parasitics 309
Conclusions on Parasitic Inductance 314
Thermal Considerations 314
Conduction 314
Convection. 315

Radiation. 315
Circuit Design Thermal Criteria 315
Thermal Analysis Computer Program 321
Thermal Testing 326

Layout Guidelines Common to Both Thick- and Thin-Film
Hybrids. 326
Preliminary Physical Layout 326
Estimating Substrate Area. 326
Final Physical Layout. 326
Assembly Aids 327
Device Placement 328
Wire Bonding Guidelines. 330
Conditions to Avoid. 331

Thick-Film Materials and Processes Description. 332
Thick-Film Substrates 333
Thick-Film Conductor Materials. 335
Thick-Film Resistors 335
Overglaze Design Guidelines 336
Solder Application. 339
Thick-Film Dielectrics 340

Thick-Film Design Guidelines 340
Artwork and Drawing Requirements. 340
Multilayer Yields. 343
Conductor Patterns—General Considerations. 344
Vias—Interlevel Conductor Connections Through Multilayer
Dielectric. 347
Wire and Die Bonding Pads 348
Thick-Film Resistor Design Guidelines. 352

Thin-Film Guidelines 360
Standard Practices. 360
Design Limitations. 361

References. 366

11. DOCUMENTATION AND SPECIFICATIONS 367
Documentation. 367
Military and Government Specifications. 369
MIL-M-38510 General Requirements for Microcircuits 370
MIL-STD-883 Test Methods and Procedures for Microelectronics . . 374
Purpose. 374
Method 5008 375
MIL-STD-1772 Certification Requirements for Hybrid Micro-
circuit Facilities and Lines. 385

12. FAILURE ANALYSIS. 389
Types and Causes of Hybrid Failures. 389
Device Failures. 389
Interconnection Failures. 391
Material Verification 391

Substrate Verification 391
Substrate Failures 392
Package Failures 392
Contamination 392
Failure Analysis Techniques 397
Electrical Analysis. 397
Chemical Analysis 397
Thermal Analysis. 397
Physical Analysis. 397
Analytical Techniques 400
Case Histories of Hybrid Circuit Failures 402
Tin Whiskers. 402
Metallic Smears. 402
Particles. 402
Flux Residues. 402
Cracked/Broken Die. 402
Collapsed Wires. 404
Package Plating. 404
Package Discoloration 405
Nickel Ion Contamination. 405
Corrosion of Aluminum Wire Bonds 405
Corrosion of Nichrome Resistors 407
Stress Corrosion of Kovar 408
Intermetallics in Wire Bonds 409
Die Bond Surface Oxidation 412
Loose Particle Short. 414
Wire Bond Short, Case 1. 415
Wire Bond Short, Case 2 417
Weak Wire Bonds, Case 1 418
Weak Wire Bonds, Case 2 420
Open Wire Bonds. 421
References 423

INDEX 424

1

Introduction

CLASSIFICATION OF MATERIALS FOR MICROELECTRONICS

Before proceeding to a discussion of hybrid circuit processes, it is important to have a general understanding of the materials most commonly used in today's electronic equipment. Indeed it is difficult to dissociate processes from materials. All electronic devices and circuits are based on combinations of three general types of materials. These are conductors, insulators, and semiconductors and are defined in terms of their abilities to either conduct or impede the flow of electrons.

Conductors

Conductivity is a function of the number of free electrons and their mobility. Conductors conduct electricity readily because they have many free and mobile electrons available to carry the current. The best electrical conductors are metals, generally elements residing in Groups I, II, and III of the Periodic Table. Elements in Group IB, typically metals having a valence of one, are most conductive. Because these elements contain only one electron in their outermost shells, the electron is easily released from the forces of attraction of the nucleus. Thus elements of Group IB (copper, silver, and gold) are among the best conductors known and the most widely used in electronics. Metals with valences of two or three are also good conductors, but not as good as those of Group IB. Among these are aluminum, palladium, platinum, nickel, and tungsten. A comparison of the resistivities (inverse of conductivities) of various metals is given in Table 1.1. The metals and alloys most widely used in the fabrication of electronic circuits and systems and their general applications are given in Table 1.2.

Metals have a positive temperature-coefficient-of-resistivity (TCR); that is, as the temperature of a conductor increases its resistivity increases or, conversely, its conductivity decreases. However, for the highly conductive metals, (Al, Au, Ag), the effect is not significant in the temperature range over which circuits are tested or used (room temperature to about 150°C). (Figure 1.1)

1

Table 1.1: Electrical Resistivities of Metal Conductors

Metal	Symbol	Resistivity (ohm-cm x 10^{-6})
Silver	Ag	1.62
Copper	Cu	1.69
Gold	Au	2.38
Aluminum	Al	2.62
Tungsten	W	5.52
Nickel	Ni	6.9
Indium	In	9.0
Platinum	Pt	10.52
Palladium	Pd	10.75
Tin	Sn	11.5
Tin-lead	Sn-Pb (60-40)	14.99
Lead	Pb	20.65

Table 1.2: Commonly Used Conductors and Their Uses in Electronics

Conductor	Symbol	Use
Gold	Au	Wire, plating, thick films, thin films, filler in epoxy conductive adhesives for device attachment
Gold-platinum	Au-Pt	Thick-film solderable conductors
Gold-palladium	Au-Pd	Thick film solderable conductors
Gold-silver	Au-Ag	Low-cost thick-film solderable conductors
Gold-silicon	Au-Si	Eutectic attachment of devices for ohmic contact
Gold-germanium	Au-Ge	Eutectic attachment of devices for ohmic contact
Silver	Ag	Low-cost conductor, thick-film plating, thin film, EMI, filler for epoxy conductive adhesive
Silver-palladium	Ag-Pd	Thick film (low silver migration)
Silver-platinum	Ag-Pt	Thick film (low silver migration)
Copper	Cu	Thick film (low cost), thin film, foil and plating for printed circuit boards, wire, filler for epoxy conductive adhesives
Aluminum	Al	Wire, thin-film metallization for devices
Tungsten	W	Thick-film conductors for co-fired tape substrates and packages
Molybdenum-manganese	Mo-Mn	Thick-film conductors for co-fired tape substrates and packages
Nickel	Ni	Barrier metal, low conductivity metal, plating
Tin-lead alloys	Sn-Pb	Solder for attachment of components, plating for circuit boards
Indium and alloys	In	Low-temperature solder for attachment of devices and components
Tin-silver alloys	Sn-Ag	Low-temperature solders

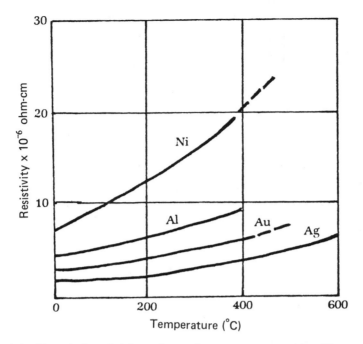

Figure 1.1: Electrical resistivity of metals versus temperature (Note: positive TCRs).

Insulators

At the other end of the spectrum from conductors are the insulators, which are very poor conductors of electricity. Their valence electrons are tightly bound and not free to move within the structure. In an insulator, the bonds that hold the atoms together, as a rule, are covalent (shared electrons) and possess high bond strengths (the energy in kilocalories per mol to break apart a bond pair). An important class of insulators are compounds based on the carbon atom—hydrocarbons, polymers and plastics fall within this group. The bond energies for C-C, C-H, and C-O bonds are 83, 99, and 110.2 kcal/mol, respectively—thus extremely difficult to break apart to create free-moving electrons. Sufficient energy, of course, could be applied in the form of extreme heat or radiation to eventually rupture the bonds, but this is done only at the expense of causing irreversible damage to the material. Free radicals and smaller molecular fractions would then be formed which are quite different in physical properties from the original material.

Insulators generally reside in Groups V, VI, and VII of the Periodic Table. Examples of excellent insulators include: plastics, rubber, ceramics, glass, sapphire (a form of aluminum oxide), other metal oxides (aluminum oxide, beryllium oxide, and silicon dioxide) and metal nitrides (silicon nitride, aluminum nitride). Insulators may be solids, liquids, or gases, but only solids are extensively used in electronics. Applications include substrates on which resistors, capacitors, and metal conductors are formed, substrates within which semiconductive, resistive, and conductive areas are generated (ICs), and passivation layers protecting active circuits from moisture and other harsh environments. Solid insulators have resistivities ranging from several megohm-cm to greater than 10^{14} ohm-cm, but caution should be used in their selection because in some cases the insulating properties can be rapidly degraded by ionic impurities, moisture absorption, or thermal and radiation exposures. In contradistinction to metals, insulators have negative TCRs (resistivity decreases with increase in temperature). For the ceramics that are most widely used as circuit substrates or as dielectric insulating layers, this decrease is significant only at the extremely high temperatures—far higher than any electronic circuit would be exposed to (Figure 1.2). At the maximum temperature that ceramic-based circuits might be exposed (about 150°C), the resistivity of the ceramic is still quite high (about 1×10^{13} ohm-cm). The situation is somewhat more critical for plastic insulating materials. It is well known that the insulation resistance of many epoxies, though very high in the dry state (10^{12} ohms or greater), decreases to 10^6 or lower when exposed to temperature/humidity cycling. This has been attributed to residues of ionic impurities (chloride, sodium, and ammonium ions) remaining in the epoxy from its manufacture. These ions become mobile in the presence of moisture and at elevated temperature, thus increasing the conductivity of the plastic. Other polymers that are inherently purer than epoxies such as silicones and fluorocarbons remain stable over the same humidity and temperature conditions (Figure 1.3).[1]

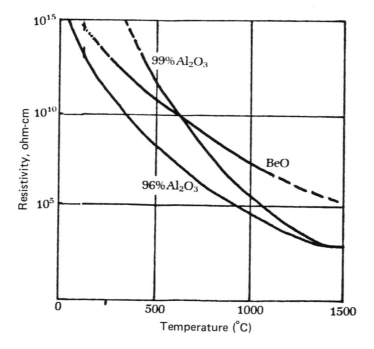

Figure 1.2: Electrical resistivities of ceramics (Note: negative TCRs).

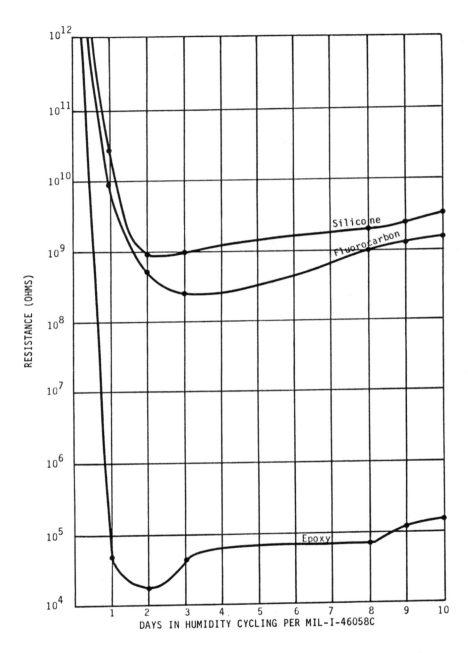

Figure 1.3: Effect of humidity on insulation resistance of polymer coatings.

Liquid insulators, when they can be used, have several advantages over solid insulators. These include: ability to self heal after discharge, better and more uniform conduction of heat, uniform electrical properties (dielectric strength and dielectric constant), and better corona and expansion control. Liquid insulators are used primarily in transformers, capacitors, circuit breakers, and switching gear. They may be used in direct contact with heat-generating devices for cooling, as impregnants for high-voltage cables and capacitors, and filling compounds for transformers and high-voltage circuits. For these applications, liquid insulators must possess high electrical insulation resistance, low dielectric constants, high dielectric breakdown voltages, and must be noncorrosive. Examples of liquid insulators currently in wide use are the hydrocarbons (mineral oil), silicones, fluorocarbons, and organic esters.[2]

Nitrogen, argon, and helium used as ambients in hermetically sealed circuits are prime examples of gaseous insulators. The permeation of moisture into these packages will, of course, degrade their electrical insulating properties. Sulfur hexafluoride (SF_6) has been shown to have excellent dielectric breakdown voltage, much better than that of nitrogen, helium, or air at the same pressure. It has been used as an insulating gas for some very high voltage applications.[3]

Semiconductors

Materials that are intermediate in their conductivities between insulators and conductors are known as semiconductors. Resistivities of semiconductors range from several ohm-cm to 10^5 ohm-cm. Two materials extensively used in the manufacture of semiconductor devices (transistors, diodes, ICs) are germanium (Ge) and silicon (Si) having resistivities of 60 ohm-cm and 6×10^4 ohm-cm, respectively. The practical significance of these materials, however, lies not so much in their initial resistivities but in the fact that the resistivities can be varied by the introduction of small amounts of impurity atoms into their crystal structure. Then, by forming junctions of the same material doped in different ways, the movement of electrons can be precisely controlled and devices having functions of amplication, rectification, switching, detection, and modulation can be fabricated.

For example, consider the crystal structure of silicon. It is similar to that of diamond, a tetrahedron with shared electron pair bonds. Both elements belong to Group IV of the Period Table and have valencies of 4. However, the Si-Si bond strength (43 kcal/mol) is far weaker than the C-C bond strength (83 kcal/mol), rendering the rupture of bonds and the release of free electrons in the silicon crystal much easier than in the diamond crystal. The breakage of bonds is a function of temperature. As a single-crystal silicon is exposed to higher temperatures, it becomes more conductive; thus it has a negative temperature coefficient of resistivity—a property of other semiconducting materials and of plastic insulators. The bonds which hold the germanium crystal together, Ge-Ge bonds, have an even lower bond strength, thus accounting for its lower resistivity and greater ease with which temperature can induce and mobilize free electrons. As one moves further down Group IV elements, it becomes so easy to detach electrons that the element may now be considered a conductor. This is the case with the gray form of tin. Thus in moving down the periodic group, one has transitioned from diamond, an insulator, to silicon and germanium, both semiconductors, to gray tin, a conductor. The energy necessary to move an electron from a low energy state to a conduction state is expressed in electron volts (ev), and the correspondence of this energy with resistivity for the Group IV elements is given in Table 1.3.

Table 1.3: Energy Requirements Versus Resistivity

Group IV Element	Energy eV	Resistivity (20°C) ohm-cm
C (diamond)	$\geqslant 7.0$	$> 10^6$
Si	1.0	$\cong 6 \times 10^4$
Ge	0.7	$\cong 50$
Sn (gray)	0.08	> 1

The semiconducting properties of a material that are induced by elevating the temperature or applying strong electric fields are referred to as intrinsic semiconductivity. However, extrinsic semiconductors are of most interest in the commercial world of electronic devices. These are rendered semiconducting by diffusing extremely small but controlled amounts of impurity atoms into the crystalline lattice. All crystals of silicon, no matter how pure, contain some atom vacancies which can be filled with atoms similar in structure. When elements of Group V are used (notably phosphorus, arsenic, and antimony), the atoms fit in the vacancies and form four covalent bonds with the surrounding silicon atoms; but, because the impurity atom is pentavalent, it will have an extra electron that cannot be accommodated as a bond. These extra unbound electrons are free and mobile and can be moved and controlled by an electric field. Semiconductors, having an excess of free electrons due to the incorporation of atoms of Group V, are known as N-types.

If atoms of Group III, notably boron, are used as the impurity atoms, only three of their electrons can enter into covalent bondage with the surrounding silicon atoms, thus leaving an electron vacancy or hole. This becames a metastable site and tends to be stabilized by drawing an electron from an adjacent silicon atom to fill the hole. No sooner is one hole filled and stabilized then another hole is created at the site from which the last electron moved. So, in essence, a movement of holes occurs in a direction opposite to the movement of the electrons, again resulting in a semiconducting material, this time referred to as P-type.

As an indication of the sensitivity of these impurity atoms, only 10 atoms of boron in one million atoms of silicon increases the conductivity of silicon by 1,000. By carefully controlling the concentration of the impurity atoms in selected regions of the silicon chip, resistors, conductors, and semiconductor devices can be formed. Figure 1.4 shows that on one extreme, heavy doping of 10^{20} atoms of carriers per cc of silicon results in a material approaching the electrical conductivity of a pure metal while, at the other end, light doping of about 10^{13} atoms/cc results in materials approaching the resistivity of insulators.

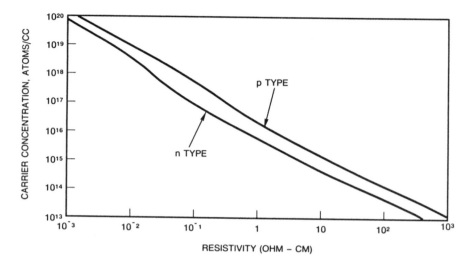

Figure 1.4: Effect of carrier concentration on resistivity of silicon.[6]

The theory of semiconductor devices is well established. The reader is referred to several books on this subject.[4,5]

CLASSIFICATION OF PROCESSES

Processes for electronics may be classified generally as either fabrication processes or assembly processes. Fabrication processes pertain to the formation of the chip devices, the integrated circuit chips, or the interconnect substrates. Generally, fabrication processes involve chemical reactions: photoresist deposition, exposure to ultraviolet light and developing, oxide formation, doping or ion implantation, metallization by vapor deposition, sputtering, screen printing and firing, etching, laser drilling, and resistor trimming. Assembly processes, on the other hand, pertain mostly to physical and physical thermal steps involved in attaching, interconnecting, and packaging the fabricated devices. Examples are adhesive attachment of devices, wire bonding, soldering, and sealing. A third group of processes are auxiliary to these two and involve cleaning, annealing, and stabilization baking.

Processes may also be classified according to the electronic products that are produced by them. The basic electronic product is the single chip which may be an active device (diode, transistor), a passive device (capacitor, resistor), or an integrated circuit (IC). A second product is the hybrid microcircuit, a packaging approach in which chip devices are assembled on a specially processed substrate which electrically interconnects the chips, creating a circuit function. The third major product is the printed wiring assembly (PWA) in which prepackaged chip devices (components) are solder attached and electrically connected to a printed wiring board (PWB), also referred to as a printed circuit board (PCB). The latter is generally an array of etched metal conductors on a reinforced plastic laminate substrate. In addition to the electronic components, hybrid microcircuits may be interconnected and incorporated in a PWA.

Processes may therefore be classified according to whether they pertain to:

- Single devices or integrated circuits in chip form or as packaged components
- Hybrid microcircuits
- Printed wiring assemblies

Two things should be noted here. First, there are other packaging or assembly formats besides the three mentioned (for example, microwave circuits and welded modules); but for the most part, these employ the same or similar processes. Thus, thin-film and thick-film deposition processes used for hybrid circuits are also used for microwave circuits. Secondly, even within the three main categories, there is commonality of processes. Sputtering or vapor deposition of metals are basically the same processes for both IC manufacture and hybrid circuit fabrication, differing only in the metals used, deposition parameters (temperature, pressure, etc.), and the conductor line dimensions.

DEFINITION AND CHARACTERISTICS OF HYBRID CIRCUITS

Types and Characteristics

Hybrid circuits are circuits in which chip devices of various functions are electrically interconnected on an insulating substrate on which conductors or combinations of conductors and resistors have previously been deposited. They are called hybrids because in one structure they combine two distinct technologies: active chip devices such as semiconductor die and batch-fabricated passive components such as resistors and conductors. The discrete chip components are semiconductor devices such as transistors, diodes, integrated circuits, chip resistors, and capacitors. The batch fabricated components are conductors, resistors, and sometimes capacitors and inductors.

Chip devices are small unpackaged components. Semiconductor chip devices, also referred to as bare chips or die, may range in size from very small 12 mil-square signal transistors to large high-density integrated circuits (ICs) about 250 mils square. Recently, with the development of VHSIC (Very High Speed Integrated Circuits) and gate arrays, chips larger than 500 × 500 mils are being produced.

Hybrid microcircuits are often classified as either thin film or thick film based on the process used to fabricate the interconnect substrate. The nomenclature refers directly to the thickness of the film and indirectly to the process used to deposit the conductors and resistors on the substrate. If the metallization is deposited by vacuum evaporation—by vaporizing a metal in a high vacuum, thin-film circuits are formed. The thickness of the metallization may be as low as 30 Å or as high as 25,000 Å. If the metallization is applied by screen-printing and firing conductive and resistive pastes onto a substrate, a thick-film circuit is formed. Thick films range from 0.1 mil (about 25,000 Å) to several mils in thickness (Figure 1.5).

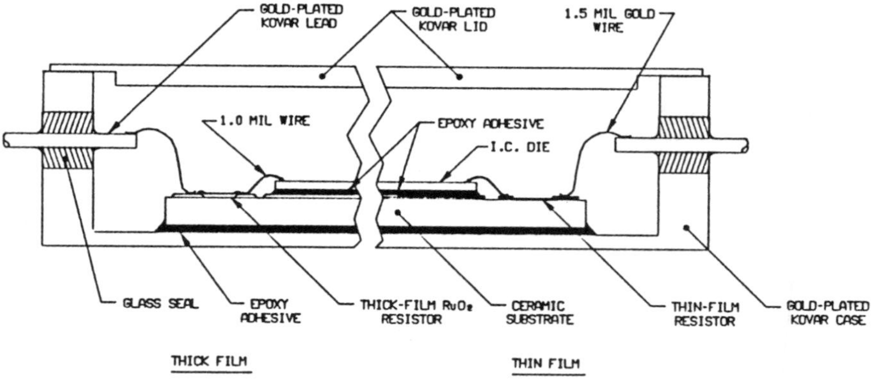

Figure 1.5: Cross-sections of thin and thick film hybrid assemblies. (Not to scale.)

Whether the interconnect substrate is thin or thick film, the assembly steps are generally the same for both. The chip devices are first attached to the substrate with a polymeric adhesive—generally an epoxy—or with a eutectic alloy or solder. After this, two sequences may be followed: (1) the devices may be interconnected to the substrate circuitry by wire bonding, then the substrate attached to the floor of the package or (2) the substrate may first be attached in the package then wire bonded. The latter sequence minimizes handling of assembled substrates, avoiding damage to wire interconnections, and avoids outgassing/contamination of both the circuit and wire bonds during the processing of the epoxy used for substrate attachment. For power circuits, the substrate must be bonded into the package prior to device attachment because of the high temperatures required in processing eutectic or alloy attachment. In all the above cases the outer bonding pads of the substrate are wire bonded to the package pins as the final assembly step. The interconnect wire may be gold or aluminum and is typically 1 mil in diameter, though it may be as thin as 0.7 mil or—for high power/amperage circuits—thicker than 3 mils. As the final step, the package is sealed by welding a lid on the package case or attaching the lid with glass or epoxy. Although the steps mentioned are the basic assembly steps, many other auxiliary steps are used in the production of hybrid circuits. Among these are numerous cleaning steps, annealing and trimming of resistors, curing of adhesives, visual inspection, electrical testing, nondestructive wire bond pull testing, and a host of mechanical/ thermal screen tests that have been imposed by the military and space agencies.

Hybrid circuits vary in size from very small, about 200 mils square containing only a few devices (Figure 1.6), to large 4 × 4 inch circuits containing several hundred devices. Typically, their sizes are intermediate, for example, 1 × 2 inches or 0.75 × 1.5 inches, and contain 20 to 30 ICs or chips (Figures 1.7 and 1.8). Integrated circuit densities of 25 chips per square inch and resistor densities of 90 per square inch with as many as 1,000 wire bonds are not uncommon. In a broader sense, circuits in which pre-packaged components are solder-attached on a ceramic substrate containing the interconnect conductor pattern may also be considered hybrid circuits (Figure 1.9).

Figure 1.6: Small hybrid microcircuit in TO package (Courtesy, Rockwell Intl.).

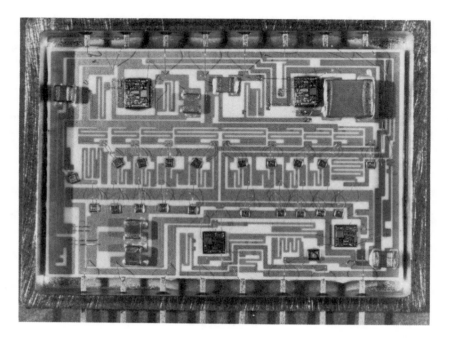

Figure 1.7: Moderate-density thin film hybrid microcircuit (Courtesy, Rockwell Intl.).

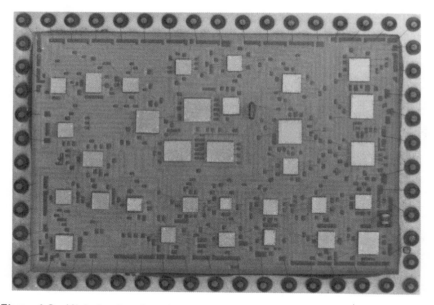

Figure 1.8: High-density three-layer multilayer thick film circuit for digital communication terminal (Courtesy Rockwell Intl.).

Figure 1.9: Ceramic printed circuits (Courtesy Rockwell Intl.).

In addition to their being classified as thin film or thick film according to the process used for the interconnect substrate, hybrid circuits may also be classified according to generic electronic functions. Thus there are digital, analog, RF microwave, and power circuits.

Comparison With Printed Wiring Boards

Hybrid microcircuits are used in applications where high density, low volume, and low weight are required for an electronic system. Without these constraints, printed wiring boards with solder-attached pre-packaged components are more economical. Because of the advantages of weight and volume which they have over printed wiring assemblies, hybrids have found extensive use in airborne and space applications. Hybrid circuits were a major contributor to the successful design and development of the Space Shuttle. Other key commercial and military applications include communication systems, high-speed computers, guidance and control systems, radar systems, heart pacemakers, other medical electronic devices, and automotive electronics. As already mentioned, hybrid circuits are much lighter and take up less space than printed wiring assemblies. They may be 10 times lighter than an equivalent printed circuit and 4 to 6 times smaller. Wire-bonded IC chips occupy only one-fourth the volume of dual-in-line IC packages used in printed circuit assemblies. Screen-printed or vacuum-deposited resistors replace pre-packaged chip resistors and avoid solder interconnections, effecting still further reductions in space and weight.

Because of the fewer interconnections that need to be made, since the resistors are batch-processed, the reliability of hybrids can also be greatly improved over printed circuit boards. Resistors can be trimmed statically or dynamically to precise values with close tracking; hence, precision circuits can be fabricated that are not possible with printed circuits. From a thermal management standpoint, hybrid circuits have many benefits. The direct mounting of high-power devices on a thermally conductive ceramic is greatly superior to mounting pre-packaged components onto a thermally-insulative epoxy or polyimide circuit board. To remove heat from a plastic printed circuit board, heavy metal heat rails must be attached with adhesive or metal-core boards must be used.

Finally, hybrid circuits are more suited to high-frequency, high-speed applications where, because of shortened distances between devices, finer, more precise conductor lines and spaces, and close tolerance resistors, parasitic capacitances and inductances can be minimized.

Comparison With Monolithic Integrated Circuits

Although many hybrid circuits can be designed and produced as monolithic integrated circuits, this is not economically feasible unless large quantities of a single type are to be produced so that the high non-recurring development costs can be amortized. The development of masks and requirements for expensive equipment and special clean-room facilities and controls make custom IC manufacture feasible only for high-volume production. The hybrid-circuit approach is more versatile and suitable for small-to-moderate production quantities of many different circuit types. Hybrids are also more suited to custom circuits than are ICs. Iterations that are required in the design are easier, faster, and less costly with hybrids than with ICs because changes in the hybrid masks can be made quickly (less than one hour per mask for thick-film circuits). Thus hybrids offer greater flexibility in making design and layout changes.

Although hybrid circuits are higher in density than printed wiring boards, the highest-density circuits can only be achieved with integrated circuits. Circuit-function densities for chip ICs are orders of magnitude greater than for hybrid circuits. VLSI (Very Large-Scale Integration) chips can have thousands of transistor functions in the space of a 100-mil-square chip, while recent advances in VHSIC (Very High Speed Integrated Circuits) have produced chips with greater than 10,000 transistors in the same area. Present gate-array ICs may have 2,000-5,000 gates with pin-counts of 100-200 on chips that are 200 mils square.[7] VHSIC chips with micron geometries may incorporate 20,000 to 40,000 gates in an area of 300 to 400 mils square. VHSIC having geometries in the submicron range are being developed with over 100,000 gates. Though these ICs may replace entire hybrid circuit functions, they, in turn, may require a hybrid-packaging approach to interconnect and integrate them with other devices. This is important not only to attain even greater density but also to enhance device performance through better heat dissipation, controlled impedances and capacitances, and shortened interconnect paths. Despite predictions of the demise of hybrid circuits because of advancements in ICs, at every stage in the development of ICs, hybrids have been found useful to effect higher levels of integration. Table 1.4 compares hybrid circuits with printed circuit boards and ICs while Table 1.5 shows a comparison of several microcircuit technologies.

Table 1.4: Hybrid Circuit Advantages (Courtesy, Tektronix)

Hybrid circuit advantages over etched-circuit boards:

- Consume less space, less weight
- Higher performance due to shorter circuit paths, closer component spacing (tighter thermal coupling), better control of parasitics, excellent component tracking
- Simpler system design and reduced system cost due to simplified assembly, and functional trimming capability
- Higher reliability due to fewer connections, fewer intermetallic interfaces, higher immunity to shock and vibration
- Easier system test and troubleshooting due to hybrid circuit pretested functional blocks

Hybrid circuit advantages over custom monolithic IC's:

- Lower nonrecurring design and tooling costs for low to medium volume production
- Readily adaptable to design modifications
- Fast turnaround for prototypes and early production
- Higher performance sub-components available (both substrate and add-on components), for example, ±0.1% resistors, ±1% capacitors and low TCR zener diodes
- Ability to intermix device types of many different technologies, leading to increased design flexibility
- Ability to rework allows complex circuits to be produced at reasonable yields, and allows a certain amount of repair

THE HYBRID CIRCUIT MARKET AND APPLICATIONS

Hybrid Circuit Market

Hybrid microcircuits have been in use for over thirty years. In the early sixties, hybrids were extensively used in weapons systems such as the Minuteman Intercontinental Ballistic Missile. The main reasons for the use of hybrids were the increased reliability and reduced size which they offered over the standard printed circuit boards. In the late sixties and early seventies when the fast-growing monolithic integrated circuits began to replace hybrid functions, many predicted the demise of the hybrid technology. These dire predictions were further fueled by the advent of large-scale integrated circuits (LSIs) as they replaced even more hybrid products. However, it soon became evident that the usefulness of ICs could only be optimized by integrating them, with other ICs, resistors, and

Table 1.5: Comparison of Microcircuit Technologies

Parameter	Thick-Film Hybrid Circuits	Thin-Film Hybrid Circuits	Monolithic Circuits
Performance	High	High	Limited
Design flexibility, digital	Medium	Medium	High
Analog	High	High	Low
Parasitics	Low	Low	High
Resistors, maximum sheet resistivity	High	Low	High
Temperature coefficient of resistance	Low	Lowest	High
Tolerance	Low	Lowest	High
Power dissipation	High	High	Low
Frequency limit	Medium	High	Medium
Voltage swing	High	High	Low
Size	Small	Small	Smallest
Package density	Medium	Medium	High
Reliability	High	High	Highest
Circuit development time (prior to prototype)	1 month	1 month	1-2 months
1:1 design transfer from bench	Yes	Yes	No
Turnaround time for design change	2 weeks	2 weeks	1 month
Part cost, low quantity	Medium	High	Impractical
High quantity	Medium	Medium	Low
Cost of developing one circuit	Low	Medium	High
Capital outlay	Low	Medium	High
Production setup and tooling costs	Low	Medium	High

capacitors in a multi-chip hybrid circuit. Hybrid circuits have thus outlived these predictions.

In the eighties, hybrids have continued to grow at a rate of 15 to 20 percent per year. According to a market survey in 1985 by Frost and Sullivan,[8] the hybrid industry was $4.6 billion in 1985 and is estimated to increase to $5.3 billion in 1986 and $6.3 billion in 1987. In 1989 the market is forecast to be $9.1 billion, which is twice the 1985 market. According to this study, linear hybrids will comprise one-half, digital hybrids one-third, and opto-electronic/RF hybrids the rest of the 1989 market. These figures comprise the total hybrid market both merchant and captive. The merchant market is defined as that in which hybrids are produced to be sold in the open market. Captive hybrids are those that are produced only for internal consumption. In the early years of hybrids, there were few merchant hybrid houses; however, that trend has been shifting. By the end of this decade, it is predicted that captive suppliers will have 53 percent of the market and merchants the other 47 percent.

Military applications will continue to dominate and are the fastest growing segment. The Frost study predicts a 21 percent growth rate for military applications and a rise from 28 percent of the market to 31 percent by 1989. A growth rate of 13 percent per year is predicted for use in computers and 18 percent per year for communications applications.

Applications

Hybrids are found in almost every military system and in most commercial products. Commercial products include: home computers, calculators, radios, televisions, aircraft equipment, heart pacemakers, communication equipment, car ignitions, watches, cameras, electronic games, VCRs, and microwave ovens. These are but a few of the commercial applications of hybrids. Hybrids are used mainly to save weight and space. However, in the commercial field, cost is usually the deciding factor, because the commercial market is very competitive and will settle only for the lowest-cost packaging approach. Only when a commercial application requires a more reliable circuit, such as a heart pacemaker, is cost overridden.

Thin-film hybrids are better suited to high-frequency applications. While thin- and thick-films can be used up to 500 MHz, above that limit thin-film technology dominates.[9]

Commercial Applications. Medical electronics is one segment of the commercial market that requires long-term reliability, along with dense circuitry. Also, irregularly shaped substrates may be needed to fit the package. Medical hybrids must pass even more stringent tests than military hybrids, and must be free of contaminants in order to be implanted in humans. The broad field of medical electronics includes instrumentation for life support, patient monitoring, and pacemakers. The pacemaker market especially has developed rapidly, and hybrid circuits along with it as the logical approach for packaging.

As stated above, the restrictions in size and materials have made medical electronics a very challenging area for the hybrid engineer.

The computer industry also makes extensive use of hybrids. Rather than wait for the semiconductor industry to design and fabricate larger-memory ICs, computer designers are packaging existing ICs in hybrid form to get the most memory in the smallest area. Micro-Technology Inc. produces two standard CMOS Memory units fabricated in hybrid form: a 32k by 8 bit and a 16k by 8 bit static RAM. These hybrids use thick-film conductors on a ceramic substrate. The company will soon be releasing a 128k by 8 bit device making it one of the largest commercial CMOS static RAM hybrid in the industry.

Analog-to-Digital (A/D) Converters are the closest thing to a "standard" product that the hybrid industry has. These hybrids are found everywhere that the analog world meets the digital world. Hybrid technology allows the very best monolithic IC to be combined with high-accuracy laser-trimmed thin-film resistors to provide the most accurate circuit. This optimizes the performance for the application. Hybrid Systems Corp. manufactures an A/D Converter (HS-9516-6) with an accuracy of 0.0008 percent linearity error (Figure 1.10). It is a 16-bit resolution hybrid, which means one part in 65,536. From the specifications of this device, it is clear that hybrid technology is used when high accuracy is a requirement. Other companies, such as Analog Devices also produce high accuracy A/D Converters like the AD1147, a 16 bit hybrid, which requires only half the space of its module counterpart.

Figure 1.10: 16-bit analog-to-digital converter (Courtesy Hybrid Systems Corp.).

Still another area in which hybrids are extensively used is in electrical-test equipment. The Tektronix 2400 series of oscilloscopes employ eleven custom hybrids. These hybrids, two of which are shown in Figures 1.11 and 1.12, replace thousands of discrete parts, rendering the instrument lighter, more reliable, and more producible. The data-acquisition hybrid is used for 100ms/sec data acquisition. It incorporates a charge-coupled device (CCD) and a clock driver. Also featured are a beryllium-oxide (BeO) substrate with plated through-holes for connection to a back-side ground plane, thin-film peaking coils, and controlled-impedance conductor runs. The hybrid dissipates 10.4 watts in normal operation when attached to a multi-finned heat sink. The A-to-D Converter is a thick-film hybrid multiplexed 300 MHz, 8-bit A/D converter with ECL input levels and a 1-volt output level into a 75-ohm load. It features through-hole connections to a 2-level ground and power distribution pattern on the back side to enhance high-speed performance. Many other types of hybrids are used in ground-support and telecommunications equipment, among which are: line filters, receivers, transmitters, A/D and D/A converters, and code/decode circuits.

Military/Space Applications. Every military weapons system uses one form of hybrid or another. Hybrids are used in missiles, satellites, aircraft, helicopters, hand-carried weapons, shipboard equipment, and submarine navigation instruments. A military system that benefited from the reduced size of hybrids was an airborne data processor built by Hughes Aircraft for the F-14 and F/A-18 aircraft. The original version of the computer consisted of eighteen 5 × 5 inch printed circuit boards. Redesigning these boards and partitioning them into hybrids condensed them to one 6 × 9 inch, ten-layer board containing eight hybrids, 13 discrete ICs, and some capacitors and resistors. Besides the reduction in space and weight, the hybrid version also increased the computer's speed by a

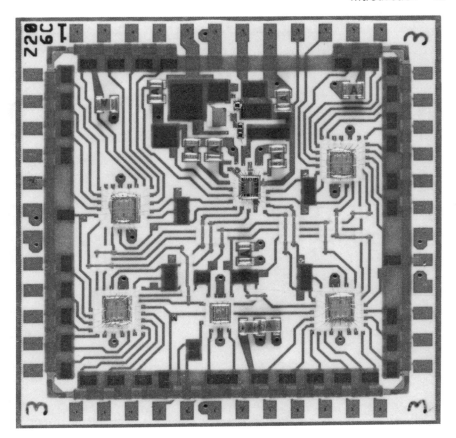

Figure 1.11: Oscilloscope hybrid—A-to-D converter (Courtesy Tektronix Inc.).

factor of 1.5 by shortening the signal paths. One of the hybrids was packaged in a 2 × 2 inch case and had 160 I/Os. It consisted of a fifteen-layer ceramic substrate with ICs in open chip-carriers that dissipated 20 watts. A later version, designed with gate arrays to replace some of the ICs, dissipated only 10 watts.

Not surprisingly, one of the biggest users of hybrids has been NASA for the Space Shuttle where every pound savings in weight was critical to the mission. Each shuttle uses over 10,000 hybrids for their reduced size, weight, and reliability. Over one thousand power hybrids are used on each orbiter. These hybrids are used to open the equipment bay doors, to release the landing gear-up-lock, and to eject the external fuel tank after the fuel has been expended.

Dramatic weight savings were achieved by converting load-controller circuits from printed wiring boards to hybrid circuits. The weight was reduced from 55 pounds to 15 pounds per load-controller. With six controllers per Shuttle orbiter, a significant weight-savings of 240 pounds per orbiter was achieved.

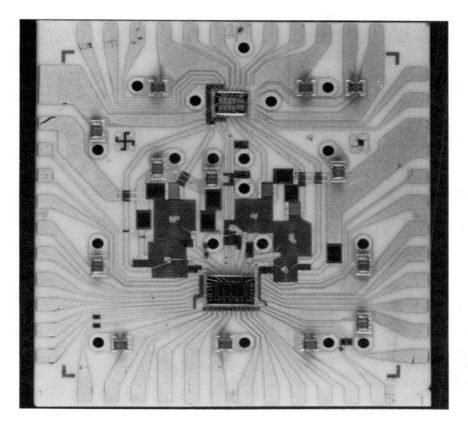

Figure 1.12: Thin film data-acquisition hybrid (Courtesy Tektronix Inc.).

A unique item that was incorporated in the load-controller hybrids was an internal fuse. The hybrid used three parallel 1-mil-diameter gold wires in series with the incoming power. The internal "fuse" protected spacecraft wiring and further decreased its overall weight. Figure 1.13 is a photograph of one of the power hybrids. Note that it was partitioned in two sections: a low-power and a high power section. The power components (transistors and diodes) were soldered to a beryllia substrate and the beryllia was then solder-attached to the Kovar case, whereas all the low-powered circuitry was attached to a standard alumina substrate with epoxy adhesive.

Power Applications. Some hybrids, such as the HS9151 power converter produced by National Semiconductor (Figure 1.14), are designed to dissipate large amounts of power.[9] This particular hybrid works directly off the ac line and supplies 5vdc at 3 amps. The second generation of this hybrid, the HS9503, supplies 5vdc and +/− 12vdc for a total power output capability of 50 watts. This hybrid uses various assembly techniques as well as several semiconductor technologies. In the hybrid, use is made of

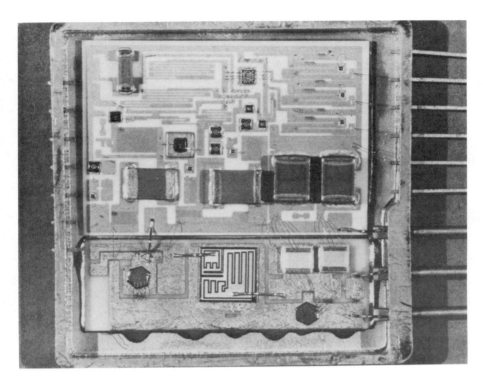

Figure 1.13: Space Shuttle power hybrid (Courtesy Rockwell Intl.).

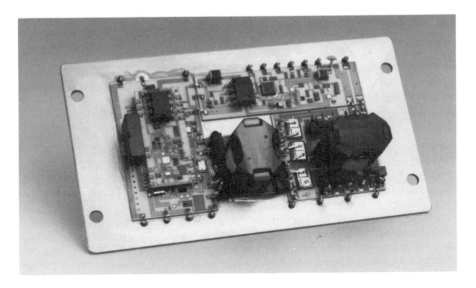

Figure 1.14: Power conditioner hybrid (Courtesy National Semiconductor).

CMOS, power MOSFETs, Schottky diodes, bipolar transistors, high-voltage rectifier diodes, several passive devices, and an optocoupler. The devices are attached by both surface-mount and chip-and-wire techniques on four separate substrates one of which is mounted "piggy-back" on another substrate. A very high heat-dissipating hybrid is an operational amplifier, the PA03 power device, manufactured by Apex Microtechnology Corp. This hybrid (Figure 1.15) dissipates 500 watts. To maximize the removal of heat, the hybrid is enclosed in a special copper package.

Figure 1.15: 500-watt operational amplifier, PA03 (Courtesy Apex Microtechnology Corp.).

Besides power conditioning and switching, power hybrids are used for motor controls. International Rectifier is a prime supplier of motor control hybrids.[9] ILC Data Device Corporation (DDC) also manufactures power hybrids including motor controls. DDC has developed an H-Bridge (Figure 1.16) which operates at a supply-voltage of 80 volts with continuous output-currents of 20 amps. The hybrid uses power MOSFETs soldered to a beryllia substrate which is then brazed to a copper case, a fabrication technique that ensures a low thermal resistance of less than 0.5°C/watt. The H-Bridge hybrid is rated for 150 watts at a case temperature of 25°C. Other power hybrids supplied by DDC are Power Drivers (Figure 1.17a) and FET Converter/Rectifiers (Figure 1.17b). The power driver has an output current

Figure 1.16: H-Bridge hybrid circuit (Courtesy ILC Data Device Corp.).

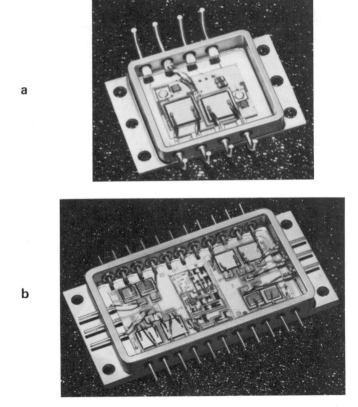

Figure 1.17: Custom power hybrid circuits. (a) Power driver (b) FET converter/rectifier (Courtesy ILC Data Device Corp.).

of 45 amps and dissipates 140 watts at 25°C. In this hybrid the power devices are mounted on beryllia substrates, then brazed to the bottom of a molybdenum case. The FET converter uses many unique assembly techniques to obtain a producible hybrid. Besides using beryllia substrates brazed to a copper case, the low-power control circuitry uses a conventional thick-film alumina substrate mounted between the two power beryllia substrates.

REFERENCES

1. Licari, J.J., *Plastic Coatings For Electronics*, McGraw-Hill, 1970.
2. Saums, H.L. and Pendleton, W.W., *Materials For Electrical Insulating and Dielectric Functions*, Hayden Book Co., 1973.
3. Von Hippel, A.R. (Ed.), *Dielectric Materials and Applications*, M.I.T. Press, 1961.
4. Glaser, A.B. and Subak-Sharpe, G.E., *Integrated Circuit Engineering, Design, Fabrication, and Applications,* Addison-Wesley Publishing Co., Menlo Park, CA, 1977.
5. Lindmayer, J. and Wrigley, C., *Fundamentals of Semiconductor Devices*, Van Nostrand, 1965.
6. Gise, P. and Blanchard, R., *Semiconductor and Integrated Circuit Fabrication Techniques*, Reston Publishers, 1979.
7. Marshall, J.F., Packaging For Performance, *Semiconductor Intl.*, Aug. 1984.
8. Frost and Sullivan, *Hybrid Microcircuits Market in North America*, No. A 1365, Sept. 1985.
9. Mennie, D., "Hybrid Circuits", *Electronic Design*, June 19, 1986.

2

Substrates

FUNCTIONS

Substrates for hybrid circuits serve three key functions:

1. Mechanical support for the assembly of the devices.
2. Base for the electrical interconnect pattern and batch-fabricated film resistors.
3. Medium for the dissipation of heat from devices.

Besides these basic mechanical, electrical, and thermal functions, substrates must meet many other electrical, thermal, physical, and chemical requirements, among which are:

- High electrical insulation resistance—to avoid electrical leakage currents between closely spaced conductor lines. Initial volume resistivities greater than 10^{14} ohm-cm and surface insulation resistance greater than 10^9 ohms are highly desirable. Other electrical properties of importance especially in high-speed, high-frequency circuits are dielectric constant, dissipation factor, and loss tangent.

- Low porosity and high purity—to avoid moisture and contaminant entrapment, arcing, tracking, and metal migration.

- High thermal conductivity—to dissipate heat produced by devices

- Low thermal expansion coefficient—to match, as closely as possible, the expansion coefficients of attached devices, thus minimizing stresses and avoiding cracking during thermal cycling.

- High thermal stability—to withstand, without decomposition or outgassing, the high temperatures involved in subsequent

25

processing. Among the highest processing temperatures are those encountered during belt-furnace sealing (295-375°C), eutectic die and substrate attachment (380°C) and, in the case of thick films, firing (850-1000°C).

- High degree of surface smoothness—to achieve stable, precision thin-film resistors and very fine conductor lines and spacings.

- High chemical resistance—to withstand processing chemicals such as acids used in etching and plating.

The combination of these requirements quickly narrows the number of available materials. Among materials that were initially used as substrates for hybrid circuits were glass and glazed-alumina ceramics. To a lesser extent, and for some special applications, beryllia, sapphire, silicon, quartz, aluminum nitride, and porcelainized steel have been used. However, high-purity alumina with an as-fired, smooth surface is the material most widely used today because it most closely meets all the requirements and also takes into consideration cost.

SURFACE CHARACTERISTICS

Surface Roughness/Smoothness

Roughness, or conversely smoothness, is defined as the average deviation, in microinches, from a center line that traverses the peaks and valleys of the surface. If a profile of the surface is taken using a moving diamond stylus (profilometer) and an arbitrary line is drawn such that the areas of the peaks above the line equal the areas of the valleys below the line, the average deviation from the line is referred to as the center-line-average (CLA). One can see that if nichrome or tantalum nitride resistor films of only 200-400 A thickness are used on a rough surface of 20 microinches CLA (5000 A) the films will contour the hills and valleys giving non-reproducible dimensions to the resistors and thus resulting in varying resistance values. When using thin films on rough surfaces there is also a greater chance for forming imperfect, discontinuous films having micro-cracks, stresses, and opens. This problem does not exist with thick films which range from 0.5 mils (127,000 A) to several mils in thickness and thus completely encapsulate even the roughest surfaces. Figures 2.1 and 2.2 compare surface profiles for a relatively smooth 99.6% alumina substrate typically used for thin film circuits with a relatively rough 96% alumina substrate used for thick films.

The smoothness of a ceramic is a function of its microstructure and density; the smaller the grain size and higher the density, the smoother the surface. Microstructure and density, in turn, are dependent on the ceramic composition and process by which the substrates are produced. Early alumina substrates were quite rough and unsuitable for thin film circuits unless the surface had been polished or a glaze applied to it.

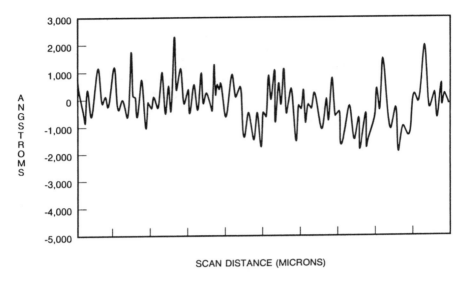

Figure 2.1: Surface roughness profile for 99% alumina used for thin-film circuits (Courtesy Rockwell International Corporation).

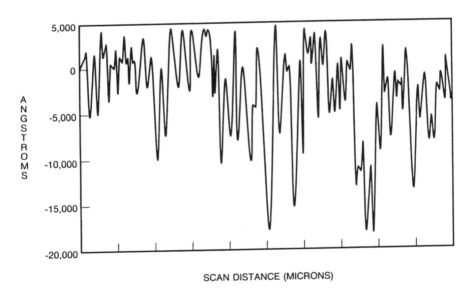

Figure 2.2: Surface roughness profile for 96% alumina used for thick-film circuits (Courtesy Rockwell International Corporation).

Camber

Camber is the warpage of a substrate or the total deviation from perfect flatness. It is normalized by dividing the total camber (in inches or mils) by the largest dimension over which the camber can occur. In the case of square or rectangular substrates this is the diagonal distance (Figure 2.3). Thus camber is expressed as inches per inch or mils per inch. Camber is an important parameter for thick film fabrication. A high camber will affect the screen printer snap-off distance and in turn affects the resolution of the pattern being printed. Alumina substrates employed in thick film circuits (96% alumina) have a camber between 2 and 4 mils per inch and are adequate for most applications.

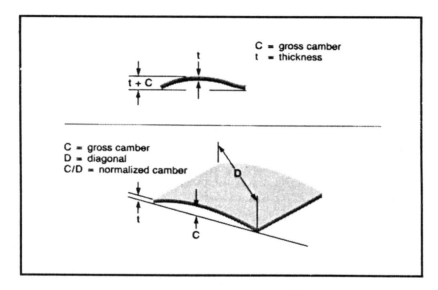

Figure 2.3: Camber.

The most expedient method for measuring camber is a "go-no go" method in which the substrates are passed through parallel plates separated by a fixed distance that is equal to the sum of the substrate thickness being tested and an allowance for the maximum camber. Though this is not a precise or accurate method, it is extensively used as a low-cost quality control and receiving inspection test that can discriminate between 1 and 2 mils of camber. Absolute camber measurements can be made by first measuring the thickness of the substrate with a ball point micrometer, then measuring the combined thickness and camber with a dial-indicating micrometer attached to a plate that is larger than the substrate. The zero-setting of the micrometer is made at the flat plate upon which each substrate is placed.

Granularity

The grain size and number of grain boundaries per unit area influence both the smoothness of the ceramic surface and the adhesion of vapor-deposited, thin-film metallization. The grain size and density are, in turn, determined by the nature and amount of impurity oxides in the ceramic composition and the firing conditions. Over-firing produces a large grain structure and a coarse rough ceramic. Impurities concentrate at the grain boundaries during firing. Some impurities, notably sodium and potassium oxides, should be avoided since they result in large grain sizes and a course porous ceramic. Other oxides such as magnesium oxide and silicon oxide are essential in promoting a fine grain structure and assisting in the chemical mechanism for adhesion of thin film metals. Such oxides also concentrate at the grain boundaries and serve as sites for nucleation and growth of the thin metal films.

ALUMINA SUBSTRATES

Alumina ceramic is almost exclusively used for both thin and thick film circuits. To a lesser extent, beryllia ceramic substrates are used but only for high power circuits where the six-fold higher thermal conductivity of beryllia is required.

The reliability of a hybrid circuit starts with the proper selection of the substrate. Though practically all ceramic substrates (alumina, beryllia) look alike, small differences in their composition and differences in their surface roughness and flatness can determine success or failure of the hybrid circuit.

Alumina is used for both thin and thick film circuits but for thin film circuits there is a significant difference in the quality and grade of alumina that must be used. Because thin films are primarily used for high-density circuits that require precision, closely-spaced conductors and close-tolerance, high-stability resistors, the purity and surface smoothness of the alumina must be of a high level. Thin film circuits require substrates with an alumina content greater than 99% by weight and a surface smoothness of 1 to 6 microinches CLA (Center Line Average) whereas substrates for thick film circuits may be of a lesser grade in both purity (94-96% alumina) and surface smoothness (15-25 microinches CLA). A comparison of the scanning electron micrographs, magnified 3100 times, shows dramatically the differences in surface finish between 99.6% alumina having a CLA of 4 microinches and a 96% alumina having a CLA of 25-35 microinches (Figures 2.4 and 2.5).

Grades of Alumina

Manufacturers of alumina substrates offer several grades of alumina ranging from those having relatively rough surfaces, used for thick film circuits, to those with extremely smooth, lapped and polished surfaces (1 microinch or less CLA finish), used for the ultimate in high resolution, precision, thin film circuits. Substrates with intermediate smoothness (4-6 microinch CLA) are available for more typical thin film hybrid applications.

Figure 2.4: SEM photo of 99.6% alumina used for thin-film cirucits.

Figure 2.5: SEM photo of 96% alumina used for thick-film circuits.

Alumina Substrates For Thick Films

Normally, thick-film circuits are fabricated on pre-fired substrates made from 96 percent alumina. The 96 percent alumina content represents only the nominal percentage; actual contents vary from slightly under 95 percent to slightly over 97 percent, depending on the supplier. All paste manufacturers use 96 percent alumina substrates to evaluate their pastes, unless specifically requested to do otherwise. The 96 percent alumina is compatible with most thick film pastes, has high flexural and compressive strength, and has good thermal conductivity, approximately one-seventh that of aluminum. Where greater thermal dissipation is required, 99.5 percent beryllia should be used because of its higher thermal conductivity— about the same as that of aluminum metal.

Substrates are available in either the "as-fired" condition or with ground and polished surfaces. Substrates with ground surfaces may cost five to seven times more than the as-fired types, but for dense, fine line circuity, high resistor density, or tight system-packaging considerations, ground and polished substrates may be necessary, at least on the surface side that is to be screened.

The approximate dimensional tolerances for "as-fired" substrates are given in Table 2.1. Dimensional tolerances for ground substrates are largely a function of cost. For a premium price tolerances of ±0.001 inch or even ±0.0005 inch are available. Cambers of 0.001 inch per inch or even 0.0005 inches per inch and nearly any surface finish are also available.

Substrates are generally between 0.025 inches to 0.060 inches thick. The use of substrates thinner than 0.020 inches is not advisable because of the fragility of handling and processing thin ceramics. A practical guideline for determining the optimum substrate thickness for multilayer thick film circuits is to allow a minimum of 0.025 thickness for each inch of length or width, whichever is longest. As an example, a 0.025 inch thick substrate should be used when the longest dimension is one inch; a 0.050 thickness should be used when the longest dimension is 2.5 inches.

Table 2.1: Dimensional Tolerances for As-Fired 96% Alumina

Outside dimensions (length or width)	±1%; not less than ±0.003 in
Thickness	±10%; not less than ±0.0015 in
Perpendicularity	≤0.004 in/in
Edge straightness	≤0.003 in/in
Camber	0.003–0.004 in/in
	0.002 in/in (special)
Hole centers	±1%
Surface finish	≤25 microinches CLA

Alumina Substrates For Thin Films

The properties of 99.6% alumina, for example, the Superstrate 996 series, the highest grade of alumina produced by Materials Research Corp. (MRC) are given in Table 2.2. The polished and low-CLA substrates provide the designer of thin film microwave and precision resistor networks

Table 2.2: Properties of Alumina Substrates for Thin Film Circuits*

Property	Test	Units	Value
Alumina content	—	wt %	99.6
Color	—	—	white
Water absorption	ASTM C373	—	nil
Gas permeability	He leakage	—	nil
Dielectric constant	ASTM D150	1MHz	9.90
25°C		10 MHz	9.90
Dissipation factor	ASTM D150	1 MHz	0.0001
Loss factor	ASTM D150	1 MHz	0.001
Volume resistivity	ASTM 1829	ohm-cm	$>10^{14}$
Thermal conductivity		cal/cm-sec-°C	
25°C	ASTM C408		0.090
300°C			0.050
500°C			0.038
Thermal coefficient of linear expansion		per °C	
25°-300°C	ASTM C372		6.3×10^{-6}
25°-600°C			7.1×10^{-6}
25°-800°C			7.3×10^{-6}
Flexural strength (avg)	ASTM F394	psi	90.000
Compressive strength (avg)	ASTM C773	psi	400,000
Hardness	ASTM E18	Rockwell 45N	81

*Properties are for Superstrate 996 Series, tradename of Materials Research Corp. (Courtesy Materials Research Corp.).

with reproducible results even for circuits with lines and spacings of 1 mil or less. Other less costly grades of alumina are suitable for the majority of hybrid applications. Lists of these grades and their properties are given in Tables 2.3 and 2.4.

As with many other materials used in electronic applications the material selected and used is often a compromise. Glass as a substrate is ideal in providing a very smooth surface, but is extremely poor in thermal conductivity. Pure alumina, though good in thermal conductivity, requires glass as a binder and does not provide a highly smooth surface. Increasing the amount of glass binder from an optimum of about 0.5% to 15% results in almost a three-fold decrease in thermal conductivity (Figure 2.6). An early solution to this problem involved the use of a composite substrate, glazed alumina. A thin glassy layer applied to the surface of the alumina provided a smooth (<1 microinch) CLA surface, yet preserved the high thermal conductivity of the bulk alumina. Over the years, ceramic materials and processes have improved, so that today as-fired alumina substrates with CLAs of 2-6 microinches are available and widely used.

Table 2.3: Alumina Substrate, Superstrate* Grades

Grade*	Description	Camber (mils/in) 0.005" Through 0.027"	0.040" and 0.050"	Dielectric Constant at 4 GHz	Specific Gravity (g/cc)	Surface Finish (CLA)	Grain Size (μ)
Micro-Rel (M)	A very fine grained, smooth ($<$3 μ in CLA) substrate with tightly controlled dielectric constant used for most demanding microwave integrated circuits (MIC's), resistivity control, and fine-line resolution	2	4	9.9 ± 2%	3.86–3.90	$<$3 μ in	$<$1.2
Hi-Rel (A)	100% inspected and selected for superior surface perfection, flatness, and dielectric constant control. Used for high-density MIC's	2	4	9.9 ± 2%	3.86–3.90	$<$4 μ in	$<$1.5
Standard (S)	Meets all requirements of the Hi-Rel grade including flatness and dielectric constant. Recommended for standard MIC's.	2	4	9.9 ± 2%	3.86–3.90	$<$4 μ in	$<$1.5
Resistor (R)	Exceptionally fine grain smooth surface ($<$3 μ in CLA). Used for nichrome resistor networks where consistency and tight tolerances are essential. 100% inspected, cost efficient equivalent of standard grade.	3	4	9.9**	$>$3.81	$<$3 μ in	$<$1.2
Hybrid (B)	A cost/quality compromise for patterns of 5-mil lines and spaces. 100% inspected for surface.	3	4	9.9**	3.83–3.91	$<$5 μ in	$<$2.0
Circuit (C)	Low-cost substrate with high degree of flatness and good surface smoothness. Not 100% inspected.	3	4	9.9**	$>$3.81	$<$6 μ in	$<$2.5

*All grades are trademarks of Materials Research Corporation, Orangeburg, NY.
**Typical.

Table 2.4: Substrates for Thin-Film Circuits

Type	Surface Roughness (CLA, Å)	Camber (mils/inch)
Glazed alumina	<250*	2–4
As-fired alumina (99.5%)	1000–1500	2–4
Polished alumina (99.5%)	250	<0.01
Polished sapphire	<250	<0.1
Soda lime glass	<250	2

*<1 μ inch.

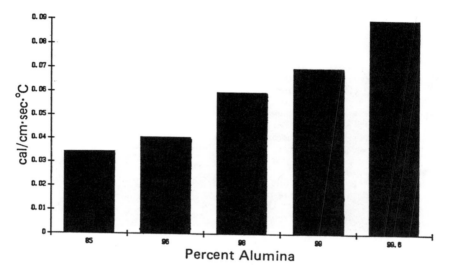

Figure 2.6: Thermal conductivity vs percent alumina.

Off-the-shelf substrate sizes range from 0.25 × 0.25 inches to 4.25 × 4.25 inches. Though larger sizes are possible, the substrate size for thin film circuits is limited by what can be processed in commercial vacuum-deposition and photoresist-processing equipment. Very small circuits are generally not processed individually but as multiple circuits on larger (about 2 × 2 inch) substrates which have been prescored on the back side so that they can be broken apart after batch processing. Pre-scoring involves cutting into the substrate to a depth of approximately one third of the substrate thickness using either laser or diamond saw cutting. Substrate thicknesses range from 0.005 inches to 0.050 inches, but standard off-the-shelf thicknesses of 0.010 and 0.025 inches are more cost effective and suitable for most thin film hybrid applications. A comparison of the properties of typical alumina substrates used for thin film and thick film is given in Table 2.5.

Table 2.5: Comparison of Alumina Ceramics*

	96% Alumina (for thick film)	99.6% Alumina (for thin film)
Grain size (μ)	11	<1.2
Surface smoothness (μ inch CLA)	10–25	1–6
Camber (mils/inch)	~4	~2
Flexural strengh (psi)	40,000–50,000	70,000–90,000
Specific gravity	3.75	3.8
Thermal conductivity (cal/cm-sec-°C)	0.04–0.08	0.09
Thermal expansion (ppm/°C, 20–300°C)	6.7–6.8	6.3
Volume resistivity (ohm-cm, 300°C)	10^{14}	>10^{14}
Dielectric constant (at 1 MHz)	9.4	9.8–9,9
Dissipation factor (at 1 MHz)	0.0002	0.0001

*Values represent ranges derived from a composite of values from several manufacturers.

Co-Fired Ceramic Tape Substrates

Alumina or beryllia in the "green" tape form (un-fired state) may be processed so that the individual layers of a multilayer substrate can be produced separately, then stack-laminated and fired together—a process analogous to fabricating multilayer printed-wiring boards using epoxy laminates. The content of the alumina in the "green" tape (about 90-95%) is lower and the glass content is higher than that for pre-fired alumina, resulting in lower thermal conductivity. Co-fired ceramic is extensively used to fabricate high-density thick film multilayer substrates and packages. A more detailed treatment of the materials and processes used for the co-fired process is given in Chapter 4.

BERYLLIA SUBSTRATES

Beryllia (beryllium oxide, BeO) is unique among electronic materials in combining high electrical insulation resistance with high thermal conductivity. Only one other material, diamond, is known to combine these two divergent properties. Generally, good electrical insulators (plastics and ceramics) have very low to moderate thermal conductivities, while electrical conductors (metals) have high thermal conductivity. The thermal conductivity of beryllia approaches that of aluminum metal yet has the electrical insulation resistivity of the best of the plastics. High purity beryllia ceramic has a thermal conductivity approximately 1,200 times greater than that of a typical epoxy plastic, 200 times greater than most glasses, and 6 times better than alumina ceramic. Figures 2.7 and 2.8 show a comparison of the thermal conductivities of beryllia with some of the most commonly used electronic materials.[1]

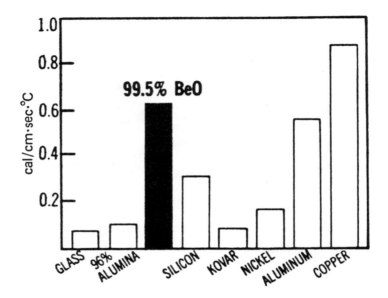

Figure 2.7: Thermal conductivity of beryllia compared with other electronic materials.

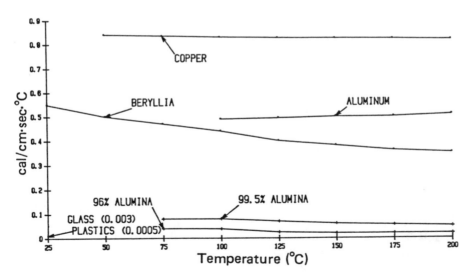

Figure 2.8: Thermal conductivities of ceramics and metals as a function of temperature.

Beryllia would therefore make an ideal substrate material were it not for its high cost and the precautions needed in processing it because of its toxicity. But even the toxicity is not a major problem for the hybrid manufacturer who purchases already fired and machined substrates. The toxicity hazard is primarily faced by the manufacturer of the beryllia who must take precautions in controlling dust from machining and vapors from high temperature processing.

Besides its use as an interconnect substrate for hybrid and microwave circuits, BeO is used as heat sink spacers (heat spreaders) beneath power devices, and as a package-construction material for discrete power transistors, integrated circuits, and multi-chip hybrids.

The electrical insulation resistance of 99.5% BeO is excellent both at room temperature and at the maximum temperatures that hybrid circuits may be expected to encounter (about 200°C). Values of 10^{17} to 10^{18} ohm-cm are reported. As with many other ceramics and inorganic and plastic materials, beryllia has a negative temperature coefficient of resistivity; that is, as temperature increases, resistivity decreases (Figure 2.9).

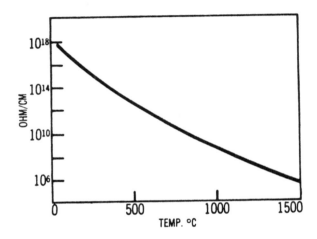

Figure 2.9: Electrical resistivity vs temperature for 99.5% beryllia (Courtesy National Beryllia, Division of General Ceramics).

The dielectric strength of beryllia ceramic (voltage required for electrical breakdown or puncturing) is more than adequate for hybrid applications; it ranges from 600 volts/mil to 800 volts/mil, depending on thickness (Figure 2.10). Again as with plastics, the dielectric strength decreases with the thickness of the sample being tested. The dielectric constant (k) and dissipation factor for beryllia are better (lower) than for alumina ceramic. The k for BeO is 6.7 compared to 9.9 for alumina ceramic of equivalent purity. Both k and loss tangents are quite stable over a wide frequency range of 1 KHz to 10 GHz (Tables 2.6 and 2.7). Other physical and electrical properties of beryllia substrates are given in Tables 2.8 and 2.9.

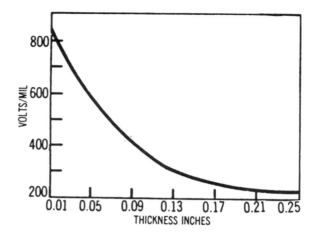

Figure 2.10: Dielectric strength vs thickness for 99.5% beryllia (Courtesy of National Beryllia, Division of General Ceramics).

Table 2.6: Loss Tangent vs Temperature and Frequency for Beryllia (Berlox K-150) (Courtesy National Beryllia, Division of General Ceramics)

Temperature ($^\circ$C)	. . . . Loss Tangent of Berlox K-150*			
	. . . . .Loss Tangent			
	1 kHz	1 MHz	1 GHz	10 GHz
25	0.0002	0.0002	0.0003	0.0003
300	0.0003	0.0003	0.0005	0.0005
500	0.0005	0.0005	0.0008	0.0008

*Trade name for 99.5% Beryllia of National Beryllia, Division of General Ceramics.

Table 2.7: Dielectric Constant vs Temperature and Frequency for Beryllia (Berlox K-150) (Courtesy National Beryllia, Division of General Ceramics)

Temperature ($^\circ$C)	. . . . Dielectric Constant of Berlox K-150*			
	. . . . .Dielectric Constant			
	1 kHz	1 MHz	1 GHz	10 GHz
25	6.7	6.7	6.7	6.6
300	6.8	6.8	6.8	6.7
500	7.0	7.0	6.9	6.9

*Trade name for 99.5% Beryllia of National Beryllia, Division of General Ceramics.

Table 2.8: Properties of 99.5% Beryllia[1]

Specific heat (cal/°C-g)	0.25
Dielectric constant (1GHz)	6.7
Dielectric loss (1 GHz)	0.0003
Compressive strength (psi)	225,000
Flexural strength (psi)	30,000
Tensile strength (psi)	20,000
Young's modulus (psi)	50,000,000
Density (g/cc minimum)	2.85
Impenetrability	Impervious to gases and liquids
Vapor pressure at 1500°C (atm)	10^{-12}
Maximum use temperature	1800°C
Radiation hardness	Excellent

Table 2.9: Average Physical Property Values as a Function of Purity
for Beryllia Ceramics[1]

	95% BeO	98% BeO	99.5% BeO	99.9% BeO
Flexural strength, psi	20,000	25,000	35,000	36,000
Electric resistivity, ohm-cm	10^{14}	10^{15}	10^{17}	10^{18}
Young's modulus, psi	44×10^6	45×10^6	50×10^6	55×10^6
Thermal conductivity, cal/cm-sec-°C	0.35	0.52	0.62	0.66
Dissipation factor at 1 GHz at 25°C	0.0005	0.0005	0.0003	0.0002

ALUMINUM NITRIDE AND SILICON CARBIDE SUBSTRATES

Aluminum nitride (AlN) and silicon carbide (SiC) are receiving considerable attention as substrate materials for hybrid circuits, particularly as low-
cost alternates to beryllia for high-power circuits. The key advantage of
both these materials over beryllia is that their thermal conductivities are
close to that of beryllia, yet they lack the toxicity and high cost associated
with beryllia. It is also anticipated that AlN may replace alumina for many
applications because of its higher thermal conductivity (5 to 7 times
greater than alumina). Reported values for the thermal conductivity of AlN
vary widely because they depend largely on the amount of impurities,
specifically oxygen. For example, the thermal conductivity of AlN has been
shown to be highly sensitive to the amount of oxygen in the formulation,
decreasing considerably with only fractions of a percent of oxygen.[2]
Aluminum nitride is similar to alumina in its electrical properties, but its
coefficient of thermal expansion (4.5 ppm/°C) more closely matches that of
silicon, making it more desirable for assemblies using silicon chip devices.

Ceramic manufacturers have developed AlN in both cast tape form and pressed form and have improved its surface smoothness to the point that it can also be used for thin-film circuits. Used as a substrate for thick-film multilayer circuits AlN requires specifically formulated thick-film pastes that have expansion coefficients similar to that of the substrate. Several paste manufacturers now provide a complete system of conductor, resistor, and dielectric pastes specifically formulated for use with AlN substrates.[3]

The development of silicon carbide substrates is not as advanced as that of aluminum nitride. A version of SiC that has been enhanced with a small amount of beryllium oxide is reported to have a thermal conductivity even higher than beryllia.[2]

CERAMIC SUBSTRATE MANUFACTURE

There are two methods for producing alumina and other ceramic substrates: pressing and tape casting. In either case, fine aluminum oxide powder (in the micron or submicron range) is mixed with binders and other oxides such as magnesium oxide, silicon oxide (glass frit), and calcium oxide. The additive oxides (4-6% in the case of thick film substrates) act as fluxes to reduce the temperature required for sintering and as agents to control grain size. In the final sintered product these oxides concentrate at the grain boundaries and play a key role in adhesion of both thin films and thick films. The ingredients are first mixed in a ball mill to produce a homogeneous mixture. Chemical purity, particle size, particle size distribution, and thoroughness of mixing must be rigidly controlled to assure the optimum microstructure of the ceramic after sintering. Sintering is effected at about 1700°C and produces a high density polycrystalline structure.

In the press process the mixed ceramic composition is compacted in metal dies at pressures of 10,000 to 20,000 psi, then sintered. All portions of the unfired pressed part must be of uniform density to avoid differences in shrinkage that result in warped parts.[4]

The tape process differs from the press process in that the ceramic composition is cast into sheets, dried to a flexible "green" state, cut to size, and then sintered. In this process solvents and polymeric thixotropic binders are added to the ceramic mix to facilitate the casting.[4] Beryllia substrates are produced in much the same way as alumina. Beryllia is extracted from several naturally occurring minerals: beryl and bertandite, both consisting of beryllia aluminosilicate ($3BeO \cdot Al_2O_3 \cdot 6SiO_2$).[5] These ores are chemically processed to remove the bulk of the aluminosilicate leaving behind high purity BeO powder. The powder is next pressed and sintered at high temperatures to yield high density substrates of 99.5% purity. As with many other electronic materials beryllia ceramics were not fully developed or utilized until after World War II. Though beryllia was known for over one hundred years it was not until the Manhattan Project that very high purity, high density beryllia substrates were fabricated.[5] It took still another decade before beryllia found applications in electronic circuits and aerospace instruments.

ENAMELED METAL SUBSTRATES

Enameled, also referred to as porcelainized or glazed-metal, substrates may be used as low-cost substrates for thick-film circuits where weight is not a significant consideration. Thick-film circuits formed on enameled-steel substrates are being used for many commercial/industrial applications such as automotive and camera electronics. They have found greater interest and applications in Japan than in the U.S. and very limited use in military and space electronics.

Enameled steel has been extensively studied and characterized. RCA conducted detailed studies on methods for depositing the enamel onto metal and on the reliability of the enamels.[6] Electro-Science Laboratories (ESL) has developed thick film conductor, resistor, and dielectric pastes and evaluated them on enameled steel substrates from three sources.[7,8] The enamels consist of materials such as quartz, feldspar, or clay reacted with a flux such as borax, soda ash, cryolite or fluorspar to form a glass. The glass is melted, fritted, and ball-milled in water. Various materials may be added to keep the particles suspended in the aqueous medium, to improve adhesion to steel, and to add color or opacity. The enamel should have a melting temperature high enough to permit firing of thick film pastes yet low enough to minimize oxidation and warping of the steel. Because the enamel consists of several glasses of relatively low melting temperatures (675-680°C), the commercially available thick film pastes that are fired at 850-1000°C cannot be used. Thus specially formulated conductor, resistor, and dielectric pastes that can be fired at 600-625°C have been introduced. The enamel should also be formulated so that it has a thermal expansion lower than that of the metal in order to keep it in compression and prevent its cracking. The chemical composition of the enamels is critical to the reliability of the final circuit. Early enamel compositions contained large amounts (>5%) of alkali metal ions (sodium, potassium, lithium). These compositions were low in bulk resistance and often displayed a discoloration, associated with screened-on thick-film silver conductors, that was termed "brown plague". Most of these problems were resolved in later formulations by reducing the alkali metal contents. Some typical properties of the fired enamel are given in Table 2.10.

Table 2.10: Properties of Porcelain Enamel Coatings*

Dielectric constant	8–22
Dissipation factor, %	0.10–0.64
Voltage breakdown, v/mil	55–675
Surface insulation resistance, ohms	1×10^7–2×10^{12}
Bulk insulation resistance, ohms	1×10^7–7×10^{11}
Thickness, mils	5–6

*Values represent ranges for several enamels.[5]

Enamel, in thicknesses of 5-6 mils, is applied to low carbon steel, <0.003% C, by dipping, spraying, electrophoretic deposition, or electro-static spraying. In electrophoretic deposition the colloidal particles of the enamel are negatively charged and attracted to and deposited onto the metal part which has been positively charged. This process is analogous to electroplating. In electrostatic spraying the enamel is sprayed as a solid powder with high voltage charges of opposing polarity applied to the steel and the powder.[9] Both the electrophoretic and electrostatic deposition processes produce more uniform coatings than spray or dip processes. The coatings are then fired in a conveyor furnace for one to three minutes at peak temperatures of 800-890°C. The enameled substrates may be purchased from several suppliers. Thick film conductor, resistor, and dielectric pastes are screen-printed onto the glazed surface and dried and fired to form single or multilayer circuits.

Enameled-steel substrates offer several advantages over ceramic substrates, among which are:

- Substrates will not break, crack, or chip; they are shock resistant.

- Substrates can be die punched or cut into complex shapes and sizes; the metal can be cut to irregular shapes and holes drilled prior to glazing.

- The core metal serves as an inherent heat sink.

- The metal, especially steel, is less expensive than ceramic.

- The metal serves as a built-in ground plane.

QUALITY ASSURANCE AND TEST METHODS

Substrates may be purchased to the manufacturer's or user's specifications. Most users generate their own procurement specifications. The key elements of such a document include: dimensions (length, width, thickness) with tolerances, allowable camber and waviness, composition (for example, alumina or beryllia content), and surface finish. For some applications, tests for microcracks and porosity should be specified. One of these, the dye-penetrant test, is a low-cost test that can be applied on a one hundred percent basis. According to this test the parts are immersed in a water or organic-solvent solution of a colored or ultraviolet-light-visible dye (Zyglo), pressurized, then lightly rinsed with the same solvent. Any residual dye that remains in cracks or fissures will be visible on inspection under normal or black-lamp light.

Many other more expensive and sophisticated tests may be used to qualify ceramic substrates or to assure their quality. Among these are:

Surface characteristics (grain size, inclusions)	Scanning Electron Microscopy (SEM)
Surface Impurities	Electron Microscopy Auger Microscopy
Bulk Qualitative and Quantitative Analysis	Emission Spectroscopy

Electrical Properties	Dielectric Constant
	ASTM D-150
	Dissipation Factor
	ASTM-D-150
	Volume Resistivity
	ASTM-1829
Mechanical Properties	Flexural Strength
	ASTM-F-394
	Compressive Strength
	ASTM-C-773
	Hardness
	ASTM-E-18
Thermal Properties	Expansion Coefficient
	ASTM-C-372
	Thermal Conductivity
	ASTM-C-408

REFERENCES

1. Fleischner, P.L., "Beryllia Ceramics in Microelectronic Applications," *Solid State Technology*, Vol. 20, No. 1, 1977.
2. Konsowski, S.G., et al, "Evaluation of Advanced Ceramics for High Power And Microwave Circuitry," *Intl. J. Hybrid Microelectronics*, Vol. 10, No. 3, 1987.
3. Cox, C.V., et al, "A New Thick Film Materials System For Aluminum Nitride-Based Power Hybrid Circuits," *Intl. J. Hybrid Microelectronics*, Vol. 10, No. 3, 1987.
4. Harper, C.A., *Handbook of Electronic Packaging*, McGraw-Hill, 1969.
5. Sidgwick, N.V., *The Chemical Elements and Their Compounds*, Vol. 1, Oxford, Clarendon Press, 1950.
6. Onyshkevch, L., "Base Metal Thick Film Inks on Porcelain Substrates," *Proc. 16th Natl. SAMPE*, Oct. 1984.
7. Stein, S.J., Huang, C. and Gelb, A.S., "Comparison of Enameled Steel Substrate Properties For Thick Film Use," European Hybrid Microelectronic Conf., May 1979, Ghent, Belgium.
8. Stein, S.J., Huang, C. and Gelb, A.S., "Thick Film Materials on Porcelain Enameled Steel Substrates," 29th Electronics Components Conference, May 1979.
9. Lacchia, A., "Electrostatic Spraying of Powder Enamels," *Ceramic Industry*, Nov. 1977; Feb. 1978; Mar. 1978.

3

Thin Film Processes

DEPOSITION PROCESSES

Having selected the substrate material, the next step in the fabrication of thin film microcircuits is the depositioin of metals or metal compounds onto the substrate. These metals ultimately provide the conductor and resistor patterns and functions. Typically, a substrate is coated sequentially with a layer of resistive material, a barrier metal layer, and a top conductor layer. These layers are relatively thin ranging from 200 Å to 20,000 Å and are deposited by either vapor deposition, sputtering or variations of these two processes. The following sections cover deposition processes specifically as they apply to the fabrication of thin-film resistor/conductor networks for hybrid circuits. Comprehensive treatments of thin-film deposition processes may be found in several other books.[1,2]

Vapor Deposition

Vapor deposition consists in heating a material in a relatively high vacuum so that its vapor pressure exceeds that of its environment, allowing it to be vaporized quickly. Substrates to be coated are placed in a vacuum chamber in the vicinity of the source material (Figure 3.1). Upon contacting the cooler surfaces of the substrate the vapor condenses by a mechanism of nucleation and growth of the film emanating from various grain boundary sites on the substrate. Most metals have very high melting and boiling points (2500-3000°C) at room ambient pressure making them impractical to deposit under these conditions. Furthermore, since the evaporation of most metals in air results in oxidation of the deposited metal, it is desirable to deposit metals in as high a vacuum as possible. Pressures of 10^{-5} to 10^{-6} Torr can reduce the vaporization temperatures of most metals to 1000°C or lower. Comprehensive tables of vapor pressures versus temperature for metals are available in the literature.[2]

A widely used method for heating the metal is to contain it in a boat wrapped with high resistance wire and to apply a current through the wire.

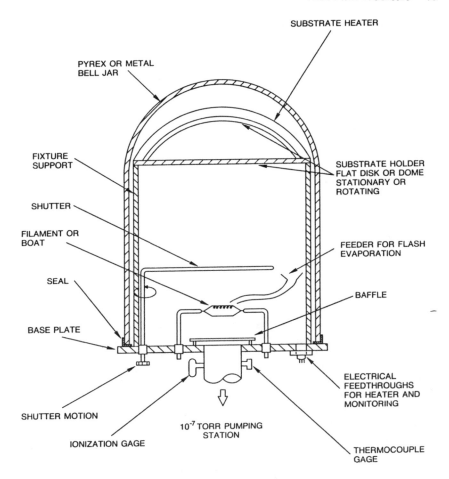

SUBSTRATE HEATER

PYREX OR METAL
BELL JAR

FIXTURE
SUPPORT

SUBSTRATE HOLDER
FLAT DISK OR DOME
STATIONARY OR
ROTATING

SHUTTER

FILAMENT OR
BOAT

FEEDER FOR FLASH
EVAPORATION

SEAL

BAFFLE

BASE PLATE

ELECTRICAL
FEEDTHROUGHS
FOR HEATER AND
MONITORING

SHUTTER MOTION

10^{-7} TORR PUMPING
STATION

IONIZATION GAGE

THERMOCOUPLE
GAGE

Figure 3.1: Schematic diagram for vacuum evaporator.

This method, referred to as resistance heating, though widely used, suffers from four problems: (1) The molten metal can alloy with the boat material thus contaminating the charge. (2) Refractory metals have such high vaporization temperatures even at low pressures that they cannot be deposited by resistance heating. (3) Alloys of metals having different vapor pressures will fractionate during deposition giving rise to variations in the composition and properties of the deposited film. (4) The rate of evaporation can fluctuate making it difficult to consistently deposit coatings of uniform thickness. These problems can be obviated to a large extent by variations in the vacuum deposition process. For example, to deposit alloys and minimize fractionation, flash evaporation should be used. This process consists in vibrating small portions of the alloy powder down an incline into a boat that is kept at a sufficiently high temperature to vaporize both constitents as soon as they contact the boat (Figure 3.2). Nickel-chromium alloy and some cermet resistors may be deposited in this manner.

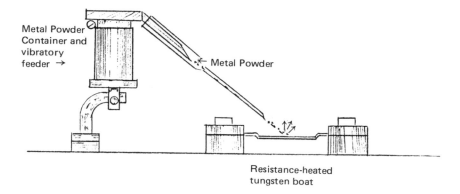

Figure 3.2: Mechanism for flash evaporation.

Electron beam evaporation is used to deposit refractory metals such as titanium, tungsten, tantalum and molybdenum. The vacuum chamber is equipped with a mechanism that permits an e-beam to be focused and impinged on the metal causing it to heat and vaporize. By controlling the energy of the electron beam, both the depth and area of the melt can be controlled and localized so that the molten metal does not contact and interact with the boat material.

Direct Current (D.C.) Sputtering

Sputtering is an electro-physical process in which a target (rendered cathodic) is bombarded with highly energetic positive ions which, by transferring their energy, cause ejection of particles of the target. The "sputtered" particles deposit as thin films on substrates that have been placed on an anodic or grounded holder. Early investigators referred to the ejection of particles as "spluttering", an anonomatopoetic word describing the sound of the process. Later in 1923 the "l" was dropped and the process became known as sputtering. Sputtering was largely a laboratory curiosity until 1960. Since then an entire industry has emerged in which there are many suppliers of equipment, targets, and accessories and many firms who are using the process for the production of electronic circuits and devices.

The simplest variant of sputtering is D.C. or direct current sputtering. The equipment consists of a diode or parallel plate system (Figure 3.3). The material to be sputtered is attached to the cathode plate while the substrates to be coated are placed on the opposite plate which is rendered anodic or grounded. A plasma (glow discharge) is generated between the plates by first evacuating to about 10^{-6} Torr to remove all air, moisture and extraneous gases then backfilling with argon gas and applying a negative bias of 1,000 to 2,000 volts d.c. to the cathode. On backfilling, the pressure increases to

10^{-3} to 10^{-1} Torr. A plasma discharge is created in which the argon is electronically activated, loses an electron, and becomes an argon cation (positively charged ion). These Ar ions are attracted and accelerated toward the cathode whereupon they bombard the target with sufficient energy, effect a transfer of momentum and cause particles of the target material to be sloughed off (sputtered). In the course of this process, highly energetic secondary electrons are emitted. These interact with neutral Ar atoms and create more positively charged Ar ions with emission of more electrons. Thus a self sustaining plasma is formed.

$$Ar \rightarrow Ar^*$$
$$Ar^* \rightarrow Ar^+ + e$$
$$e + Ar \rightarrow Ar^+ + 2e$$

A simple sputtering apparatus consists of a vacuum chamber in which are assembled two parallel electrodes: a cathode target and an anode plate or holder on which the substrates to be coated are placed. The cathode has provision for cooling because of the heat generated during sputtering. The anode has provision for both heating and cooling since temperature control is important in obtaining adequate adhesion. The cathode is typically 5 to 50 cm in diameter and spaced 1-12 cm from the anode.

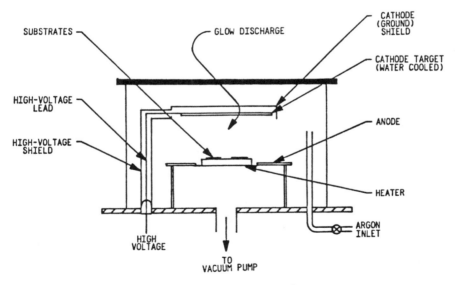

Figure 3.3: Schematic of a dc sputtering apparatus.

Deposition rates range from less than 100 Å/min to several thousand Å/min and depend on many variables including the nature of the target material (binding forces of the atoms), the atomic mass of the bombarding gas ions, the ion current, pressure of the gas, and the interelectrode distance. The deposition rate (G) is proportional to the ion current (I) and the sputtering yield (S) and is given by the equation:

$$G = CIS$$

where C is a proportionality constant based on the characteristics of the particular sputtering chamber used. The ion current may be increased by increasing the power or the pressure of the gas. However, there is a compromise between pressure and power: too high a pressure will result in diffusion of sputtered atoms back to the cathode while the maximum power that can be used is limited. The sputtering yield (S) is a function of the target material and the gas that are used. Argon, the bombarding gas most commonly used, is actually a compromise between cost and efficiency. The heavier inert gases, xenon and krypton, would provide higher sputtering yields than argon but would also be too costly to use in commercial applications. Helium, the least costly gas, would result in a very low sputtering yield (Figure 3.4). A detailed discussion of these factors has been given by Vossen and Kern.[3]

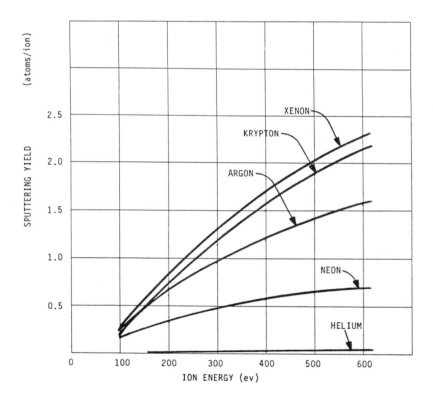

Figure 3.4: Sputtering yield for platinum as a function of noble gas and its ion energy.

Radio-Frequency (RF) Sputtering

The DC sputtering process, just described, is limited to electrically conductive targets (metals and metal alloys); it cannot be used to deposit inorganic materials such as oxides or dielectric insulators. When an insulator is used as the cathodic target in DC sputtering, positive charges quickly accumulate on its surface during ion bombardment. These charges impede the initiation and maintenance of the glow discharge. To overcome this situation, radio frequency or RF sputtering is used where the target is subjected alternately to positive ion and electron bombardment. Thus the accumulation of positive charges during the negative portion of the cycle is neutralized by the electrons formed during the positive cycle. During the first half (negative cycle), sputtering of the target material occurs; while during the second half, electrons neutralize the positive charges, effectively rendering the target ready for sputtering again.

RF sputtering is much more versatile than DC sputtering. In addition to metals and alloys it can be used to deposit almost any dielectric material at relatively low temperature and pressure. Examples include: silicon oxides, silicon nitride, glasses, alumina, refractory oxides, and some plastics such as Teflon, polyethylene, polypropylene. Among the plastics, the thermoplastic polymers are better suited to sputtering than the thermosetting types. Though some polymer chain scission occurs with thermoplastics because of the high kinetic energy imparted to the target, the deposited film is essentially the same structurally as the starting material. Thermosetting, highly cross-linked polymers, however, decompose and recombine in somewhat different molecular structures and display different electrical and physical properties than the starting material.

The equipment used for RF sputtering is essentially the same as for DC sputtering except that an RF generator of 13.56 MHz and 1-2 KW power has been added and provision has been made for cooling the target (Figures 3.5 and 3.6). The dielectric material to be sputtered is attached to a metal backing plate with conductive adhesive.

Reactive Sputtering

Reactive sputtering is still another practical variation of sputtering. Here a reactive gas is introduced along with the inert argon to form the plasma. The reactive gas becomes activated and chemically combines with the atoms that are sputtered from the target to form a new compound. Reactive sputtering is therefore a combined physical, electrical, and chemical process. Generally, the amount of reactive gas used is small compared to that of the inert gas, but by varying the ratio, films ranging in properties from almost a metal to a semiconductor, insulator, and resistor can be produced. Two widely used reactive gases are oxygen (producing oxides of metals) and nitrogen (producing nitrides of various elements). Reactive sputtering has become a valuable commercial process for depositing dielectrics, resistors, and semiconductors. For example, the reactive sputtering of tantalum nitride is one of the two most widely used processes for depositing thin film resistors. A list of some films that may be deposited by reactive sputtering and their applications is given in Table 3.1.

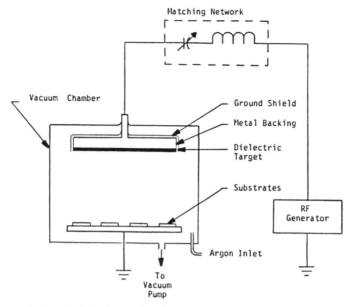

Figure 3.5: Schematic of an RF sputtering apparatus.

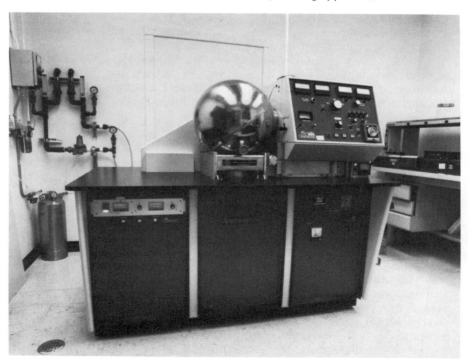

Figure 3.6: Commercial RF sputter equipment (Sputter-Sphere, Courtesy of Materials Research Corp.).

Table 3.1: Reactive Sputtering

Sputtering process in which a reactive gas is introduced alone or along with the Ar such that the sputtered particles react to form a new compound.

. . . . Examples . . . .

Target	Reactive Gas	Film
Ta	N_2	Ta_3N_2
Ta	O_2	TaO
Al	N_2	AlN
Al	O_2	Al_2O_3
Si	N_2	Si_3N_4
Si	O_2	SiO_2
Ti	O_2	TiO_2

Though the exact mechanism for the formation of the reactively sputtered films is not known, it is believed that the films are formed on the substrate surface where the energy of formation can be easily dissipated without decomposing the newly formed compound. When the reactive gas is used in high concentration, the reaction is believed to occur at the cathode and the resultant compound is then carried to the grounded substrate. Reaction in the gaseous phase is deemed unlikely because the energy of formation and kinetic energy of the atoms cannot be dissipated, resulting in spontaneous decomposition of any compound as it is formed.

Comparison of Evaporation and Sputtering Processes

The sputtering of thin films has significant advantages over vapor deposition and is rapidly supplanting vapor deposition as a commerical production process Among these advantages are:

1. Sputtered films adhere more strongly to substrates than vapor deposited films. This is attributed to the high kinetic energies with which the sputtered atoms impinge upon the substrate.

2. Sputtered films are denser and more uniform.

3. The sputtering process is more versatile. The target may be composed of alloys or composite materials in addition to pure metals. For example nickel-chromium targets of various ratios may be used to deposit resistor films of various sheet resistivities. Unlike vapor deposition, little or no fractionation occurs.

4. Both conductive (metals, alloys) and non-conductive (dielectrics, insulators) may be deposited.

5. The process may be used in a reverse mode just prior to sputtering to clean the surface of the substrates or to etch fine lines. In these cases the substrates are rendered cathodic for a brief period.

6. The rate of deposition, film thickness and uniformity of films are better controlled.

A more detailed comparision of the two processes is given in Table 3.2.

Table 3.2: **Comparison of Evaporation and Sputtering Processes**

	Vacuum Evaporation	Sputtering
Mechanism	Thermal energy	Momentum transfer
Deposition rate	Can be high (to 750,000 Å/min)	Low (20–100 Å/min) except for some metals (Cu = 10,000 Å/min)
Control of deposition	Sometimes difficult	Reproducible and easy to control
Coverage for complex shapes	Poor, line of sight	Good, but non-uniform thickness
Coverage into small blind holes	Poor, line of sight	Poor
Metal deposition	Yes	Yes
Alloy deposition	Yes (flash evaporation)	Yes
Refractory metal deposition	Yes (by e-beam)	Yes
Plastics	No	Some
Inorganic compounds (oxides, nitrides)	Generally no	Yes
Energy of deposited species	Low (0.1–0.5 eV)	High (1–>100 eV)
Adhesion to substrate	Good	Excellent

THIN FILM RESISTOR PROCESSES

Thin Film Resistors

Important considerations in selecting thin film resistors are:

- controlled and reproducible sheet resistance (ohms/sq) for a practical thickness range,
- low TCRs,
- close resistor tracking
- long term stability (low resistance drift on powering or on thermal aging).

Thin film resistors may be classified as pure metals, metal alloys, metal compounds, or cermets (combinations of ceramics and metals). Examples are given in Table 3.3

Table 3.3: Thin-Film Resistor Types

Type	Examples
Metal	Tantalum
	Chromium
	Nickel
Alloy	Nickel-chromium (nichrome)
	Cobalt-chromium
	Tantalum-tungsten
Cermet	Silicon monoxide-chromium
Metal compounds	Tantalum nitride

Though there are many resistor materials available, only three are in wide use today for thin film hybrid circuits: nichrome, tantalum nitride, and chromium-silicon oxide cermets. All employ gold terminations with a barrier metallization. Barrier metals are often required to separate the resistors from the gold conductors and prevent interdiffusion. In the case of the nichrome/nickel/gold system the nickel, deposited in thicknesses of 400-1,500 Å prevents diffusion of chromium (from the nichrome) into the gold. This diffusion causes two potential reliability problems—unstable resistors because of the change in the nickel to chromium ratio of the nichrome and wire bond degradation due to contamination of the gold bonding pads with chromium. Besides functioning as the resistor element, nichrome functions as an adhesion layer, a so-called "tie layer" to improve the adhesion of the gold to the ceramic substrate. In the tantalum nitride process, titanium is the tie layer and palladium serves as a diffusion barrier. Gold is seldom deposited directly onto ceramic because of its poor adhesion.

Both nichrome and tantalum nitride offer similar resistor capabilities: sheet resistances of 25-300 ohms/square and low TCRs (0 ± 50 ppm/C). Cermet resistors, however, provide an extended capability for sheet resistances from 1000 to several thousand ohms/square thus offering the design engineer the freedom to lay-out very high valued resistors (over 100 Kohms to megohms) without taking up too much surface area.

The Nichrome Process

Regardless whether nichrome, tantalum nitride, or cermet resistors are used, the process steps are quite similar. First, a resistor/conductor "sandwich" structure is formed on the ceramic substrate by sequentially depositing thin layers of the resistor material, the barrier metal, and the top conductor metal. Then through precision photolithography involving a series of photoexposures of a photoresist coating and selective etching steps, an intricate pattern of conductor lines and spacings (from 1-10 mils) and thin-film resistors (from 5-20 mils wide) is formed. The sequential process steps are shown in Figure 3.7. The nichrome/nickel/gold sandwich structure may be deposited either by vacuum evaporation or by sputtering (discussed above). It is desirable to deposit the metals sequentially in the

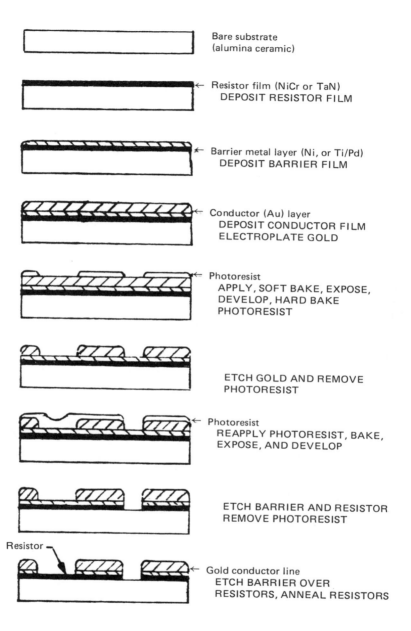

Bare substrate
(alumina ceramic)

← Resistor film (NiCr or TaN)
DEPOSIT RESISTOR FILM

← Barrier metal layer (Ni, or Ti/Pd)
DEPOSIT BARRIER FILM

← Conductor (Au) layer
DEPOSIT CONDUCTOR FILM
ELECTROPLATE GOLD

← Photoresist
APPLY, SOFT BAKE, EXPOSE,
DEVELOP, HARD BAKE
PHOTORESIST

ETCH GOLD AND REMOVE
PHOTORESIST

← Photoresist
REAPPLY PHOTORESIST, BAKE,
EXPOSE, AND DEVELOP

ETCH BARRIER AND RESISTOR
REMOVE PHOTORESIST

Resistor —

← Gold conductor line
ETCH BARRIER OVER
RESISTORS, ANNEAL RESISTORS

Figure 3.7: Process sequence for deposition and photodelineation of thin-film conductor/resistor network (not to scale).

same vacuum pump down. If the vacuum is interrupted or if the parts are exposed to air between depositions, oxidation or contamination of the films can occur, degrading the adhesion of subsequent layers and the stability of the resistors. It should be noted that the two key processes for depositing thin-film gold, vacuum evaporation and sputtering, result in very thin layers 3,000 Å to 10,000 Å, insufficient to provide enough electrical conductivity or enough metal thickness for subsequent wire bonding. To increase the gold thickness the thin films are electroplated with an additional 50-100 microinches of gold.

An alternative nichrome process used by some manufacturers involves depositing the nichrome layer and photoetching it to form the resistors prior to depositing the gold layer. After the resistors are formed, titanium, palladium, and gold layers are sequentially sputtered over the entire surface, the gold is electroplated to increase its thickness and conductivity, then photoetched to form the termination pads and circuit lines. The resistors are then annealed and laser trimmed, if necessary. This process is reported to offer very close tolerance, precision resistors necessary for high frequency circuits. Although nichrome resistors are more susceptible to electrolytic and chemical corrosion and oxidation than are tantalum nitride resistors, they are generally not passivated if the circuit is hermetically sealed in nitrogen. For commercial non-hermetic applications passivation has been accomplished by selectively coating the nichrome resistors with organic coatings or with inorganic sputtered coatings such as silicon monoxide-dioxide. The organic coatings must be of high purity, thermally and chemically stable, and possess stable electrical insulative properties. Examples include the polyimides, silicones, and polyxylylenes (Parylenes).

Nichromes used for thin film resistors are alloys of nickel and chromium. Various ratios of these metals may be used to achieve different sheet resistance values. Two commonly used compositions are 50% nickel, 50% chromium with 1-5% silicon and 80% nickel, 20% chromium. Nichrome resistors are difficult to deposit consistently in the same thickness and sheet resistance when resistance heating is used. The composition of the deposited film differs from that of the starting material because of differences in the vaporization temperatures and vapor pressures of the nickel and chromium (Figure 3.8). Further, the composition of the starting material in the crucible or boat changes continuously during evaporation, becoming richer in nickel because of its lower vapor pressure. Studies performed by Alderson and Ashworth[5] showed that the composition of the deposited nichrome film approximated that of the source material (80% Ni/20% Cr) only at temperatures above 1600°C (Figure 3.9). But deposition at these high temperatures resulted in such a rapid evaporation rate that the quality of the film was degraded. Now, films that approximate the composition of the starting material can be deposited at lower temperatures either by flash evaporation or sputtering. By these processes little fractionation of the two metals occurs and reproducible, stable resistors are produced. In flash evaporation, nickel-chromium powder is slowly dropped into a preheated boat, constructed of tungsten or other refractory metal. The powder is vibrated down an incline into the boat. The nichrome vaporizes immediately on contact with the boat. The vacuum in the chamber is maintained at 10^{-6} to 10^{-7} Torr and the temperature of the boat is kept between 1000°C and 1100°C. For best adhesion of the film to the substrates, the latter are preheated at 150-300°C.

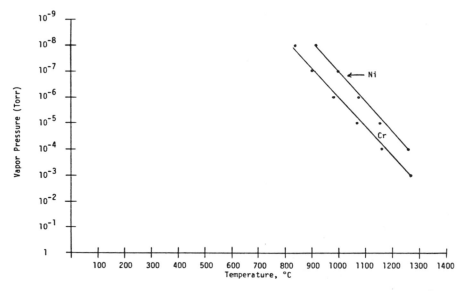

Figure 3.8: Vapor pressure–temperature curves for nickel and chromium.[4]

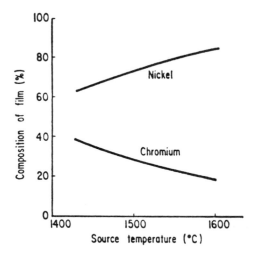

Figure 3.9: Composition of vacuum deposited nichrome film vs source temperature starting with 80% Ni, 20% Cr mixture.[5]

Characteristics of Nichrome Resistors

Typical properties of nichrome resistors are given in Table 3.4. As discussed in the chapter on Substrates, thin film resistor characteristics are highly dependent on the surface characteristics of the substrate onto which they are deposited, the smoother the surface finish the more stable the resistor values. However many other factors contribute to stable resistors, chief among which are the annealing, stabilization bake, and trimming conditions.

Table 3.4: Characteristics of Nichrome Resistors

Sheet resistance	25–300 ohms/sq, 100–200 ohms/sq (typical)
Sheet resistance tolerance	± 10% of nominal value
TCR	0 ± 50 ppm/$^\circ$C, 0 ± 25 ppm/$^\circ$C (with special anneal)
TCR tracking*	2 ppm
Resistance drift	<2,000 ppm after 1,000 hr @ 150°C
	<1,000 ppm with special anneal
	<200 ppm, sputtered films with 350°C anneal
Ratio tracking	5 ppm
Resistor tolerance after anneal and laser trim	±0.1%
Noise (100 Hz to 1 MHz)	–35 db (max)

*–55° to 125°C.

Nichrome resistors are annealed in air at temperatures from 225°C to 350°C for several hours. Annealing in air involves a chemical reaction in which the more reactive chromium is oxidized forming a passivating layer that slows further oxidation and stabilizes the resistor values. Generally, during annealing, sheet resistances change (increase) about 10-20%. The long term stability, resistance drift after aging 1000 hrs at 125-150°C, depends largely on the annealing schedule used; the higher temperature, longer anneal schedules provide more stable resistors, lower TCRs, and closer resistance tracking (Table 3.5). Some sputtered nichrome resistors annealed at 350°C for 4-5 hrs have drifted less than 100 ppm after the 1,000 hrs aging at 150°C and exhibited TCRs of 0 ± 3 ppm/°C.

Though nichrome resistors are among the most precise, thermally stable resistors, they are also among the most susceptible to chemical and electrolytic corrosion. Chloride ions transferred from a fingerprint, combined with moisture, will corrode and etch the nichrome film. Nichrome resistors as part of integrated circuits that have been overcoated with silicon dioxide have been reported to disappear. This has been attributed to acid etching in which phosphorus contained in the silicon dioxide passivation layer interacts with moisture to form phosphorous acid. Some organic coatings will also interact with nichrome resulting in resistance changes. Any coating used, be it organic or inorganic, must therefore be evaluated on nichrome and resistance values measured as a function of time and temperature. Generally, for military and space applications, nichrome resistors are left uncoated because the hybrid circuits containing them are hermetically sealed with nitrogen in metal or ceramic packages.

Table 3.5: Nickel-Chromium Resistor Stability and TCR Data
(Courtesy Rockwell Intl.)

Deposition Method	Sheet Resistance of NiCr (ohms/sq) as Deposited	Annealing Conditions . . (Air) . . (°C)	(min)	Resistance Change (ppm) . . . . . . After Aging . . . . . . 50 hr, N$_2$ 150°C	168 hr, N$_2$ 125°C	500 hr, N$_2$ 125°C	Final TCR
Flash evaporated without Ni barrier	165	225	120	2,000	500	700	–20
Flash evaporated with 400 Å Ni barrier	165	225	120	1,700	350	630	–16
Flash evaporated with 800 Å Ni barrier	165	225	120	1,400	280	400	–12
Flash evaporated with 800 Å Ni barrier	165	300	120	1,310	260	300	–10
Flash evaporated with 800 Å Ni barrier	165	350	120	460	70	110	+34
Sputtered with 800 Å Ni barrier	125	350	235*	71	41	–	+3
Sputtered with 800 Å Ni barrier	105	350	315**	96	–96	+80	–3

*In 4 steps.
**in 6 steps.

A further limitation of nichrome resistors is that the practical upper limit for sheet resistance is low, about 200-300 ohms per square. Higher sheet resistances are possible by depositing ultrathin layers (much less than 100 Å) since sheet resistance is inversely proportional to thickness. However, it is not practical to deposit such extremely thin layers because discontinuous, imperfect films are formed, giving rise to opens and unstable resistors. Figure 3.10 shows the variation of sheet resistance with thickness for two sputtered nichrome compositions. These plots show that a sheet resistance of 100 ohms/sq is a good compromise. Although using high sheet resistances would reduce the size of most circuits, the slope of the curve is so steep in this range that reproducibility and resistor stability would be jeopardized.[6] High-value resistors using low sheet resistance materials such as nichrome or tantalum nitride can only be designed by using a large number of squares having small widths and serpentining the lines to conserve on space. Even so, the design of a 100 Kohm resistor with a 200 ohm/sq nichrome takes up considerable space (Figure 3.11). Fortunately, for many thin-film circuits, the number of high valued resistors is small compared to the low and intermediate values and these can be accomodated by using chip resistors.

The Tantalum Nitride Process

The processes for fabricating tantalum nitride resistor/gold conductor

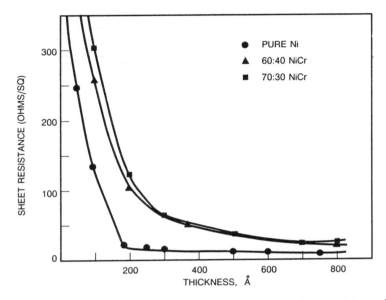

SHEET RESISTANCE (OHMS/SQ)

THICKNESS, Å

● PURE Ni
▲ 60:40 NiCr
■ 70:30 NiCr

Figure 3.10: Nichrome sheet resistance as a function of film thickness.[6]

circuit patterns on ceramic are generally similar to those for fabricating nichrome resistors. In both cases photolithography involving selective etching of a multilayered metal structure is used. Process differences include: 1. tantalum nitride (TaN) is deposited by reactive sputtering of tantalum in a partial nitrogen atmosphere instead of by direct sputtering or flash evaporation, 2. titanium and palladium are used as a tie layer and barrier layer, respectively, between the TaN and the gold; nickel is used as the barrier layer with nichrome resistors, 3. the titanium and TaN are etched with hydrofluoric acid-nitric acid solution or may be sputter-etched for very fine line definition.

The partial pressure of nitrogen gas introduced during reactive sputtering affects both the sheet resistivity and the TCR of the deposited resistors (Figure 3.12) since tantalum nitride goes through several crystallographic forms as the concentration of nitrogen is increased. Besides this batch process for depositing TaN on alumina substrates, TaN resistors may be obtained in chip form. Films are sputtered onto highly polished silicon, quartz, or sapphire wafers then batch processed with aluminum/nickel barrier terminations and sputter-etched to yield fine line resistors of high resistor values (to 12 Megohms).[7] Cross-sections of tantalum nitride chip resistors showing two constructions are given in Figure 3.13.

Characteristics of Tantalum Nitride Resistors

Tantalum nitride resistors are similar to nichrome resistors in their electrical properties including sheet resistance range, TCRs, resistor tracking, and long term resistance drift after elevated temperature aging. Some electrical characteristics are given in Table 3.6.

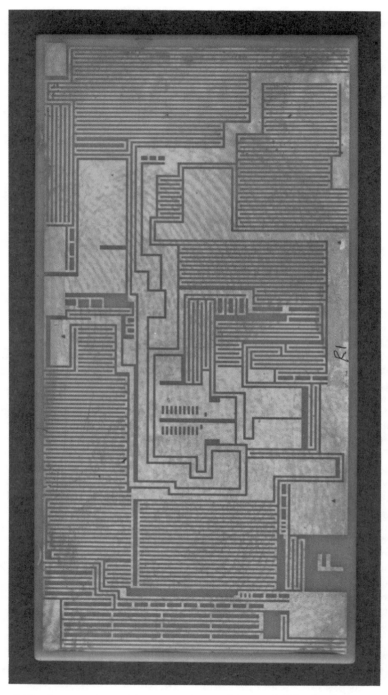

Figure 3.11: Large value nichrome resistors (about 100K ohms) (note large surface areas required).

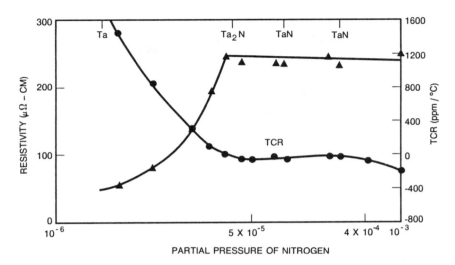

Figure 3.12: Effect of N_2 partial pressure in the sputtering chamber on the resistivity and TCR of Ta films.[2]

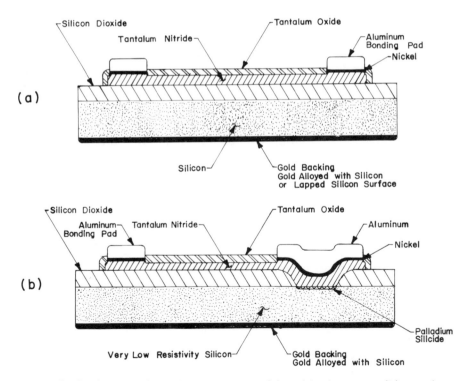

Figure 3.13: Cross-section of a top contact (a) and back contact (b) tantalum nitride resistor chip (Courtesy National Micronetics, Inc.).

Table 3.6: Properties of TaN Resistors

Sheet resistance	20–150 ohms/sq, 100 ohms/sq (typical)
Sheet resistance tolerance	± 10% of nominal value
TCR	–75 ± 50 ppm/°C (typical), 0 ± 25 ppm/°C with vacuum anneal
TCR tracking*	<2 ppm
Resistance drift (1,000 hr at 150°C in air)	<1,000 ppm
Ratio tracking	5 ppm
Resistor tolerance after anneal and laser trim	± 0.10% standard, ± 0.03% bridge-trim
Noise (100 Hz to 1 MHz)	<–40 dB

*–55° to 125°C.

Tantalum nitride resistors are more rugged and more chemically and thermally resistant than nichrome resistors. A passivating tantalum pentoxide layer (Ta_2O_5) is formed by annealing the resistors in air or in oxygen or by introducing controlled amounts of oxygen during the reactive sputtering process. Annealing is performed at 450°C or higher. This inherent oxide is quite resistant to moisture and other hostile environments.

Cermet Thin Film Resistors

Cermet, a word coined from the words ceramic and metal, is a composition of metal oxides and metals. They may be thick-film pastes (see Chapter 4) or evaporated thin films and are useful in forming resistors having high resistance values. The most widely used thin-film cermet resistors are compositions of silicon monoxide and chromium (SiO-Cr) and are used primarily because of their high sheet-resistances and their ability to form high-valued resistors in a minimum of substrate area. By varying the ratio of silicon monoxide to chromium, a wide range of sheet-resistances is possible. Silicon oxide is inherently an insulator, so as the amount of silicon monoxide in the composition is increased, the resistivity increases. Sheet-resistances of several hundred ohms/sq to tens-of-thousands of ohms/sq may be produced. However, it is difficult to control the stability and reproducibility of the very high resistor values. The most practical range is 1,000 to 3,000 ohms/sq. The relationship between composition and resistivity for cermet films of 900 to 1750 Å thickness is shown in Figure 3.14.[8] Thus, for the thicknesses indicated, compositions containing about 60% SiO are required to produce resistor films of 1,000 ohms/sq.

Cermet films are best deposited by flash evaporation or by sputtering. In flash evaporation, the cermet powder is dropped onto a tantalum boat maintained at a temperature higher than the evaporation temperature of either of the cermet constituents. The process for fabricating the cermet resistor-conductor pattern for a hybrid circuit substrate is similar to that for nichrome or tantalum nitride resistors. In one sequence, a "sandwich" or layered structure of SiO-Cr, palladium, and gold is first formed. The palladium serves as a barrier layer. Photoresist is then applied, exposed and devel-

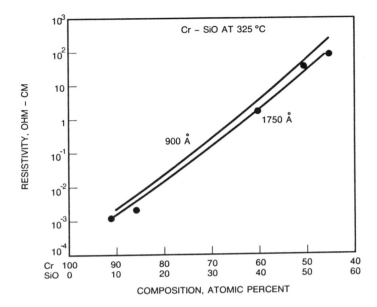

Figure 3.14: Chromium-silicon monoxide resistivity versus composition.[8]

oped to define the conductor pattern; the unwanted gold is chemically etched using the standard potassium iodide-iodine solution. A composite photoresist pattern is then applied. The palladium is etched with an aqua regia solution then the cermet with a nitric acid solution. In the last step, a photoresist is applied to mask and protect all areas except those that are to become resistors. Again, aqua regia is used to selectively remove the palladium leaving the exposed cermet resistors. Etching the SiO-Cr cermet has always been somewhat difficult, especially for compositions high in SiO. Two etchants that have been used are postassium ferricyanide/sodium hydroxide and ceric ammonium nitrate/nitric acid solutions. Sputter-etching as a dry-etching process has also been found to be effective.

Stabilizing cermet resistors by annealing requires much higher temperatures than for nichrome. Typical annealing conditions are 450-500°C for 2-4 hours. At these temperatures a nickel barrier was found to be ineffective in preventing interdiffusion of the chromium and gold, but palladium at about 3,000 Å thick was effective.

The TCRs for cermet resistors are generally negative, becoming more negative as the oxide content increases. For a cermet composition of 70% Cr/30% SiO, TCRs of −60 to −40 ppm/°C have been reported.

PHOTORESIST MATERIALS AND PROCESSES

Photoresists are organic compositions consisting of light-sensitive polymers or polymer precursors dissolved in one or more organic solvents.

They are of two types: those that on exposure to light are further polymer-
ized or crosslinked forming a hardened coating which is resistant to
etching solutions (negative types) and those that on exposure to light are
decomposed, break down and can be dissolved (positive types). In the
latter case, the unexposed portions become hardened coatings resistant
to etching solutions. Ultraviolet light in the 2000-4000 Å wavelength, is
generally used. The photoresist coating is applied over the entire surface
of the substrate, baked at a low temperature to remove solvent (soft
baked), then exposed to ultraviolet light through a separate mask which
may consist of either a Mylar film or a glass plate having opaque and
transparent areas corresponding to the image to be produced. This mask
is often referred to as the artwork or the photo-tool; it is based on either a
Mylar transparent film or glass. After exposure to light, the photoresist is
developed. It is immersed in a chemical solution, called the developer,
which dissolves the unexposed portions of the photoresist (in the case of a
negative resist) or the exposed portions in the case of a positive resist. The
remaining photoresist pattern may then be hard baked to render it more
resistant to the subsequently used etching chemicals. In using a negative
photoresist, the mask must have a negative image of the pattern to be
produced. Figure 3.15 illustrates the steps in etching a thin film of gold on
an alumina ceramic substrate using a negative photoresist and a negative
image mask. The pattern produced is the opposite image of the mask used.
In a positive resist the areas exposed to ultraviolet light degrade or
decompose and are then readily dissolved and removed. The remaining
areas become resistant to the etching solutions; hence a positive image
on a mask results in the same pattern on the substrate. The steps in etching
metallization using a positive photoresist are shown in Figure 3.16. In all,
four combinations are possible depending on whether a positive or negative
resist is used with a positive or negative mask. These combinations are
given in Figure 3.17.

Photoresists are essential not only in the fabrication of microelectronic
devices and circuits but also in fabricating thin film hybrid microcircuits.
The process by which photoresists are used to etch intricate and precise
patterns in metal or dielectrics known as photolithography or photoetching,
was a key factor in the rapid development of microelectronics in the 1950s
and 1960s. Photoresists allow selected areas of a surface to be removed
leaving other areas (protected by the photoresist) as defined patterns of
metal conductors, resistors, dielectrics, or inorganic passivation layers-all
essential in the fabrication of monolithic integrated circuits and hybrid
microcircuits.

Chemistry of Negative Photoresists

The chemical reactions that occur during the processing of negative
photoresists are simpler and easier to understand than those for positive
resists. The photoresist compositions now used in electronics evolved
from technology that already existed in the printing industry in the 1940s
and early 1950s. During this period hundreds of patents were issued for
materials for printing and copying applications. A key patent was that of

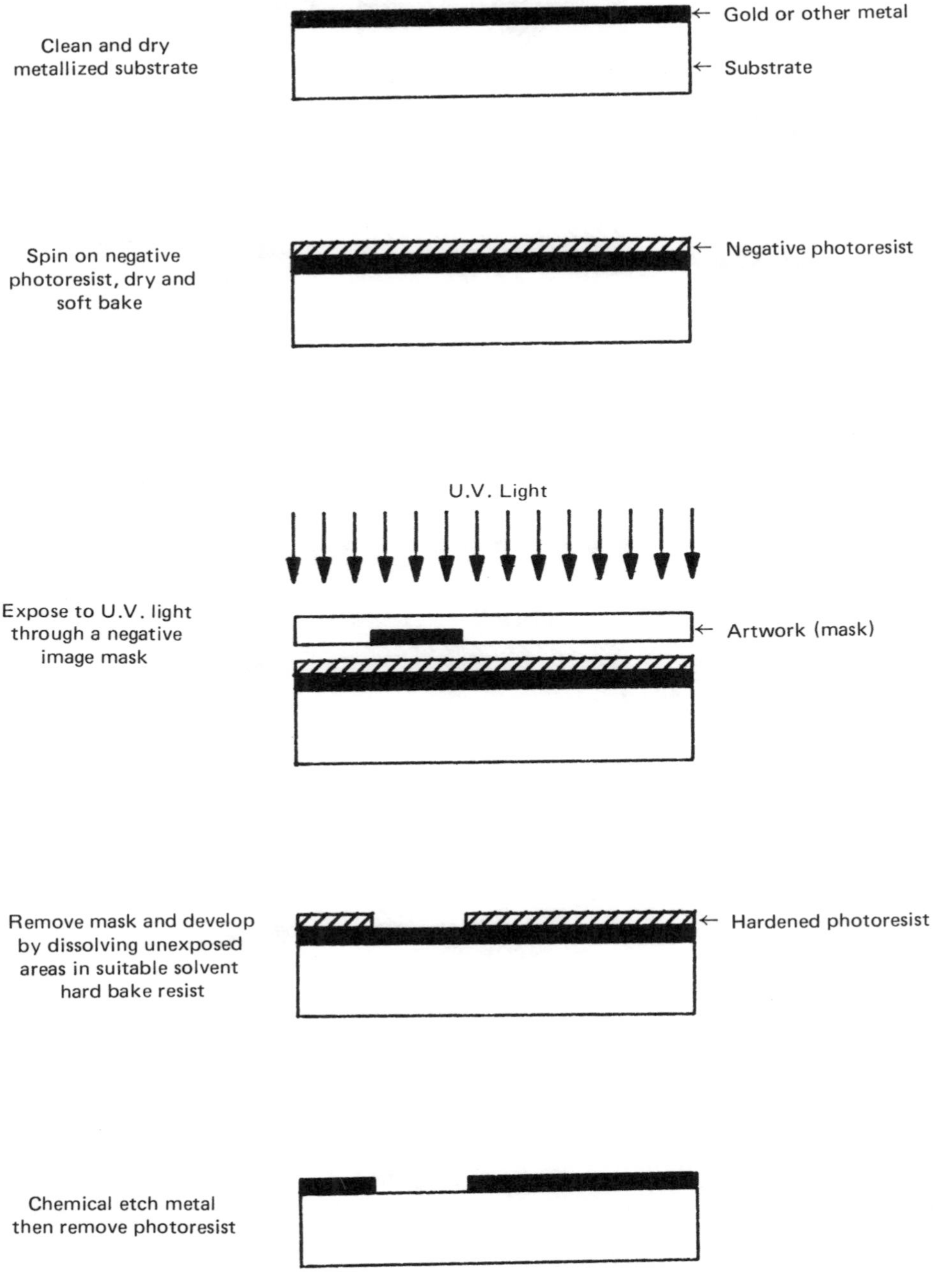

Clean and dry
metallized substrate

← Gold or other metal

← Substrate

Spin on negative
photoresist, dry and
soft bake

← Negative photoresist

U.V. Light

Expose to U.V. light
through a negative
image mask

← Artwork (mask)

Remove mask and develop
by dissolving unexposed
areas in suitable solvent
hard bake resist

← Hardened photoresist

Chemical etch metal
then remove photoresist

Figure 3.15: Steps for photolithography using negative photoresist (side view).

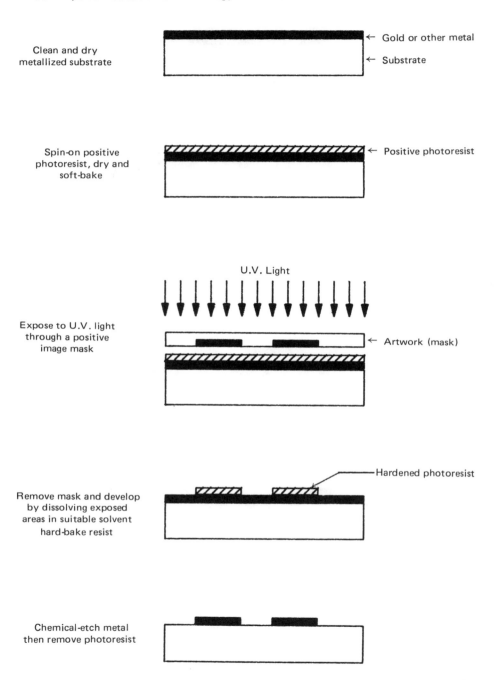

Figure 3.16: Steps for photolithography using positive photoresist.

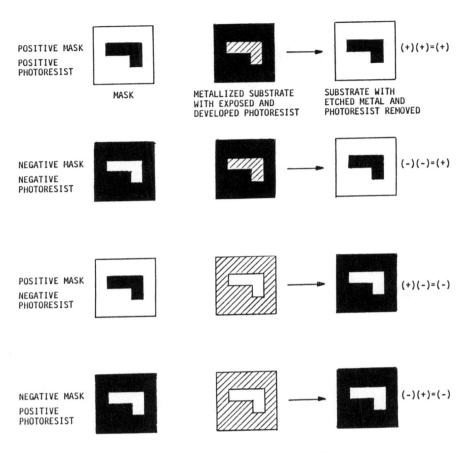

Figure 3.17: Mask/photoresist combinations.

Minsk issued to Eastman Kodak in 1954[9] that described a photosensitive composition for fabricating printing plates. This patent provided the basic technology for the subsequent development of negative photoresists to be used in fabricating electronic devices and circuits. Minsk described a cinnamic ester of polyvinyl alcohol (Figure 3.18). The double bonds of the cinnamic acid portion are quite sensitive to ultraviolet light, opening up into diradicals. These diradicals being unstable species quickly join with other free radicals forming new carbon to carbon bonds and tying together the linear molecules into highly crosslinked macromolecules. In general, negative photoresists are based on compounds having ethylenic or double bonds which decouple to diradicals on being energized with ultraviolet light. These free radicals, being unstable, quickly join head-to-tail forming long chains or cross-linked polymers which are insoluble and chemically resistant compared to the original unexposed coating (Figure 3.19). The exact structure of these cross-linked molecules is not known but it is theorized that coupling occurs through the formation of truxillic acid structures as in Figure 3.20. The solubility of the exposed portions is low and chemical resistance high compared to the unexposed portions. This allows the unexposed portions to be dissolved and removed in an organic solvent (developer). This is the basic chemistry for KPR (Kodak Photoresist) and for many other negative photoresists. Numerous variations of the cinnamic acid ester compound based on chemical variations of the parent molecule have been made. Formulations can be optimized by adding other ingredients such as photosensitizers, solvents, flow-control agents, and stabilizers.[10]

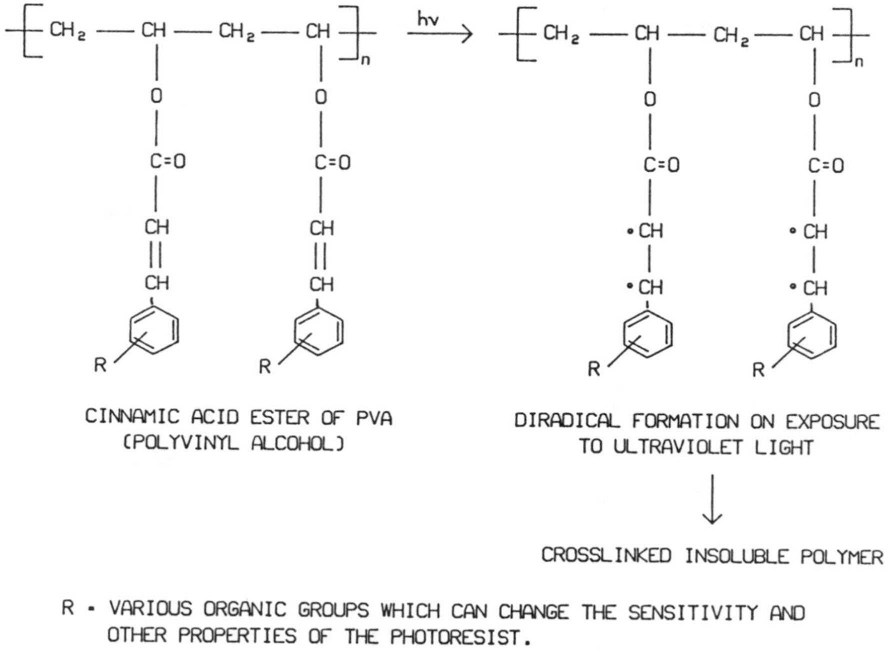

CINNAMIC ACID ESTER OF PVA
(POLYVINYL ALCOHOL)

DIRADICAL FORMATION ON EXPOSURE
TO ULTRAVIOLET LIGHT

CROSSLINKED INSOLUBLE POLYMER

R • VARIOUS ORGANIC GROUPS WHICH CAN CHANGE THE SENSITIVITY AND OTHER PROPERTIES OF THE PHOTORESIST.

Figure 3.18: Negative photoresist chemistry based on cinnamic ester of polyvinyl alcohol.

Chemistry based on decoupling of a double bond with ultraviolet energy to form unstable free radicals which then join head-to-tail to form long chains or cross-linked polymers.

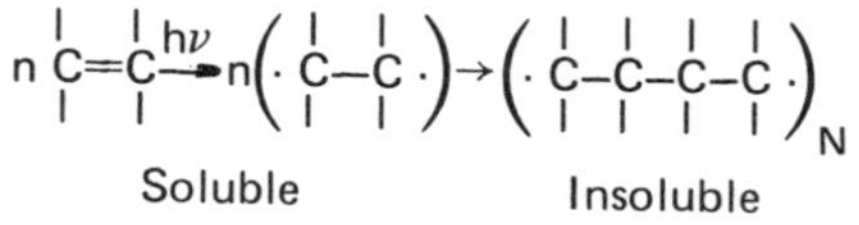

Soluble   Insoluble

Figure 3.19: General mechanism for negative photoresists.

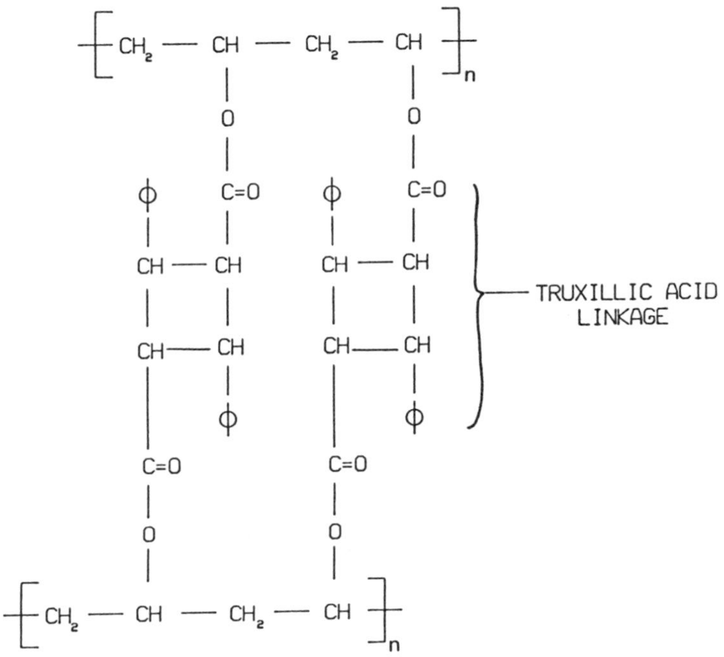

Figure 3.20: Mechanism for cross-linking (negative photoresists).

Chemistry of Positive Photoresists

Positive photoresists are based on chemistry that was initially developed in Germany for the azo-dye, printing, and copying industries. Prior and during World War II the Germans had a very advanced industry that was based on the azo dyes for the coloring of textiles and other materials. The chemistry of positive photoresists is based on two reactions of diazonium salts or diazides. First, diazonium salts or diazides react quickly and almost quantitatively with a coupling agent (a phenolic compound) under alkaline conditions to form various colored insoluble azo dyes, the color and solubility depending on the structure of both the diazide and the coupling compound (Figure 3.21). The reaction can be inhibited or prevented by formulating the two ingredients in a buffered slightly acid solution. The developer provides the alkalinity to cause the coupling reaction to occur. Secondly, diazonium compounds are unstable when exposed to ultraviolet light; the diazo-group decomposes and releases nitrogen. Once exposed to ultraviolet light, the resulting compound is no longer capable of coupling and forming the insoluble dye.

Photolithography that employs positive photoresists is based on both these reactions (Figure 3.22). Most commercially available positive resists contain naphthoquinone diazide. The group designated R may be a polymer group optimized for adhesion, solubility, or other characteristics. Exposure

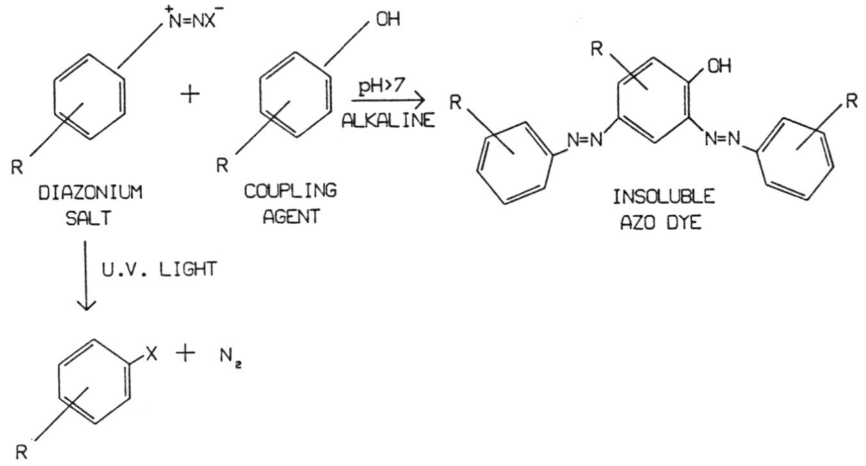

Figure 3.21: Diazo reactions involved in positive photoresists.

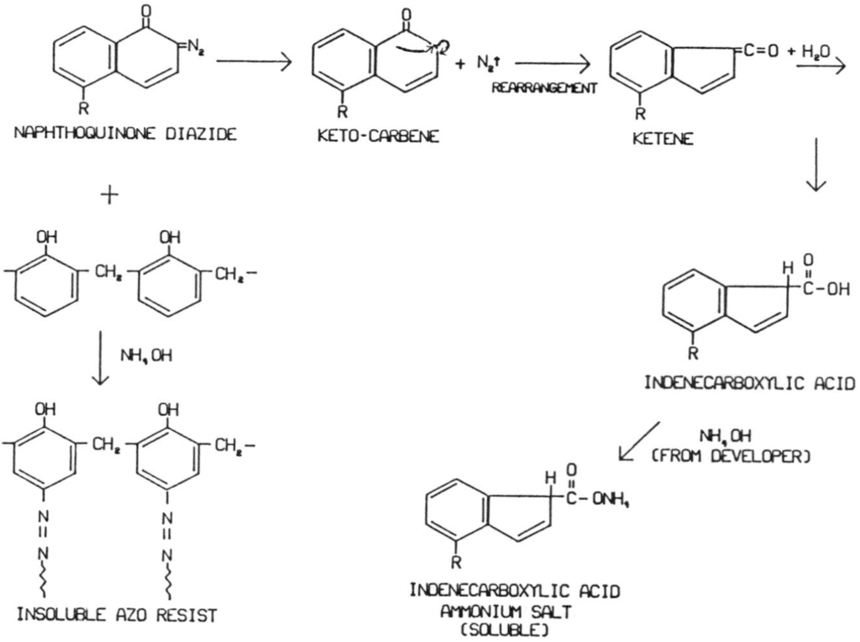

Figure 3.22: Chemical reactions for positive photoresist.

of the diazide to ultraviolet light results in a rearrangement and evolution of nitrogen. The resulting ketene reacts with some moisture which is invariably present and converts to an indene carboxylic acid. On developing the image with an alkaline solution, the carboxylic acid dissolves. The alkaline conditions which dissolve the exposed portions of the resist also provide the alkalinity to cause coupling between the unexposed azide with itself or with a phenolic polymer such as a Novolac that had previously been formulated with the diazide. This coupling reaction results in crosslinking of the chains and decreased solubility.

Processing

The key steps in the processing of photoresists are:

- Application to the substrate
- Pre-baking
- Alignment and exposure
- Development
- Post-baking
- Removal or stripping.

Liquid photoresists are generally applied to substrates by spinning, a process wherein the photoresist coating is applied to the center of a substrate and the substrate rotated on a circular table. Substrates are held down by a vacuum chuck. The liquid photoresist is first applied to a stationary substrate and allowed to spread. The spreading may be enhanced by spinning the substrate at about 500 rpm for 3 to 4 seconds. The spin speed is then increased to 3,500 to 5,000 rpm for about 30 seconds. This forces the resist outward by centrifugal force, coating the substrate evenly. The thickness of the deposited resist is largely a function of its viscosity and speed at which it is spun. Figure 3.23 shows the relationship of viscosity of several positive resists to thickness while maintaining a constant spin-speed while Figure 3.24 shows the relationship of spin-speed to thickness for several positive photoresists. Some thickness variation occurs at the outer edges of the substrate since there is always a build up of material in that area. However, several new "striation-free" resists are now available that reduce edge build-up. Besides viscosity and spin speed, other factors that affect thickness and uniformity are acceleration, pre-bake temperature and time, amount of resist dispensed, and temperature of the ambient.[11]

To obtain optimum results, especially in fabricating very fine geometries, as for microwave or high-density hybrid circuits, many process details must be followed meticulously. The spin, bake, and exposure parameters must be experimentally determined and followed precisely to obtain reproducible fine geometries. Any blemishes, pinholes or other imperfections in the hardened resist become sites for the penetration of etchants, resulting in opens or shorts in the circuitry. Two main causes of imperfect

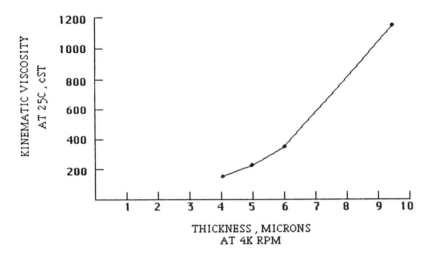

Figure 3.23: Photoresist viscosity vs thickness.

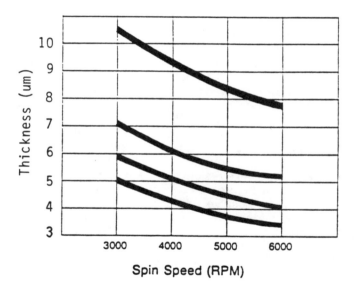

Figure 3.24: Relationship of spin speed to thickness for Microposit S1600 series positive resists (Courtesy Shipley).

resists are air-borne particulates that deposit on the coating during spinning, alignment, or exposure and particulate contaminants contained in the liquid resist. These particle sources can be minimized or eliminated completely by processing the photoresists in a laminar flow station of Class 100 or better and to use liquid resist that has been filtered through a

Millipore superfine filter. Most photoresists can be purchased pre-filtered to remove 0.45 micron or 0.2 micron particles or larger, depending on the filter. If the resist is transferred from its original container, for example, to an in-line coater further filtering at the point of dispensing may be necessary because particles may be picked up during the transfer.

Adequate adhesion of the photoresist to the substrate is another important factor in obtaining high yields. Substrates should be free of ionic salts, organic residues, moisture, and particles, all of which degrade adhesion. Substrates should be cleaned and dried. Special cleaning methods such as plasma cleaning and the use of primers are sometimes beneficial in improving adhesion.

Soft baking, also called pre-baking, allows solvents from the photoresist to escape prior to complete polymerization or hardening. Soft baking involves heating the coated substrates to 80-100°C in a forced air convection oven for about 30 minutes. This is important in achieving line width control. Hard baking, also called post-baking, at 110-115°C for 30 minutes, after the resist has been developed, further improves adhesion and increases its hardness and resistance to the etchant solutions.

ETCHING MATERIALS AND PROCESSES

Etching methods for thin films are classified as either "wet" or "dry" types. Wet methods use chemical solutions, usually aqueous solutions of acids or alkalis. These react with thin films forming salts that can be dissolved and washed away. Dry etching involves either molecular removal of the film by reverse sputtering, ion etching, or by a gas phase chemical reaction in which a gas is activated in a plasma in contact with the film. The activated gas reacts with metal film forming a compound that is easily volatilized. This process, referred to as plasma etching, is rapidly supplanting wet chemical etching for many of the more complex high density circuits. Some reasons for the increased popularity of plasma or dry etching are:

- The use of highly corrosive and dangerous chemicals such as acids and alkalis are avoided.
- Disposal and safety problems associated with chemical solutions are avoided.
- Chemical and ionic contamination of surfaces are avoided.
- Etching rate is better controlled providing finer definition of conductor lines, resistor patterns, and interconnect vias.
- Etching is anisotropic, avoiding undercutting.

Chemical Etching of Gold Films

Chemical etchants that selectively remove a conductor film or a barrier layer in the presence of a resistor film are essential in the manufacture of hybrid microcircuits. Aqua regia (a 3 to 1 mixture of concentrated nitric and hydrochloric acids) is the age-old solution for dissolving gold, but not widely used today for thin film processing. Besides the danger in handling this highly corrosive acid mixture, the photoresists that are used in defining the conductor pattern are not resistant to it. Further, the acid is not selective for gold; it will attack nickel, nichrome, and many other metals.

Solutions that will selectively etch gold in the presence of nichrome or tantalum nitride consist of inhibited aqueous solutions of potassium iodide (KI) and iodine (I_2). Uninhibited KI/I_2 solutions are not useful because they etch nichrome as well as gold. By adding inhibitors such as dibasic ammonium phosphate [$(NH_4)_2HPO_4$] or dibasic potassium phosphate/ phosphoric acid (K_2HPO_4/H_3PO_4), the etching of nichrome and nickel can be suppressed or even avoided. Etchant solutions may be used at room temperature or at elevated temperatures, about 90°C. The optimum temperature may be determined empirically based on the rate of etching and the line definition desired. Reaction equations for the etching of gold are as follows:

$$KI + I_2 \rightarrow KI_3 + KI \text{ (excess)}$$
$$3KI_3 + 2Au \rightarrow 2KAuI_4 + KI$$

To completely dissolve the iodine, a two to six molar excess of KI should be employed.

Chemical Etching of Nickel and Nickel-Chromium Films

Etchants that react with and dissolve nickel-chromium (nichrome) without attacking the gold consist of aqueous solutions of ceric sulfate, ammonium nitrate, and nitric acid. The mechanism generally involves oxidation of the nickel and chromium to salts that are soluble. These etchants have no effect on the gold. To remove the nickel barrier without etching the gold and with minimal etching of the underlying nichrome, a ferric chloride solution can be used. Immersion times must be established empirically to avoid excess removal of nichrome.

Etching is a semi-empirical process at best. In each case optimum conditions (exposure times, temperature, concentration of solution, and degree of agitation) must be experimentally established. Visual monitoring for end points and measurements of resistor values, conductor line widths and spacings, and thicknesses must be made before a process can be implemented in a production line.

Dry Etching

The removal of metal, dielectric, or semiconductor thin films by methods other than dissolution in chemical etchant solutions (wet chemical method) is referred to as dry etching. Dry etching, because of the extremely fine geometries that can be produced (in the micron and even submicron range), is widely used in the manufacture of devices and integrated circuits and is also finding increasing interest and more applications in hybrid circuit fabrication. Dry etching has many significant benefits over chemical etching, chief among which are:

- Fine line geometries, less than 0.1 mil and down into the micron range can be produced because the process is anisotropic, thus avoiding undercutting (Figure 3.25).

- Ionic contaminants can be avoided thus preventing chemical corrosion.

- The process is safer—no possibility for chemical spills and better control of toxic materials.

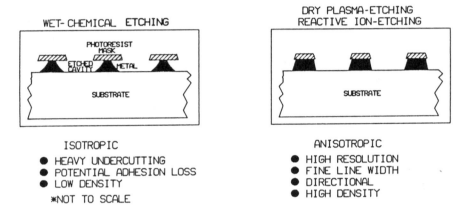

Figure 3.25: Comparison of wet versus dry etching.

Although many dry etching variations exist, there are basically two processes that are widely used in microelectronic fabrication: sputter-etching (a physical process) and reactive plasma etching (a physical-chemical process). In both cases a gaseous plasma is employed.

The purely physical process, sputter-etching, is the reverse of sputter deposition. The parts to be etched, appropriately masked with photoresist or a mechanical mask in the areas to be preserved, are attached to the cathode and bombarded with argon ions produced from the plasma as in d.c. or r.f. sputtering.[12] The key advantage of sputter-etching over chemical etching is that it is anisotropic, that is, the rate of vertical etching is faster than lateral etching. Thus dry etching is uni-directional and avoids undercutting. For this reason it is used extensively in device and integrated circuit fabrication where very fine line circuits (less than 0.1 mil) are required. Sputter-etching, however, is not selective. It does not discriminate between the material to be removed and the resist. Thus the resist must be applied in sufficient thickness so that a protective layer will remain and protect the underlying surface after the desired etching process has been completed. Sputter-etching is a slower process than reactive plasma etching and much slower than wet chemical etching, the etch rate often being in the low Angstrom per minute.

A second variant of dry etching is reactive plasma etching, a physical-chemical process. Here a chemically reactive gas (free radicals, ions) react with the surface to be removed to form a volatile gaseous compound, one that is easily pumped away from the surface. Reactive plasma etching is selective, preferentially reacting with the film to be removed but not with the mask or resist material. Depending on the plasma conditions employed, plasma etching may be omni-directional (isotropic) to directional (aniso-tropic).[13] Compared to sputter-etching, chemical plasma etching is a more rapid process. Typically, etch rates are several thousand Angstroms per minute.

The greatest amount of work in reactive plasma etching has been done with aluminum, silicon, and silicon oxide in connection with wafer process-ing of integrated circuits. Detailed studies have been conducted on hun-dreds of reactive gases, gas mixtures, and plasma conditions. Generally, chlorine and chlorinated compounds will etch aluminum by forming alumi-num trichloride, $AlCl_3$, which is volatile. Carbon tetracholoride (CCl_4), CCl_4/Cl_2, BCl_3, BCl_3/Cl_2, and $SiCl_4$, in a helium or argon plasma have been success-fully used. Etch rates of 1,000-2,000 A/min are typical. Carbon tetrachloride used alone produced especially low etch rates and inhibited reactions because of the build-up of carbonaceous deposits formed through poly-merization of carbon free radicals. The addition of oxygen to the mixture increased the etch rate, presumably by oxidizing the carbon free radicals to CO_2 before they could combine with themselves to form polymers. Also adding chlorine to the CCl_4 or BCl_3 mixtures increased the etch rate to as high as 5 micrometers/min.[14]

Fluorine and fluorinated compounds (Freons, CF_4, fluorocarbons) have not been found useful in etching aluminum because the resultant aluminum trifluoride is not volatile. The fluorinated compounds are, how-ever, very effective in etching silicon, silicon oxides, tantalum nitride, nickel, and nickel-chromium films.

THIN FILM MICROBRIDGE CROSSOVER CIRCUITS

Thin film hybrid microcircuits consist essentially of a single layer of conductor and, if applicable, single resistor layers photolithographically defined on a ceramic substrate. Numerous attempts have been made to develop a multilayer thin-film interconnect substrate, the counterpart to multilayer thick film, but these have not met with success largely because of the high probability of pinholes that exist when vacuum depositing or sputtering a thin dielectric film over a large area. However, a unique process that approaches a two conductor layer circuit is the air gap microbridge process, first introduced by Lepselter of the Bell Labs in 1968.[15] In this process, after the first conductor layer has been patterned (photoetched) a layer of copper is vacuum deposited, then electroplated to about 1-mil thick. Vias are then photoetched in the copper in those sites that will become the posts (pillars) for the microbridges. An additional exposure and development of photoresist defines spaces in the photoresist (spans) that will connect the posts. Gold is then electroplated, simultane-ously filling in the vias to form the posts and forming and connecting the spans. As a final step the copper is selectively etched and removed in all areas including the areas beneath the spans, thus creating numerous microbridges with air as an insulator.[16]

Over the years, several variations of this process have been developed to improve reliability, increase yields, extend the process to metals other than gold, and to incorporate resistors. A reliability risk existed with the original process because some of the spans, especially the longer ones, (greater than 50 mils) sagged or collapsed onto the underlying conductors, thus shorting them. Studies correlating the length of the bridge span with bridge failures showed that a high incidence of sagging bridges occurred for spans longer than 48 mils.[17] Cerniglio[18] improved the process by adding an insulative layer over the bottom conductor layers while others[19] resorted to encapsulation of the air gap with soft silicones. Licari, et al[20] extended the process to the production of aluminum crossovers, incorporated thin-film nichrome resistors, and introduced polyimide as a permanent insulator and support for the crossovers. Micro-ramps instead of microbridges were thus formed which obviated the problems of collapsing spans and edge cracking at the posts (Figure 3.26).

Finally, Burns and DiLeo developed a batch process for forming the crossovers separately on a metallized polyimide film as a temporary carrier, then aligning them to the thin film substrate, bonding the ends of the spans to corresponding pads on the substrate by a batch process, and peeling away the polyimide film. This separate, parallel process is reported to result in higher crossover test yields, less handling damage, and lower costs.[21]

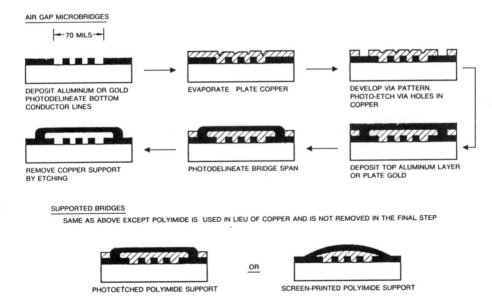

Figure 3.26: Fabrication steps for air gap microbridge and supported bridge interconnect substrates.

REFERENCES

1. Bunshah, R.F., et al, *Deposition Technologies For Films and Coatings*, Noyes Publications, Park Ridge, N.J., 1982.
2. Maissel, L.I. and Glang, R., Eds., Handbook of Thin Film Technology, McGraw-Hill, 1970.
3. Vossen, J.L. and Kern, W., Eds., *Thin Film Processes*, Academic Press, 1978.
4. Honig, R.E., *RCA Rev.* 23, 1962.
5. Alderson, R.H. and Ashworth, F., "Vacuum-Deposited Films of Nickel-Chromium Alloy," *Brit. J. Applied Physics*, 8, 205, 1957.
6. Bhatt, A.P., Luck, C.A., and Stevenson, D.M., "D.C. Sputtering of Ni-Cr Thin-Film Resistors," *Proc. ISHM*, 1984.
7. Puri, N. and Yaser, T., "Tantalum Nitride Chip Resistors For High Reliability Hybrid Microelectronics," *Proc. ISHM*, Sept. 1978.
8. Jones, D.E.H., "Electrical Properties of Vacuum and Chemically Deposited Thin and Thick Films," *Microelectronics and Reliability*, Vol. 5, No. 4, Nov. 1966.
9. Minsk, L.M., et al, U.S. 2,670,285, "Photosensitization of Polymeric Cinnamic Acid Esters," Feb. 23, 1954.
10. DeForest, W.S., *Photoresist Materials and Processes*, McGraw-Hill, 1975.
11. Eastman Kodak Co., Product Bulletin, *Kodak Micropositive Resist 820*, 1982.
12. Mogab, C.J., "Dry Etching," Chapter in *VLSI Technology*, Sze, S.M., Ed., McGraw-Hill, 1983.
13. Fonash, S.J., "Advances in Dry Etching Processes—A Review," *Solid State Technology*, Jan., 1985.
14. Reichelderfer, R.F., "Single Wafer Plasma Etching," *Solid State Technology*, April 1982.
15. Lepselter, M.P., "Air Insulated Beam-Lead Crossover For Integrated Circuits," *Bell System Technical Journal*, 48, Feb. 1968.
16. Basseches, H. and Pfahnl, A., "Crossovers For Interconnections On Substrates," *Proc. Electronic Components Conf.*, 1969.
17. Piacentini, G.F. and Minelli, G., "Reliability of Thin Film Conductors and Air Gap Crossovers For Hybrid Circuits," *Microelectronics and Reliability*, Vol. 15, Pergamon Press, 1976.
18. Cerniglio, N.P., "Multi-chip Integrated Circuit Package," *AFML-TR*-71-167, 1971.
19. RCA Solid State Div., "Manufacturing Methods for COS/MOS Memory," *AFML-TR*-74-24, April 1974.
20. Licari, J.J., Varga, J.E., and Bailey, W.A., "Polyimide Supported Micro-ramps For High Density Circuit Interconnection," *Solid State Technology*, July 1976.
21. Burns, J.A. and DiLeo, D.A., "Batch Bonded Crossovers For Thin Film Circuits," *Solid State Technology*, July 1976.

4

Thick Film Processes

FABRICATION PROCESSES

Thick film circuits are produced by the screen-printing process. Screen-printing using silk as the mesh material was an art used by the ancient Greeks and Egyptians to produce signs, designs, and works of art. In the last twenty years screen-printing has been rediscovered by the electronics industry and is now competing strongly with vapor deposited thin films as the main production method for microelectronic circuits. Silk mesh, however, is not used in electronics manufacture because of its dimensional instability and poor abrasion resistance. The mesh of choice is stainless steel, though sometimes synthetic fibers such as Dacron (polyester) or Nylon (polyamide) are used. The three key processes used to fabricate thick-film circuits are:

Screen-printing

Drying

Firing

Screen-Printing

The basic concept in screen-printing is to force a viscous paste through apertures of a stencil screen in order to deposit a pattern onto a substrate. A rubber blade called a squeegee is used to force the paste through the screen. To produce the stencil screen, a stainless steel wire mesh is stretched, then either mechanically or adhesively attached to a metal frame, normally consisting of cast aluminum. A negative mask must then be generated on the mesh so that the conductive, resistive, or dielectric pastes can be squeegeed (forced through selective openings of the mesh by applying pressure), thus producing a positive pattern on the substrate. The negative image on the mesh is formed by applying a photosensitive emulsion on the entire screen surface by either spraying a photoresist coating or applying a photosensitivie solid film. The emulsion is next exposed to ultraviolet light through artwork which has the desired pattern and the image is developed using vendor-specified solvents. Two options exist for producing a screen having the desired negative pattern. These depend on whether positive or negative artwork and positive or negative photoresists are initially used to make the screen (Figure 4.1).

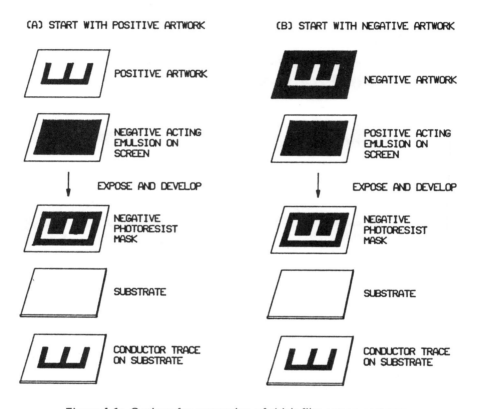

Figure 4.1: Options for generation of thick film screen patterns.

Many factors contribute to the success of producing quality thick-film circuits. Prime among these is the screen itself.[1] The screen consists of woven mesh stretched and attached to a frame on which is applied a photoemulsion. The woven mesh material may consist of Nylon, polyester, or stainless steel, but stainless steel is the material of choice when mechanical durability and high resistance to chemical solvents are required. Because of its excellent dimensional stability, stainless steel is used for the most precise fine line printing. The screen mesh count, the number of openings in the screen per linear inch, largely determines the dimensions of conductors and resistors and their tolerances, spacings between conductor lines, and via sizes. A screen with a high mesh number such as 325 or 400 results in finer lines and spacings (in the 3-5 mil range) than a coarser screen with lower mesh count. Low mesh count screens (80-200) are more suitable for screening the coarser pastes where exacting defini- tion is not required. Screens having a low mesh count are appropriate for applying sealing glasses, dielectrics, glazes, solder pastes, and some resistors. The width of the opening of the screen is related to the mesh count and to the diameter of the wire by the following equation:

$$W_o = \frac{1 - DM}{M}$$

where W_o = width of the opening in inches
D = diameter of the wire in inches
M = mesh count

The diameter of the mesh wire may range from 0.0008 to 0.0037 inches. Popular sizes for the 325 and 200 mesh screens are 0.0011 and 0.0016 inches, respectively.

Closely associated with the screen and also playing a key role in thick film dimensions and reproducibility of geometries is the photoemulsion that is used to prepare the screen pattern. There are three methods for the application of emulsion to the mesh: direct, indirect, and indirect-direct, also referred to as a hybrid emulsion. In the direct emulsion process the photoresist in liquid form is applied to the mesh, dried, and then exposed to ultraviolet light through the patterned artwork and developed. In the indirect emulsion process the pattern is exposed and developed on a separate photosensitive film, then adhered to the screen mesh. The direct- indirect process involves a combination of both processes. A separate photosensitive film is applied to the mesh then exposed and developed on the mesh. The key advantages of the indirect processes are that they are cleaner and easier to handle (no spraying or handling of liquids) and permit the application of various thicknesses. Emulsion thicknesses of 0.0015 to 0.020 inches are available.

A second key factor in producing quality thick-film circuits is the rheology or the flow behavior of the thick-film paste. Rheology, especially thixotropy, is extremely important in achieving small, well-defined vias and conductor/resistor geometries and thicknesses that are predictable and reproducible. Thixotropy is a variable flow property in which the viscosity of a paste at rest is high and has little flow, behaving more like a solid; but on applying a shear force, viscosity decreases sharply, allowing rapid flow. Then as the shear force is withdrawn, the viscosity increases again. This flow behavior, which is rather complex, is important in screen printing.[2] The flow changes that occur in the paste during the screen-printing process are shown in Figure 4.2.

Though considerable knowledge of the flow properties of thick-film pastes has been gained, screen-printing is still largely an art. A skilled operator sets the screen-printing parameters by experimenting with a number of trial runs before screening the actual parts. Screen tension, screen angle, squeegee speed, squeegee pressure, angle of squeegee,

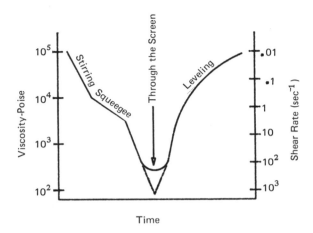

Figure 4.2: Viscosity changes in thick film pastes during screen printing.

and snap-off distance are variables that must be established empirically. The operator develops a "feel" for the equipment parameters and paste properties and makes changes until the optimum results are obtained. Manual, semiautomated, and fully automated screen printers are available (Figure 4.3). Computerized equipment is also available where optimum screening parameters, once established, can be programmed and reproduced exactly from run to run.

Figure 4.3: Microprocessor-controlled screen printer, Model 1505 (Courtesy AMI, Affiliated Manufacturers Inc.).

Drying

The drying operation consists of removing the organic solvents from the screen printed paste by moderate heating. The substrate with the freshly screened paste is first air-dried for 5 to 10 minutes to allow the paste to settle, then conveyorized through a belt furnace maintained at 120-150°C to fully dry it. In 10-20 minutes all of the organic solvents are removed by evaporation. In some cases better results are obtained if the freshly screened patterns are allowed to air dry at room temperature from one hour to several hours before drying at elevated temperature. This allows slow removal of solvent and leveling which minimizes the flow of paste and dimensional variations and avoids blistering.

Firing

After drying, the substrates are conveyorized through a high-temperature furnace comprised of several zones of increasing temperature (Figures 4.4 and 4.5). The temperature profile that should be used is specified by the paste manufacturer but must often be verified or optimized experimentally by the user. In the first portion of the furnace (Zone A, 200-500°C), the temporary organic binder is decomposed by air oxidation and removed. At these temperatures, the binder disintegrates into small, gaseous fragments that are quickly exhausted through the furnace vents. Incomplete removal of the organic binders leaves carbonaceous deposits that become entrapped in the paste and may alter both the electrical and physical properties of the final product. Complete burnout requires efficient air flow so that the organic materials can be quantitatively oxidized and decomposed into small easily volatilized species. In the intermediate temperature zone (Zone B, 500-700°C), the permanent binder (glass frit) melts and wets both the surface of the substrate and the particles of the functional material. Some softening and melting of the glassy constituents of the substrate also occurs, causing it to fuse with the glass in the paste. In the third temperature zone (Zone C, 700-850°C), the functional particles are sintered and become interlocked with the glass frit and the substrate. The parts are kept at the peak temperature of 850°C for approximately 10 minutes. The last zone provides for rapid cooling from peak temperature to slightly above room temperature. Overall, the total cycle takes approximately one hour using a conventional convection/conduction furnace. This time can be shortened by using an infrared furnace. A typical temperature profile for an air-firing paste is shown in Figure 4.6. In general, the peak-firing temperature should be approximately 100°C below the melting temperature of the metal constituent of the paste. Thus, the maximum firing temperature for a silver paste should be approximately 850°C since the melting point of silver is 960°C. Firing schedules should be established experimentally to optimize adhesion to the substrate, electrical values, wire bondability, and, if applicable, solderability.

Fritless pastes require a slightly different firing profile than fritted or mixed bonded pastes. A higher peak temperature (900-1000°C) is necessary to effect a chemical reaction between the copper oxide contained in the fritless paste and the alumina of the substrate. This adhesion mechanism is more fully discussed in another part of this chapter.

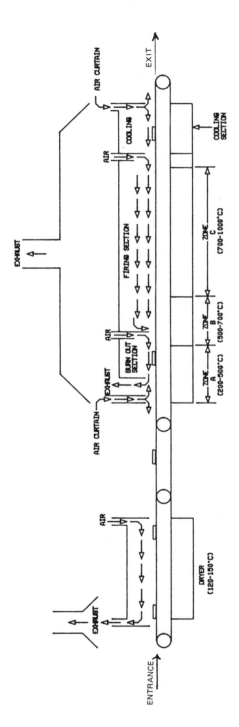

Figure 4.4: Cross-sectional view of dryer and furnace.

Figure 4.5: Thick-film engineering facility showing screening machine (center) and dryer (front right) and two furnaces. (Courtesy Rockwell International Corp.)

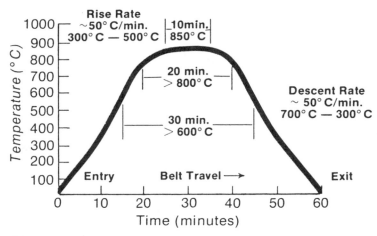

Figure 4.6: Typical firing temperature profile for thick films (Courtesy Du Pont).

Multilayer Thick-Film Process

A key advantage of the thick-film process over the thin-film process is the ability to form multilayer interconnect circuits. The thick-film process permits multiple layers of conductors, dielectrics, and resistors to be screen-printed and fired sequentially onto a ceramic substrate, resulting in high-density interconnections. Interconnect substrates having conductor lines and spacings of 5 to 10 mils, vias of 10 to 20 mils diameter, and up to 7 conductor layers are now fairly common and important in producing digital and analog circuits used in compact electronic systems. The conductor lines on the first layer are purposely laid out as the most detailed and densest. Any imperfections can thus be touched up by applying fresh conductor paste to the affected area and refiring. Next, a dielectric layer is screen-printed and fired over the entire surface except for interconnect vias, apertures where resistors will be subsequently applied, and apertures where power devices will be inserted. Vias are holes left in the dielectric which are subsequently filled with conductive paste and serve as electrical interconnections between conductor layers. Because there is always the probability of pinholes in the dielectric, it is customary to screen-print a second identical dielectric layer over the first. The double layer reduces the probability of pinholes becoming continuous and shorting out conductors between layers. The two dielectric layers may be dried and fired separately or dried separately and then co-fired.

The next step in the multilayer process involves screen-printing and firing the second level conductors on top of the dielectric. In this step the vias are simultaneously filled with conductive paste thus forming the z-direction interconnects. However, vias may also be filled in a separate step using another screen mask. Separate via-filling is beneficial when fabricating multilayer substrates having a large number of layers so that flat, level, top surfaces are assured. The screen-printing and firing steps may be repeated to increase the number of conductor and dielectric layers and hence the "wiring" density of the substrate.

During the screen-printing process, apertures are left in the dielectric extending down to the ceramic substrate so that resistor pastes may be screened and fired directly to the substrate. Applying the resistors directly to the substrate, as the last step, provides improved resistor stability because the resistors are not subjected to repeated high-temperature firings. However, as the multilayer structure becomes thicker with the increased number of dielectric layers, screen-printing through the deep apertures results in distorted resistor geometries and resistor values that are hard to control. There is thus a tradeoff in using this approach. An alternate approach involves screen-printing the resistors on the top dielectric. This process allows the design engineer greater flexibility in laying out the circuit. It increases the circuit density by about 20%, but some problems in the compatibility of the resistor with the dielectric may be encountered. Some dielectrics soften during the firing of the resistor and, in so doing, interact with it chemically or physically to change resistor values. Furthermore, laser-trimming the resistors to tolerance requires special care because the laser may cut into the underlying dielectric.

A wide range of resistor values extending from very low values (1-10 ohms) to the megohm range are possible with thick films because two, three, or even four resistor pastes having different sheet resistances may be used. This is a major advantage over thin films where generally only one sheet resistance material of low value (100-300 ohms/sq) can be used. The sequential steps and process flow for producing multilayer thick-film interconnect substrates are depicted in Figures 4.7, 4.8, and 4.9.

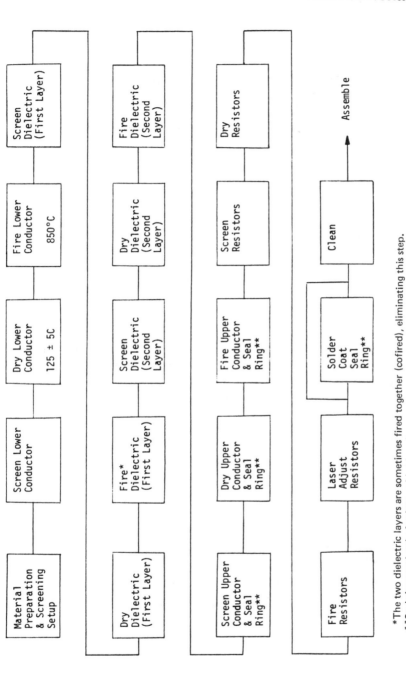

Figure 4.7: Multilayer thick film process flow diagram (two conductor layers).

Material Preparation & Screening Setup → Screen Lower Conductor → Dry Lower Conductor 125 ± 5C → Fire Lower Conductor 850°C → Screen Dielectric (First Layer)

Dry Dielectric (First Layer) → Fire* Dielectric (First Layer) → Screen Dielectric (Second Layer) → Dry Dielectric (Second Layer) → Fire Dielectric (Second Layer)

Dry Resistors → Screen Resistors → Fire Upper Conductor & Seal Ring** → Dry Upper Conductor & Seal Ring** → Screen Upper Conductor & Seal Ring**

Fire Resistors → Laser Adjust Resistors → Solder Coat Seal Ring** → Clean → Assemble

*The two dielectric layers are sometimes fired together (cofired), eliminating this step.
**Seal ring used only for integral-lead packages.

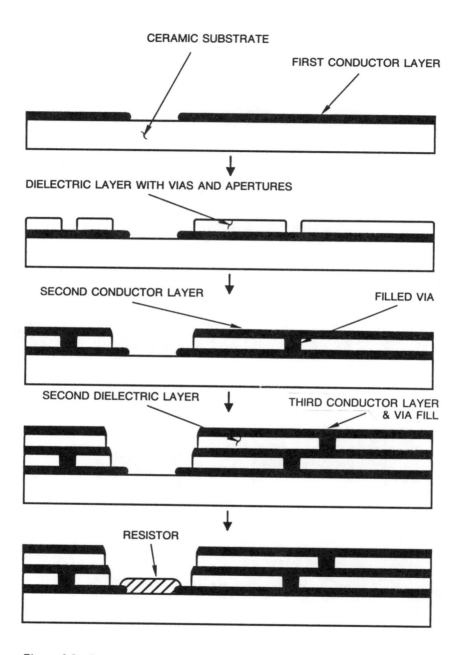

Figure 4.8: Representation of thick-film multilayer fabrication steps (not to scale).

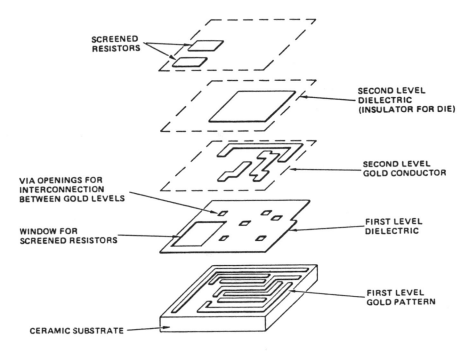

SCREENED RESISTORS

SECOND LEVEL DIELECTRIC (INSULATOR FOR DIE)

SECOND LEVEL GOLD CONDUCTOR

VIA OPENINGS FOR INTERCONNECTION BETWEEN GOLD LEVELS

FIRST LEVEL DIELECTRIC

WINDOW FOR SCREENED RESISTORS

FIRST LEVEL GOLD PATTERN

CERAMIC SUBSTRATE

Figure 4.9: Simplified thick film multilayer interconnection model.

Multilayer Co-Fired Ceramic Tape Process

The multilayered co-fired tape process differs from the thick-film process already described in that the base substrate and all the dielectric layers are initially in the unfired condition. This unfired tape is called "green" ceramic or "green" tape. It is formed by blending and milling mixtures of oxides, glasses, organic binders, solvents, and plasticizers, casting this mixture (slurry) into large sheets, and drying. The slurry is poured onto a Mylar film, then passed under a doctor blade to produce a uniform sheet of specified thickness. Colored oxides may be added to the slurry if opacity is required to protect light-sensitive devices. Black ceramic, for example, is available for these applications. The tape, so formed, is soft and somewhat rubbery. It is punched to form the z-direction vias, then cut to size or may be cut first, then punched. Thick-film conductor pastes consisting of refractory metals are screen-printed to form the conductor lines and to fill in the vias. The layers of a multilayer co-fired module are processed separately, then aligned, and stack laminated in a press to join them together. After removal of the stack from the press it is co-fired as a single unit in a high temperature furnace at about 1600°C in a hydrogen atmosphere. In the standard thick film process, one starts with a pre-fired ceramic substrate then builds on this by sequentially screening and firing the conductor and dielectric pastes.

The properties of the fired alumina tape are given in Table 4.1 along with those for fired beryllia tape which is also available for selected applications. The properties of white and black alumina ceramic tape from another leading manufacturer are given in Table 4.2.

Table 4.1: Physical and Electrical Properties of Co-fired Ceramic Tape
(Courtesy of Ceramic Systems Division of General Ceramics Inc.)

	Alumina	Beryllia
COMPOSITION	92-96% Alumina	99.5% BeO
COLOR AVAILABLE	White or Black	White
FLEXURAL STRENGTH	55,000 psi	35,000 psi
COEFFICIENT OF THERMAL EXPANSION 25°C to 200°C	$6.3 \times 10^{-6}/°C$	$6.4 \times 10^{-6}/°C$
THERMAL CONDUCTIVITY AT 100°C	0.035 cal/sec-cm°C	0.48 cal/sec cm°C
VOLUME RESISTIVITY AT 25°C	10^{14} ohm-cm	10^{14} ohm-cm
DIELECTRIC CONSTANT AT 25°C	8.9	6.6
CAMBER	0.002 to 0.004 in/in	0.002 to 0.004 in/in
TAPE THICKNESS (BEFORE FIRING) (AFTER FIRING)	0.005 to 0.025 in 20% Shrinkage	0.005 to 0.025 in 20% Shrinkage

Table 4.2: Properties of Alumina Tape Ceramic (Courtesy of Interamics)

COLOR	White	Black
ALUMINA CONTENT	92%	90%
SPECIFIC GRAVITY (gm/cc)	3.6 min.	3.6 min.
FLEXURAL STRENGTH (psi)	50,000 min.	50,000 min.
SPECIFIC HEAT (cal/gm × °C)	0.2	0.2
THERMAL CONDUCTIVITY (cal/cm × sec × °C)	.04	.04
THERMAL EXPANSION (25-400 °C) (inch/inch/°C)	6.6×10^{-6}	6.5×10^{-6}
DIELECTRIC CONSTANT (@ 1 mHz)	9.5	10.5
LOSS ANGLE (@ 1 mHz)	.0046	.0051
VOL. RESISTIVITY (ohm/cm)	1.4×10^{15}	2.0×10^{15}
THERMAL SHOCK RESISTANCE (°C)	200	200

Tungsten, molybdenum, molybdedum-manganese, or other refractory metal conductor pastes must be used in the co-fired tape process because of the high firing temperatures needed to sinter the tape. Because these refractory metals oxidize in air at high temperatures, a hydrogen atmosphere must be used during firing. The use of refractory metals as conductors, though compatible with the process, inherently exhibit higher sheet resistances compared with noble metal pastes. Tungsten and molybdenum pastes have sheet resistances of 15 milliohms/square while moly-manganese is 35 milliohms/square. These high sheet resistances can be compensated to some extent by designing wider conductor lines to increase their cross-sectional areas and hence their electrical conductance. The tape conductor layers require plating with nickel followed by gold to increase their electrical conductance and to provide a surface that is amenable to hermetic sealing, brazing of leads or lead frames, wire bonding, soldering, and device attachment.

The general steps for the co-fire tape process are shown in Figure 4.10. The co-fired tape process requires the user to have a large investment in facilities, equipment, and skilled personnel for blending and casting the tape, punching, drilling and cutting, and for the high temperature, hydrogen ambient firing. Thus, few users of co-fired tape products (substrates and packages) have an in-house capability. They will purchase these parts from established ceramics manufacturers by providing them with designs and layouts.

The key advantage of the multilayer co-fired tape process is the ability to design and fabricate very high density multilayer interconnects, of twenty or more conductor layers. These conductor layers may consist of signal lines, striplines, ground planes and power planes. Because of the large number of circuit layers and interconnect vias that can be produced, the co-fired tape process is extensively used for producing ceramic packages having large numbers of input-outputs (100 to over 300) as required for VHSIC (Figure 4.11) and large area substrates required for surface mounting components. A cross-section of a multilayer co-fired circuit is shown in Figure 4.12. The thick-film process has an upper practical limit of about seven conductor layers, above which yields begin to drop and is generally not used for producing complex ceramic packages. A further advantage of the co-fired tape process is that, once the initial investment is capitalized and the non-recurring paid for, packages and substrates are lower in cost on a production scale. There are however a few design limitations which must be considered:

- Screen-printed resistors cannot be incorporated as part of the co-fired process

- The higher amount of glass contained in the alumina tape reduces the thermal conductivity

- The use of refractory metal conductors reduces the electrical conductivity of the inner conductor lines

- The outer conductors must be plated with nickel and gold to increase the surface conductivity and provide a metallization that will permit reliable wire bonding or solder attachment of components.

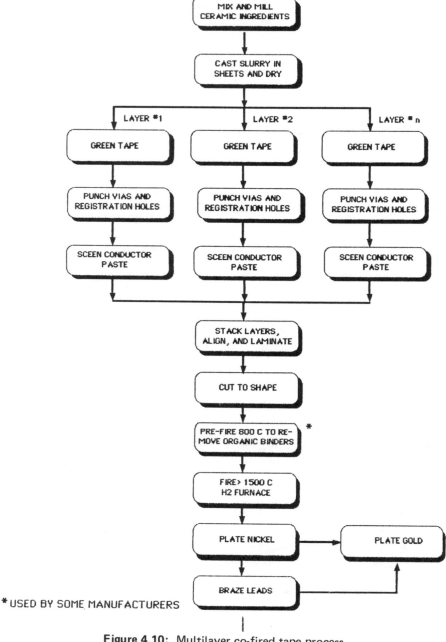

Figure 4.10: Multilayer co-fired tape process.

Figure 4.11: Co-fired alumina package with high I/O count for VHSIC.

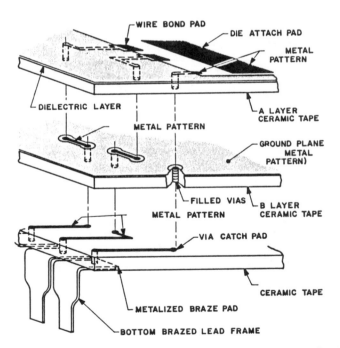

Figure 4.12: Co-fired tape multilayer structure (Courtesy Ceramic Systems Div., General Ceramics Corp.).

A detailed comparison of the thick-film process with the multilayer co-fired tape process is given in Table 4.3.

Table 4.3: Comparison of Thick-Film Process with Co-fired Tape Process

Thick Film	Co-fired Tape
Sequential screening, drying, firing	Stack lamination and batch firing
Air-fired at 850-1000 C	Fired in hydrogen at >1500 C
Permits wide variety of conductor and dielectric pastes	Limited to refractory metal conductors and high glass content dielectrics
Permits wide range of batch fabricated resistors	No screen/fired resistors
Practical dielectric thickness: 2 to 5 mils	Dielectric thickness may be as high as 25 mils
Vias: 7 to 15 mils diameter	Vias may be as small as 4 mils
No shrinkage problem	High shrinkages during firing (20-30%) must be taken into account
Separate via fill steps to provide flat top surface and hold via tolerances for multilayer circuits	Controlled dimensional tolerances once shrinkage factors are taken into account
Limited adhesion of conductors to alumina ceramic	Excellent conductor adhesion; good for brazing on lead frames
No auxilliary plating required to augment conductivity	Requires gold plating of top tungsten layer
Yields drop with conductor layers >7	Excellent for high number of conductor layers, >20
Quick turn-around for design iterations	Costly and long lead times for design iterations
Low non-recurring cost, excellent for low to moderate production runs	High non-recurring cost; more suitable for high production

Low-Temperature Co-Fired Tape Process

A low-temperature co-fired tape process has recently been introduced that offers significant advantages over the high temperature co-fired tape process. It provides a new dimension for engineers to design both high density multilayer interconnect substrates and advanced packages for high speed integrated circuits.[3] The key feature of this process is that the "green" tape can be fired at 850°C in a conventional air ambient belt furnace instead of at 1600°C in a hydrogen ambient furnace as required by the conventional co-fired tape process. The low-temperature process permits the use of air fired resistors and precious metal thick film conductors such as gold, silver, or their alloys. In the high-temperature process, screen-printed resistors cannot be used and only refractory metal pastes can be used as conductors. Furthermore, the high temperature process requires a hydrogen or reducing atmosphere during firing. From a production standpoint, the low-temperature process opens up many new opportunities to a user who can now, with little investment, produce his own interconnect substrates and even ceramic packages. If a user has an established thick-film processing line he can now purchase the green tape and process it himself using existing screen printers and furnaces. Only two additional pieces of equipment are needed: a lamination press and a precision hole-drilling or punching machine. Low-temperature co-fired taped may be purchased in blanked or roll form. They are being produced and used in Japan and in the U.S. In the U.S., DuPont has been the main developer and promoter of this technology.

The steps for the production of the low-temperature tape are similar to those already described for the high-temperature tape. A ceramic slurry is cast onto a support film of Mylar (about 0.005 inches thick) and dried. After removing the Mylar film, blanks are cut and registration holes and orientation marks are simultaneously made using a die tool. Vias are then punched or drilled in each layer using precision computer-controlled machines. Via diameters as small as 0.004 to 0.006 inches have been achieved, though a minimum of 0.008 inches is typical. In the next step, the vias are filled with conductive thick-film paste by screen-printing, then conductor patterns are screen-printed onto the tape and dried at 120°C for 5 minutes. The conductor compositions are specially formulated to shrink at the same rate as the tape when they are co-fired. The dielectric tape layers are then aligned, stacked, and laminated in a press at 70°C and 3,000 psi. The laminated circuits are trimmed to remove the outer borders containing the registration holes and pre-fired at 350°C for one hour to burn out approximately 85% of the organic binders. This burn-out phase also prevents blistering of the layers during firing. The final step involves firing in a conventional belt furnace at a peak temperature of 850°C for 15 minutes with an overall temperature profile of about two hours.

As with the high-temperature co-fired tape process, shrinkage occurs during firing and must be accurately measured and taken into account in the initial design and layout of the circuit. Shrinkage for the low-temperature tape is about 12% but varies somewhat with lamination conditions (Figure 4.13). The properties of the fired dielectric tape are given in Table 4.4.[4]

PASTE MATERIALS

Thick-film pastes, also referred to as inks, are thixotropic screenable compositions used in forming the conductor, resistor, and dielectric patterns of a circuit. As with many other technologies, the use of thick films in the manufacture of electronic circuits emerged after World War II and intensified and proliferated in the sixties and seventies. Initially, very few pastes were commercially available. Users had to formulate and produce their own compositions. Among the pioneers were Beckman Instruments (resistors for potentiometers) and Sprague Electric. Today there is little need for a user to produce his own pastes. At least a dozen firms specialize in the manufacture and sale of high-quality thick-film pastes.

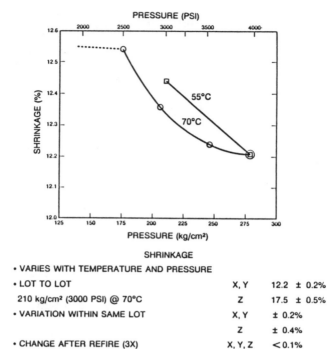

SHRINKAGE

- VARIES WITH TEMPERATURE AND PRESSURE
- LOT TO LOT   X, Y   12.2 ± 0.2%
  210 kg/cm² (3000 PSI) @ 70°C   Z   17.5 ± 0.5%
- VARIATION WITHIN SAME LOT   X, Y   ± 0.2%
                                                   Z   ± 0.4%
- CHANGE AFTER REFIRE (3X)   X, Y, Z   < 0.1%

Figure 4.13: Shrinkage reproducibility control of cofired ceramic system.[4]

Table 4.4: Physical Properties of Low-Temperature Co-fired Dielectric Compared with 96% Alumina[4]

Thermal expansion (25° to 300°C)
  Fired dielectric   7.9 ppm/°C
  96% alumina   7.0 ppm/°C
Fired density
  Theoretical 3.02 g/cm^3
  Actual $>2.89 \text{ g/cm}^3$ (>96%)
Camber
  Fired $\pm 75 \mu m$ (± 3 mil)
  86 mm x 68 mm (2.7 x 2.7 inch)
Surface smoothness
  Fired dielectric $0.8 \mu m$/50 mm
  50 mm x 50 mm (2 x 2 inch)   (peak to peak)
Thermal conductivity
  Fired dielectric   15 to 25% of 96% alumina
Flexural strength
  Fired dielectric $2.1 \times 10^3 \text{ kg/cm}^2$ (3.0×10^4 psi)
  96% alumina $3.8 \times 10^3 \text{ kg/cm}^2$ (5.6×10^4 psi)
Flexural modulus
  Fired dielectric $1.8 \times 10^6 \text{ kg/cm}^2$ (2.5×10^7 psi)
  96% alumina $0.9 \times 10^6 \text{ kg/cm}^2$ (1.3×10^7 psi)

Types and Compositions

The three general types of thick-film pastes are based on their electrical functions: conductors, resistors, and dielectrics. Dielectric pastes may be of two types: insulators or capacitors, depending on whether a filler with a low dielectric constant or high dielectric constant, respectively, is used. Regardless of their electrical function, all pastes consist of four generic ingredients: the functional material, a solvent or thinner, a temporary binder, and a permament binder.

1. *Functional Material.* The functional material is the ingredient that imparts to the paste its conductive, resistive, or dielectric properties. It is usually the constituent used in the greatest amount in the paste formulation. Functional materials may be metal or metal-oxide powders. To formulate conductors, metals such as silver, gold, platinum, palladium, copper, nickel, tungsten, or molybdenum are widely used. Alloys of silver or gold with platinum or palladium are often used to modify soldering or wire bonding characteristics. Resistor pastes consist largely of combinations of metal oxides, metals, and glasses. Dielectric pastes are composed of one or more metal oxides and glass. Aluminum oxide and silicon oxide are widely used.

2. *Solvent or Thinner.* Organic liquids of various boiling points are used as solvents to disperse the solid ingredients and to adjust the viscosity which is helpful during milling and screen-printing. Some widely used solvents are pine oil, terpineol, isomers of terpineol, butyl Carbitol acetate (Union Carbide), and esters such as dibutyl phthalate and trimethyl-pentanediol isobutyrate. Solvents are removed during the drying and initial firing stages.

3. *Temporary Binder.* Binders are organic polymers or organic compounds of moderate molecular weight that provide rheological (flow control) properties, holding the other ingredients together during the screen-printing and firing cycles. Polyvinyl acetate, polyvinyl alcohol, and ethyl cellulose are examples of some binders used. Temporary binders are removed by oxidation and decomposition during the early stages of firing.

4. *Permanent Binder.* The permanent binder is the material that fuses the particles of the functional material together and to the substrate. It remains with the functional material after the solvents and temporary binders have been removed, thus becoming an integral part of the final fired film. Permanent binders are glasses, also referred to as frit, which melt and resolidify thus wetting the particles and fusing them together. The most widely used binders are lead borosilicate, bismuth silicate, and aluminosilicate glasses. A typical lead borosilicate glass composition consists of 63% lead oxide, 25% boron oxide, and 12% silicon dioxide by weight. Glass comprises approximately 2-3% by weight or 10-15% by volume of the paste formulation. Some conductor pastes (fritless types) do not contain glass; they adhere and fuse by a chemical mechanism described below.

Conductor Pastes

Functions of Thick-Film Conductors. Like thin-film conductors, thick-film conductors serve a number of functions in a hybrid microcircuit.

Primarily, they are used to conduct electrical current in signal lines and to form an electrical path between conductor layers and between devices and circuits lines. A list of uses is as follows:

Signal layers

Ground planes

Voltage planes

Wire bonding pads

Ohmic contact pads (for metallurgical or adhesive attachment of die)

Via interconnections

Seal rings

The functionsl materials in conductor pastes are metals of high electrical conductivity. Unfortunately, most of these metals are expensive noble or precious metals such as gold, platinum, palladium, silver, or combinations of these. Low-cost conductor pastes such as those based on copper or nickel are available and may eventually replace the precious metals for some applications. Nickel, however, does not have the high electrical conductivity of copper, silver, or gold and its applications to electronic circuits are limited. Table 4.5 provides a comparison of the electrical conductivities of various metals. Sheet resistances and adhesion values for paste conductors after firing are to be found in Table 4.6. It is difficult to give absolute values because of differences in commercial formulations. For the silver or gold alloy types, there can be large differences depending on the concentrations of platinum or palladium in the formulation; the higher the platinum or palladium content, the higher the sheet resistance.

Table 4.5: Electrical Conductivities of Metals Used in Thick Film Pastes

Metal	Conductivity (micro ohm-cm)$^{-1}$
Silver	0.616
Copper	0.593
Gold	0.420
Aluminum	0.382
Rhodium	0.220
Iridium	0.189
Tungsten	0.181
Nickel	0.145
Ruthenium	0.10
Platinum	0.095
Palladium	0.093
Chromium	0.078

Note: The electrical conductivities of alloys of the above metals vary depending upon the ratio of the two metals used.

Table 4.6: Typical Thick Film Conductor Characteristics*

	Au	Au-Pt	Au-Pd	Ag	Ag-Pt	Ag-Pd
Adhesion to 96% alumina						
In tension (psi)	1,000–3000	600–1600	1500	800–1200	500–1200	800–1700
In peel (lb/in)	5–13	10–20	10–30	12–15	11–15	8–27
Thickness						
Dry (μm)	20–25	25–30	25–30	25–30	25–30	25–30
Fired (μm)	7–13	13–19	13–15	15–17	15–18	10–17
Sheet resistance						
(milliohms/sq/mil)	2–5	50–100	50–100	2–10	2–7	3–18

*Data represent a composite of values reported by several leading thick film suppliers.
 Adhesion values represent specification values; actual values are generally higher.
 Values for alloy compositions vary widely depending on the ratios of the metals used.

Besides high electrical conductivity and low cost, other desirable characteristics of conductor pastes are high adhesion to substrates, fine line resolution, good wire bondability (to both aluminum and gold wire) and good solderability.

Adhesion Mechanisms. Conductor pastes may be classified into three types depending on the adhesion mechanism that is involved. These are:

Fritted Pastes

Fritless Pastes

Mixed Bonded Pastes

The early compositions were all fritted types, that is, they contained frit, a low melting glass, as the permanent binder. Because there is always some glass in the ceramic substrate, a fusion or interlocking of the glasses occurs at the melt temperature. Thus the mechanism of adhesion is essentially mechanical. A key drawback in using fritted pastes is the difficulty in obtaining a homogeneous film. Chunks of frit are exposed on the top surface and can interfere with wire bonding and soldering. To resolve these problems, a second generation of conductor pastes was introduced about fifteen years ago.[5] These are fritless types. They contain no glass and depend on a purely chemical mechanism for adhesion. The majority ingredient in the fritless pastes is gold (98-99%) and the remainder (1-2 weight percent) is copper oxide, cadmium oxide, chromium oxide, or mixtures of these. At temperatures of 950-1000°C these oxides react chemically with the aluminum oxide of the substrate to form copper aluminate, or cadmium or chromium aluminate spinel compounds (Figure 4.14). These spinel structures are believed to form molecular layers at the interface, chemically bonding the two materials. Because of the method of adhesion and the high amount of gold that they contain, fritless pastes are also called molecularly bonded, reactively bonded, or "99+" types.

Though fritless gold pastes are better suited to producing fine-line, high density circuits, improved adhesion to alumina substrates, and improved wire bondability, a key disadvantage is the criticality of the high temperature range required for firing. Maintaining a narrow-temperature firing range of 950-1000°C is necessary to obtain best adhesion results. If temperatures below 950°C are used, the necessary spinel compounds cannot form and marginal adhesion results. Firing temperatures above 1000°C are undesirable because the gold will melt. This limitation led to the development of still a third generation of pastes—the "mixed-bonded" types which, in essence, are a compromise combining some of both mechanisms. As the name implies, mixed-bonded pastes are composed of ingredients that permit both adhesion mechanisms to occur. Copper oxide or mixtures of spinel-forming oxides are added to the paste formulation to achieve molecular bonding along with some glass frit to produce mechanical bonding. A key advantage of the mixed-bonded pastes is that the more conventional firing temperature of 850°C can be used.

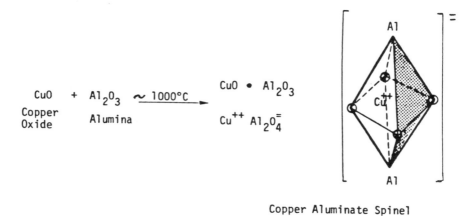

Copper Aluminate Spinel

Figure 4.14: Fritless gold adhesion mechanism.

Though significant improvements have been made by using either the mixed-bonded or fritless conductor pastes, the problem of wire bondability has not been entirely resolved. Riemer,[6] for example, reports that the top surface of a fired fritless gold conductor is still not homogeneous; it contains an oxide layer of about 2000 Å which interferes with wire bonding. The top surface of mixed-bonded conductors also contain oxides and some glass frit. Research and development programs are therefore still in progress to formulate the ideal paste that will combine excellent wire bondability with high adhesive strength.

Adhesion Tests. There are two widely used tests for measuring the adhesion strength of thick film conductors to ceramic substrates or to dielectric layers—tensile adhesion and peel adhesion. Conducting these tests under controlled conditions and obtaining reproducible values that lie in a narrow range is difficult because subtle variations in producing the test specimens result in large differences in values. Nevertheless, averages of many values are indicative of differences in adhesion strength and therefore significant in ranking thick film conductors.

The test specimen for the tensile test consists of small square or circular pads of the thick film conductor that are screen-printed and fired onto a ceramic substrate. The size of the pads may be as small as 50 mils square but some firms use 100 mil square pads or circular pads of 0.01 in² so that multiplying the pull force by 100 conveniently gives pounds per square inch. A wire is then solder attached to a copper plug which in turn is soldered to the conductor pad. The wire is pulled in tensile (normal to the substrate) using an Instron or other tensile testing machine. The maximum pull force, in grams or pounds, at detachment or prior to detachment of the conductor is reported as the pull strength. A sufficiently thick ceramic substrate should be used for the test specimen (generally 40 to 60 mils thick is adequate) so that the breaking strength of the ceramic is greater than that of the conductor film. Tensile strength values for a fritless and a mixed-bonded gold conductor showing the range of values obtained are given in Table 4.7. These values, generally averaging about 5,000 psi, exceed the 3,000 psi requirement specified in most procurement specifications for thick film conductors.

Table 4.7: Tensile Adhesion Test Data (psi)

Fritless Gold	Mixed-Bonded Gold
5,687*	5,072
6,168	5,669
5,854	4,518
6,311	4,930
4,867	6,746
3,882	5,705
Average 5,461	5,440
(24 data points)	

*Each value represents an average of 4 pull test values.

Peel and shear forces are more representative of the stresses that a hybrid circuit might encounter during assembly or operation than tensile forces. Conductors that have marginal or poor adhesion may detach as a result of peel or shear forces from pulling, bending, or twisting the attached wire leads. Large diameter wire, stiff wire, or lead frames attached to the outward pads of a thick film circuit can transfer shear stresses to the pads and lift them. This is a costly situation because, at this stage in the assembly, the substrate metallization cannot be repaired. In spite of the greater significance of the peel test, the tensile test is still widely used as a comparative test to evaluate and select conductor pastes, to establish firing parameters, and to monitor quality.

The peel test utilizes a test specimen similar to that for the tensile test. Copper wire (about 20-gage) is solder attached to square conductor pads

(80 mils square). Both the wire and conductor pads are solder dipped prior to attachment. The wire is laid over a row of pads so that a straight length of wire contacts the pads prior to solder dipping. The wire is fastened to one end of the substrate by hooking it around the edge of the substrate. At the other end it is bent at a 90-degree angle at a distance of approximately 50 mils from the edge of the conductor pad (Figure 4.15). This bend distance is critical in obtaining reproducible results. The specimen is then clamped to a tensile tester and the force necessary to detach the wire is measured. Samples should be aged 16 to 24 hours at room conditions prior to pull testing to allow the solder to anneal and obtain reproducible values.[7] A second variation of the peel test consists of fabricating the test specimens with 100-mil-square conductor pads, then solder-attaching and pulling copper ribbon that is 100 mils wide and 3 mils thick. Multiplying the pull force by 10 then gives the peel strength in pounds-per-inch width.

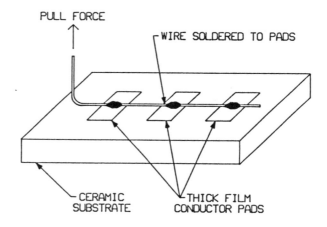

Figure 4.15: Peel test specimen.

Metal Migration. Metal migration, especially silver migration, can occur between closely spaced conductor lines and can result in reduction of insulation resistance, increase in leakage current and eventual electrical shorting, arcing, or dielectric breakdown. Metal migration has been known and studied for over thirty years and has been the cause of many cata-strophic microcircuit failures.[8-12] Though metal migration can occur in either thick- or thin-film conductors, it is treated here under thick films because silver conductors are more widely used in thick-film circuits. Metal migration has been dominant with silver and silver-bearing alloys but may occur within almost any metal. Even gold under the appropriate condi-tions of bias, ionic contaminants, and moisture has been reported to migrate.[13] It is generally reported that three conditions are necessary for metal migration to take place. These are: a dc or ac potential which may be as small as one or two volts; some ionic contaminants on the surface of a dielectric or within the dielectric, and moisture as a liquid layer, albeit even a mono-layer. There are reports that even moisture may not be necessary as long as the dielectric medium allows ionic migration, thus acting as a solid electrolyte.[14] Besides the three conditions mentioned, the nature of the metal, the solubility of the metal hydroxide or oxide that is formed, the porosity of the dielectric, and the spacing between conductors also have a pronounced effect on metal migration.

Basically, metal migration is an electrochemical phenomenon. Silver, the metal having the highest propensity for migration, is oxidized at the anode (positively biased conductor line) to positively-charged silver ions. The silver ions then are attracted to the opposite, negatively biased-line (cathode). On their journey across or through the dielectric the silver ions are accelerated by the presence of moisture and ionic contaminants. A chemical reaction usually occurs between the ions and the hydroxyl ions of water; for example, silver will form a silver hydroxide which appears as a milky-white smudge between the conductors. This reaction can either accelerate or suppress the migration depending on the solubility of the hydroxide that is formed. If the hydroxide is soluble in the medium it will quickly dissociate and the metal ions will continue their migration to the cathode, as in the case of silver. If, however, the hydroxide is insoluble, the migration of the metal ions will be arrested or slowed, as in the case of copper. The insoluble hydroxide forms a passivating barrier that can essentially stop the migration. If the positively charged ions are unobstructed and arrive at the cathode they pick up electrons and are reduced back to the metal. The ions move along field lines and are deposited at points of high field strength. The metal builds on itself and grows as filaments backward toward the anode. Thus in essence one has a two dimensional electrolytic cell, a solid plating bath. Because these filaments have the appearance of tree branches, they are called dendritic growth. A model for silver migration on a dielectric surface is given in Figure 4.16.

It is good practice not to use metals that are known to migrate readily, such as silver. However, since there is a propensity for almost any metal, even gold, to migrate, some proven ground rules should be followed to suppress or prevent migration. Among these are:

- Design the circuit to avoid biases between critical components

- Avoid the use of pure silver. Use alloys of silver and noble metals such as platinum or palladium; noble metals slow down silver migration by reducing the rate at which silver ions are formed at the anode. The larger the concentration of Pd or Pt in the alloy, the slower the silver migration but it must be understood that this is achieved at the expense of electrical conductivity. Thus the sheet resistance of silver may be increased from less than 5 milliohms/sq. to over 100 milliohms/sq. by adding large amounts of Pd or Pt. This lower conductivity may render them unsuitable for some electrical circuits.

- Use low-porosity, hydrophobic dielectrics and substrates. Test the substrates for insulation resistance at elevated temperature (250°C) or under moisture—conditions under which ions are most likely to be mobilized. A high insulation resistance of $>1 \times 10^{11}$ under these conditions assures low ionic mobility within the dielectric.[14]

● Handle, store, and seal the circuits in a moisture-free environment

● Overcoat the circuits with an organic coating or a glassivation layer to prevent moisture from condensing on the surface

● Assure that the surfaces are ultra-clean and free of ionic contaminants

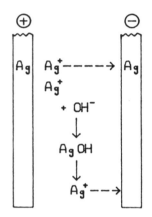

ANODE REACTIONS

$$Ag - e \longrightarrow Ag^+ \quad OXIDATION$$

INTERELECTRODE REACTIONS

$$Ag^+ + OH^- \longrightarrow Ag\,OH \downarrow$$

$$Ag\,OH \longrightarrow Ag^+ + OH^-$$

CATHODE REACTION

$$Ag^+ + e \longrightarrow Ag \quad REDUCTION$$

Figure 4.16: Model for silver migration.

It is difficult to predict to what extent a metal conductor will migrate. This is especially true for thick film conductor pastes because they are heterogeneous mixtures of many materials. Therefore, it is good practice to test thick-film conductors for metal migration prior to qualifying them for use. An effective and widely used test is the water-drop test. This simple accelerated test involves forming a pattern of lines pointed at the ends that are separated at several distances representative of the spacings that will be used in a circuit (Figure 4.17), applying a small bias (1-2 volts) across the gap, and bridging the gap with a drop of deionized water. For those metals that have a strong tendency to migrate, dendritic growth can be observed in a matter of seconds at the negative terminal.[15] The rapid dendritic growth from a silver-palladium thick film conductor after only 10 seconds at 4 volt bias is shown dramatically in Figure 4.18. For slower migrating conductors, the time required to develop a 10 mA-current with a 10 vdc applied bias across a 10 mil gap has been used to rate quantitatively the migration tendencies of various thick film conductors. Results for three thick film conductors, silver-platinum, silver-palladium, and gold, are shown in Figures 4.19 and 4.20.

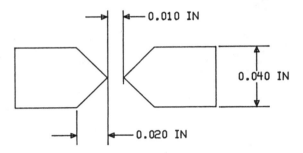

Figure 4.17: Migration test pattern.

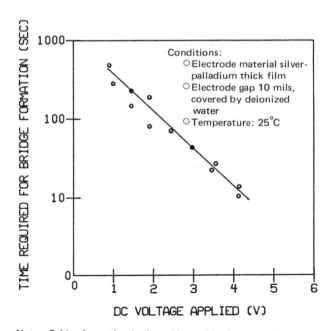

Note: Bridge formation indicated by sudden increase of current.

Figure 4.18: Time for bridge formation by silver migration in water as a function of applied voltage.[14]

**Silver-Platinum
Thick Film**

Complete bridging after
25 minutes with 4 volts
dc applied

Cathode (-)

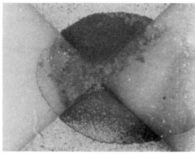

Anode (+)

**Silver-Palladium
Thick Film**

Some metal migration
evident after 1 hour
with 4 volts dc applied

Cathode (-)

Anode (+)

Figure 4.19: Metal migration water drop test results (Courtesy Rockwell International Corp.).

Gold Thick Film Conductor

No evidence of metal
migration after 1 hour
with 4 volts dc applied

Cathode (-)

Anode (+)

Figure 4.20: Metal migration water drop test for gold thick film (Courtesy Rockwell International Corp.).

Soldering and Solder Attachment. Tin/lead solder interconnections and attachments are generally avoided in hybrid circuits where chip (uncased) devices are used because of the high potential for tin/lead contamination of wire bonds, contamination of devices and circuits with flux, and solder splatter. Entrapment of these contaminants in a sealed package can cause degradation of wire bonds, corrosion, changes in device electrical parameters, and electrical shorting. However, solder is very much used in attaching and interconnecting packaged components to ceramic printed circuit boards. Thick film multilayer ceramic circuit boards are finding extensive applications for interconnecting surface mount components, both the leaded and leadless types. A popular sequence is to screen-print solder paste onto the thick film conductor pads, attach the components, and reflow the solder by vapor phase, infra-red, or wave reflow. In the use of solder on thick films there are two critical considerations: solderability and solder leach resistance.

Solderability. Solderability is the ability of a solder to quickly and uniformly wet a surface. Generally, all precious metal conductors and many non-noble metals (copper, nickel) are solderable. Poor solder wetting occurs when the metal surface has become contaminated externally or internally as with organic residues, inorganic salts, or through oxidation. Thick film pastes that contain large amounts of glass frit are also difficult to solder to. Commonly used techniques for optimizing solderability are:

Use of a flux, to remove surface oxides

Burnishing the surface mechanically

Plasma cleaning the surface, to remove organic residues

Chemically cleaning or etching the surface

Controlling the time-temperature schedule and the furnace atmosphere in the firing of the thick films

Employing thick film pastes that have a low glass content or no glass (fritless)

Solder Leach-Resistance. Solder leach-resistance is the reluctance of the thick film conductor to be absorbed, alloyed or otherwise removed by the hot solder. Controlled alloying of the solder with the top metal layer of the thick film conductor is important in forming a sound electrical and mechanical joint. This process, however, must not be allowed to go too far since the thin layer of precious metal can be completely absorbed (leached) by the solder. The commonly used techniques for minimizing or obviating solder leaching include:

Controlling and minimizing the time and temperature that the molten solder remains in contact with the thick film conductor

Increasing the thickness of the thick film conductor

Using thick film conductor pastes that contain a second metal that is known to suppress solder leaching.

The addition of platinum or palladium to gold or silver pastes improves their solder leach-resistance, but at the expense of reduced electrical conductivity. Thus, in the fabrication of multilayer substrates, these alloy pastes are used only for the top conductor layer or for the solderable pads. The inner conductor layers are generally formed from the unalloyed gold or silver paste.

Resistor Pastes

The early thick film resistors, introduced about 1960 to 1965, were based on combinations of silver, silver oxide, palladium, and palladium oxide with compatible binders and solvents. Because of the ease with which palladium oxide is reduced to free palladium, these early resistor formulations were very unstable. Small amounts of hydrogen generated from the chemical reaction of metals with moisture were sufficient to reduce the palladium oxide and reduce the resistor values. Today, high stability resistors, based on the thermodynamically more stable ruthenium (Ru^{+4}) oxides are available and have essentially supplanted the palladium-silver compositions. In these compositions, ruthenium dioxide (RuO_2), ternary oxides such as barium ruthenate ($BaRuO_3$) and bismuth ruthenate ($Bi_2Ru_2O_7$) are used. These oxides are not easily reduced or oxidized. Besides the ruthenium oxides other stable oxides of iridium, rhodium, and osmium can be used, but at a somewhat higher cost.

Thick film resistor pastes are available in sheet resistances as low as 1 ohm per square to as high as 10^9 ohms per square per mil thickness. Tolerances and stability, however, are much better for the intermediate ranges than for the very low or high values. Pastes of different sheet resistances can be formulated by varying the amount of glass that is blended with the metal or metal oxide phase; the higher the glass content the higher the sheet resistance. TCR values are also largely a function of the glass content. For low-value resistors, where the metal phase dominates, TCR values are positive and relatively low, whereas for the high sheet resistances, the glass phase dominates and TCRs are negative and higher. Glass content also has a significant effect on resistor noise; the high value resistors, because of their high glass content, will be noisier than low value resistors. The geometry of the resistor also affects noise. As the resistor increases in volume by either an increase in its thickness or area, the noise increases. Typical thick film resistor characteristics are given in Table 4.8.

Table 4.8: Typical Thick Film Resistor Characteristics

Tolerances, as fired	±10–±20%
Tolerances, laser trimmed	±0.5–±1%
TCRs	
5 to 100K ohms/sq (-55° to +125°C)	±100–±150 ppm/°C
100K to 10M ohms/sq (-55° to +125°C)	±150–±750 ppm/°C
Resistance drift after 1,000 hr at 150°C, no load*	+0.3 to –0.3%
Resistance drift after 1,000 hr at 85°C with 25 watts/in² *	0.25 to 0.3%
Resistance drift, short term overload (2.5 times rated voltage)	<0.5%
Voltage coefficient	20 ppm/(V) (in)
Noise (Quan-Tech)	
100 ohm/sq	–30 to –20 db
100K ohm/sq	0 to +20 db
Power rating	40–50 watts/in²

*Resistance drift curves for various sheet resistances and power loads are given in Figures 4.21 and 4.22.

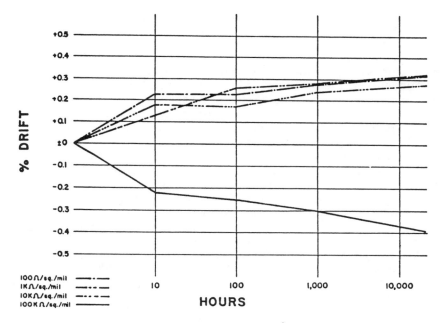

Figure 4.21: Resistor drift at 150°C, no load.

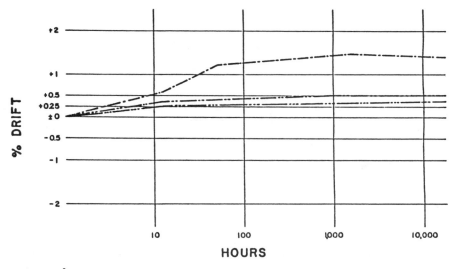

Figure 4.22: Resistor drift at 85°C under various loads for low value sheet resistance (100 ohms/sq/mil).

There are many excellent commercially available thick film resistor systems. Quantitative performance data for each resistor series may be obtained from the manufacturer, but, for high reliability applications, these data should be verified by the user. Each resistor system should be characterized experimentally by testing samples that simulate the values and geometries that will be used in a specific application. Besides these inherent material properties, several processing variations affect resistor values and performance. The nature of the conductor thick film that is used to terminate the resistors can have a profound effect on sheet resistance, noise, TCR, and drift. The characteristics of a resistor system terminated with gold thick film will differ from those terminated with silver, silver alloys, or other metallization. Even within formulations of the same metal, such as fritted gold versus fritless gold, differences will exist. This termination effect is more pronounced in small resistors (less than 50 mils in length) where resistor-conductor interactions become significant.

Resistors may be pre-terminated or post-terminated. In *pre-terminated resistors*, the conductor layer is screen-printed and fired onto the substrate first, forming the termination pads; then resistors are screen-printed such that their ends overlap the conductor pads. In *post-terminated resistors*, the resistors are screen-printed first, then the conductor circuit is screen-printed so that the terminals overlap the ends of the resistors (Figure 4.23).

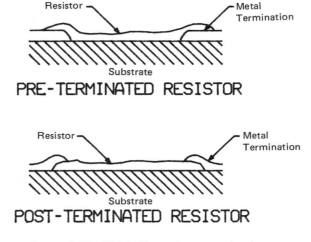

Figure 4.23: Thick film resistor terminations.

Fairly reproducible resistor values are obtained when resistors are pre-terminated; that is, screen-printed last through several layers of dielectric. However, when the thickness of the dielectric becomes too high, as with multilayer circuits having greater than three conductor layers, it is difficult to obtain reproducible dimensions and uniform thicknesses of the resistors because the screen must be pressed through deeper cavities. This creates a situation in which the paste tends to slump and run thus changing the desired aspect ratio and the resistor value. This effect is especially pronounced in small resistors. In these cases either greater allowance should be made in designing resistors so that they can subsequently be laser-trimmed to value or the resistors should be screen-printed on the top of dielectric. Post-terminated resistors hold their dimensions better and give more reproducible values than pre-terminated resistors. However, in fabricating multilayer circuits, post-terminated resistors are subjected to multiple high-temperature firings, which also can change their values. To maintain high stability, most manufacturers screen-print the resistors last, either on the top dielectric or onto the ceramic substrate through apertures left in the dielectric. Screen-printing the resistors on the top dielectric should be verified experimentally because of possible physical and chemical interactions that can occur between the resistor paste and the dielectric during firing.[16-18]

Dielectric Pastes

Thick film dielectrics are used as electrical insulating layers to separate conductor layers in a multilayer circuit, as insulation for crossovers, and as a protective overcoating for resistors and conductors. Generally, dielectric pastes consist of a mixture of alumina or other ceramic powder, a devitrifying glass (one that converts to a crystalline structure after firing but does not reflow upon reheating), a thixotropic organic binder, and organic solvents. Dielectrics may also contain temporary (fugitive) or permanent colored dyes. These dyes (usually blue) aid in registration and inspection. The temporary dyes are "burned out" during firing and become colorless, whereas the permanent dyes remain after firing and add some aesthetics to the circuit.

Most commercially available dielectric pastes have been formulated so that they can be fired between 850 and 950°C and, as such, are compatible with the furnace profiles used for the majority of resistors and conductors. Thus the same furnace settings can be used for all three functional pastes.

Some of the more important engineering parameters for thick film dielectrics are electrical values (insulation resistance, dielectric constant, dissipation factor, breakdown voltage), integrity of the film (pinholes and porosity), and via resolution.

Insulation Resistance. The insulation resistance of commercially available dielectric pastes is generally higher (greater than 10^{10} ohms) and remains high even under conditions of high humidity and elevated temperature. These values are excellent for most hybrid circuit applications where hermetically sealed packages are used. For thick-film ceramic printed circuits (unsealed), moisture and ionic contaminants can degrade (lower) the insulation resistance especially of the more porous dielectrics.

Dielectric Constant. Dielectric constants of dielectric pastes may range from 6 to 14 but typically are 9 to 10. These values are acceptable for most hybrid circuit applications. However, for very high frequency, high speed circuits (GHz range) a low dielectric constant of 2-4 is desirable and often mandatory. Dielectric pastes based on ceramic fillers that meet this requirement are not available. Consequently, quartz, polyimide, Teflon or other low dielectric constant materials must be used for these applications.

Dielectric Breakdown Voltage. Breakdown voltages for thick-film dielectrics are high (greater than 500 volts/mil) and adequate for most applications in which 1.5-2.0 mils-thick dielectric is used. For very high voltage circuits thicker (3-4 mils) dielectric layers can be screen-printed, if necessary.

Pinholes/Porosity. The physical integrity of the dielectric layers that separate conductor layers in thick-film ceramic multilayer boards is critical to both yield and reliability, particularly for large, high density ceramic multilayer boards. Reports of interlayer conductor shorting and failure of boards to survive bias and humidity demonstrate the importance of conducting tests on commercially available dielectric materials to determine their integrity before final selection and use in production.

Shorting can occur between successive conductor layers for two reasons: (1) the dielectric contains pinholes which become filled with conductor paste during the screen-printing process thus causing shorts to the underlying conductor layer and (2) under bias and moisture conditions some conductors, notably silver and copper, oxidize, form cations, then migrate through a porous dielectric, form metal dendrites, and again cause interlayer shorting. Dielectric porosity which allows moisture and ions to permeate results in decreased reliability because of decreased insulation resistance and may lead to catastrophic shorting.

Several tests may be used to evaluate and compare the integrity of dielectrics so that reliable materials and processing conditions may be selected. Among these tests are: (1) a pinhole identification test—detects pinholes or discontinuities in the dielectric (2) a water drop test—measures the porosity or permeability of the dielectric to water and ions and (3) a metal migration test—evaluates the susceptibility of dielectrics to metal migration. The test specimen that is used for all three tests consists of a capacitor structure as shown in Figure 4.24. Specimens are fabricated by screen-printing the dielectric between metal electrodes.

The *pinhole identification test* is an adaptation of a test previously developed to determine the integrity of silicon oxide passivation layers on semiconductor devices.[19] For the pinhole test, the top electrode is omitted. The apparatus consists essentially of an electrolytic plating bath. As shown in Figure 4.25, a d.c. potential (40 volts) is applied between the bottom electrode of the capacitor pattern (cathode) and a copper ring (anode), while the sample is immersed in a dilute aqueous solution of copper sulfate. If pinholes are present, metal dendrites will form at the pinhole sites. This results in electrophoretic decoration of these sites due to the electrolytic reduction of the copper ions. Thus the pinholes become visible, are easily identified, and can even be counted.

The *water drop test*[20] is used to determine the porosity or permeability of dielectrics to water and impurity ions by measuring the change in insulation resistance. According to this test, a drop of tap water is carefully placed on the capacitor pattern at the edge of the top electrode. Voltage is then applied between the capacitor electrodes and the inter-electrode

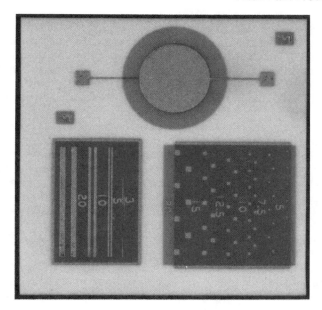

Figure 4.24: Test specimen for dielectric evaluation. Top: capacitor pattern to evaluate pinholes, porosity, dielectric constant and dissipation factor; bottom right: via pattern; bottom left: line-definition pattern.

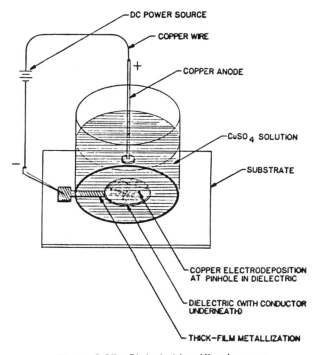

Figure 4.25: Pinhole identification test.

resistance is measured. With an applied voltage of 200 volts d.c., resistance is monitored for several minutes. Under these conditions, if the resistance drops below 10^9 ohms, the dielectric is considered to be porous and fails the test. In addition to selecting low-porosity dielectrics, most manufacturers screen-print two and sometimes three dielectric layers in order to reduce the probability of pinholes extending through the thickness of the dielectric.

The *metal migration test* is an extension of the water drop test. The capacitor is fabricated using silver or copper thick film conductors as the electrodes. The samples, containing a drop of deionized water, are exposed to a d.c. bias of about 100 volts for an extended time of about six hours. The specimens are biased such that the top electrode is rendered anodic and the bottom electrode cathodic. With this polarity, oxidation occurs at the top electrode and any resulting metal cations are reduced to free metal at the bottom electrode. This test is useful not only to evaluate the porosities of various dielectrics but also to assess the protection afforded by various organic coatings.[21]

Via Resolution. Typical via resolution is 10-15 mils, though with special pastes and screens, 7-mil vias have been achieved. Because the paste flows to some extent after screen-printing, the diameter of the via will be smaller than that defined in the screen pattern. Hence it is very easy for a small diameter via of 5 mils or less to close up completely. The resolution of vias should be experimentally determined for each dielectric paste by screen-printing a pattern having vias of different sizes (Figure 4.24).

Thick-Film Capacitors

Capacitors may be batch fabricated by screen-printing and firing a dielectric paste, much as is done with resistors. Capacitor pastes are composed of ferroelectric materials having high dielectric constants (several hundred to ten thousand). Titanates are extensively used as fillers in capacitor pastes. A typical paste consists of barium titanate formulated with the normal organic binders (ethyl cellulose), solvents (terpineol), and glass frit. The dielectric constant of barium titanate is about 1600 at room temperature, but abruptly increases to 6000 at the Curie temperature of 120°C (Figure 4.26).[22] Significant changes in dielectric properties occur at both the Curie temperature (120°C) and at 5°C because these are transition temperatures at which the crystal structure of barium titanate changes from tetragonal to cubic and from rhombic to tetragonal, respectively.[23] Ceramic capacitors based on barium titanate offer a wide range of dielectric properties. By varying the microstructure and chemical composition, different Curie temperatures and dielectric constants can be obtained. Additives such as lead titanate shift the Curie temperature above 120°C while strontium titanate will lower the Curie temperature. Compositions of barium titanate with small amounts of rare earth oxides (such as La_2O_3) or blends with other titanates ($SrTiO_3$) have been reported to shift the Curie temperature of barium titanate close to room temperature and even below 0°C. Other additives called depressers will also lower the dielectric constant. An excellent review of thick film capacitors based on the titanates has been given by Ulrich.[24]

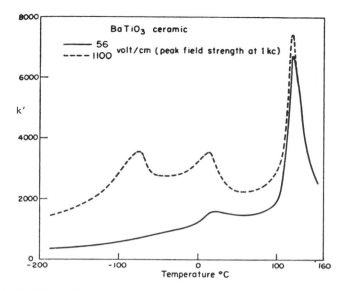

Figure 4.26: Dielectric constant of barium titanate ceramic as function of temperature. (Measurements of W.B. Westphal,[22] Laboratory for Insulation Research.)

Low value capacitors are formed by sandwiching the dielectric between a bottom and top conductor, in essence forming a parallel plate capacitor (Figure 4.27). The process sequence for batch fabricating capacitors is given in Figure 4.28. Where high value capacitors are required, the process sequence may be repeated so that a multilayer structure is formed. The numerous steps involved in screen-printing, drying, and firing such high capacitance capacitors must be weighed against simply buying the chip capacitors and assembling them. In fact, generally, discrete capacitors are used in hybrid circuits. Compared with screen-printed and fired capacitors, chip capacitors have closer tolerances, are more stable to temperature variations, take up less substrate area, and are less costly.

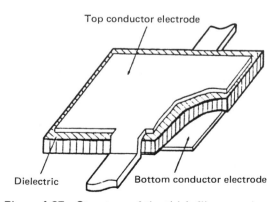

Figure 4.27: Structure of the thick film capacitor.

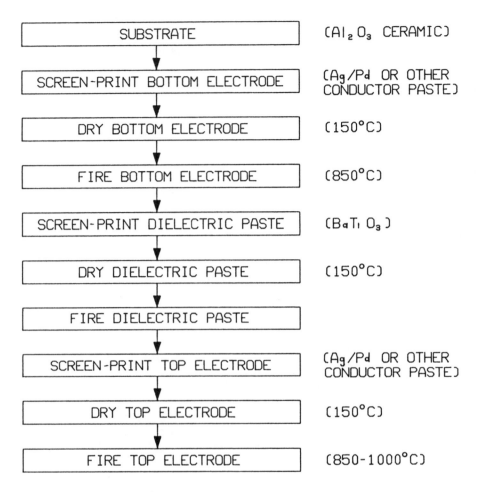

Figure 4.28: Process steps for thick film capacitor.

NON-NOBLE-METAL THICK FILMS

There has always been an interest in developing low cost non-noble-metal thick film pastes as replacements for gold pastes, but it wasn't until the price of gold approached 1,000 dollars per ounce (1979-1980) that industry became serious about developing non-noble metal pastes, in particular copper pastes. Besides its lower cost, copper has several other financial and technical benefits. Because of its abundance, the price of copper does not fluctuate as does that of gold, thus making it easier to price a product. Strict inventory controls and secure facilities are also not required. From a technical standpoint, adhesion of copper to ceramic substrates is better than for gold, often twice as high; the electrical conductivity is slightly higher; and solderability and solder leach resistance are much better. On the negative side: copper pastes require firing in a nitrogen atmosphere and some of their cost advantage is offset by the cost of nitrogen; special, higher priced dielectric pastes are required to prevent warping or bowing of the substrate; a full range of resistor pastes fireable in nitrogen and compatible with the copper is not available; and copper is susceptible to corrosion and must be protected in handling, storage, and in the final product.

Of the three main non-noble metal options—copper, nickel, and aluminum—the greatest amount of development work has been performed on copper. Thick-film aluminum pastes, though they would be ideal in keeping all interconnections mono-metallic, are not commercially available. The industry has not been successful in developing an aluminum paste stable enough to withstand the multiple firings that are necessary to produce a multilayer circuit, even when fired in an inert atmosphere. Nickel pastes have been developed and are commercially available but are not used as direct replacements for gold because of their much lower electrical conductivity. Nickel pastes, however, are widely used as electrical contacts in display panels.

Extensive research and development have been performed on copper thick-film pastes, notably DuPont during the period 1975-1983. Copper paste formulations were optimized and firing conditions established so that copper conductors with high electrical conductivity and high adhesion to substrates were obtained. Concurrently, dielectric and resistor pastes, also fireable in nitrogen and compatible with the copper, were developed and furnace manufacturers designed and produced special inert gas ambient furnaces suitable for processing the new pastes. Though copper thick-film systems were thoroughly evaluated as replacements for gold in the fabrication of hybrid microcircuits there has been a reluctance to implement them because the cost savings relative to the total cost of the hybrid has not been considered significant. Moreover, some anomalies were discovered in aluminum wire bond reliability, and resistor pastes with a wide range of resistor values have not been available. However, copper thick-film conductors have found greater applicaton in the manufacture of large area ceramic printed circuit boards in which components or leadless chip carrier devices are solder-attached (Figure 4.29).

Processing of Copper Thick Films

Copper thick film pastes and their associated dielectric and resistor pastes are processed similar to their precious metal counterparts except that the high temperature firing step must be performed in a controlled inert atmosphere, commonly nitrogen. The pastes are first screen-printed through 325 mesh screens, then dried for 10 minutes at 120°C. Drying may be conducted in air but temperatures above 120°C should be avoided, again because of oxidation. After drying, the films are fired for 6-10 minutes at a peak temperature of 900°C, though temperatures between 850-950°C have also been successfully used. A typical temperature profile is shown in Figure 4.30.

Figure 4.29: Large (4 x 4 inch) ceramic printed circuit with leadless chip carriers solder-attached to copper thick-film conductor pads (Courtesy Rockwell International Corp.).

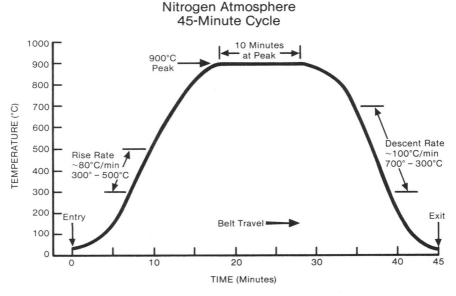

Figure 4.30: Furnace temperature profile for firing copper thick-film conductors (Courtesy DuPont).

An inert (nitrogen) atmosphere is essential in firing copper pastes but a compromise has to be achieved in the purity of the nitrogen used. Pure nitrogen (no air/oxygen) is important in preventing oxidation of the copper; yet some oxygen must be present to oxidize and remove the organic binders in the paste. Oxidation of copper is manifested in discoloration of the films, from a copper-pink color (no oxidation) to gray or purple (oxidation); poor wire bondability, poor solderability, and loss of film adhesion. The firing furnace should be equipped with an oxygen analyzer to monitor the amount of oxygen in various zones of the furnace. Oxygen concentrations of 6-8 ppm in the burn-out zone and less than 15-20 ppm in the sintering zone are considered satisfactory for copper processing. Extremely pure nitrogen presents a problem in the burn-out zone since some oxygen is necessary to oxidize and remove the organic binders. If incompletely removed, these organics remain an integral part of the conductor resulting in a sooty appearance. Thus, in some cases, controlled amounts of oxygen must be introduced. Data on the effects of varying quantities of oxygen on the electrical and physical characteristics of copper conductors have been reported.[25]

Characteristics of Copper Thick Film Conductors

The main advantage in using copper thick-film conductors is their much lower cost compared with the precious metal compositions. Copper ores are among the most abundant in the earth's crust. In fact, there has been such an over-supply that many mines in the Southwest United States have been closed. Besides its low cost, copper has several engineering advantages over gold thick films, among which are:

Higher electrical conductivity

Improved adhesion to alumina ceramic substrates

Better solder wetting and solder leach resistance

Copper thick-film conductors have electrical sheet resistances in the range of 1.5 to 2 milliohms/square/mil compared to 2 to 5 milliohms/square/mil for gold conductors. This higher conductivity is of benefit in some microwave circuits. The resistivity of copper conductors remains fairly constant on aging at 150°C for 1000 hours (Figure 4.31) and on storage at 90% relative humidity at 40°C (Figure 4.32). The adhesive strength of copper conductors to 96% alumina substrates in both peel and tensile is also better than for gold. Adhesive strength remains fairly stable after temperature cycling (Figure 4.33). There is some loss of adhesion on aging at 150°C for 1,000 hours. The bulk of this loss occurs during the first 100 hours, but the remaining adhesive strength is still high and satisfactory for most applications (Figure 4.34).

There has been some concern about the reliability of aluminum wire ultrasonically bonded to copper thick film. Extensive studies have been carried out and reported by DuPont showing that no problems exist even after accelerated temperature aging (2000 hrs at 150°C) (Figure 4.35), temperature cycling (−55 to 150°C), and storage at 90% relative humidity.[26]

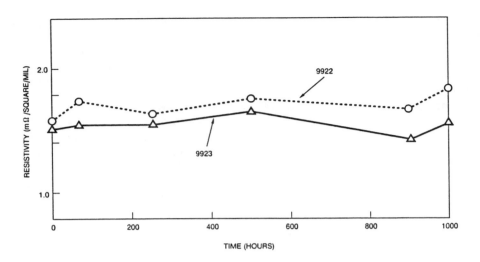

Figure 4.31: Resistivity of copper conductors as a function of storage at 150°C (Courtesy DuPont).

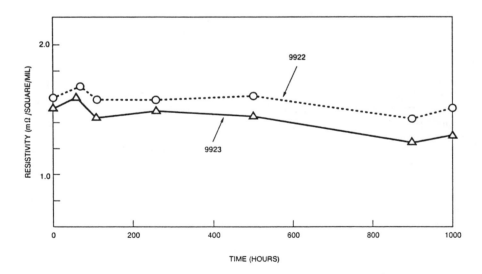

Figure 4.32: Resistivity of copper conductors as a function of storage at 40°C and 90% RH (Courtesy DuPont).

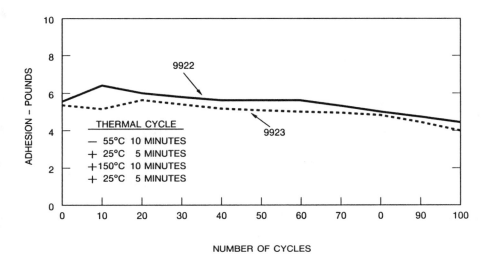

Figure 4.33: Effect of multiple temperature cycles on the adhesion of copper conductors (Courtesy DuPont).

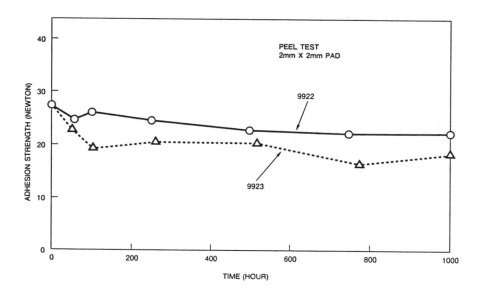

Figure 4.34: Adhesion of copper conductors aged at 150°C (Courtesy DuPont).

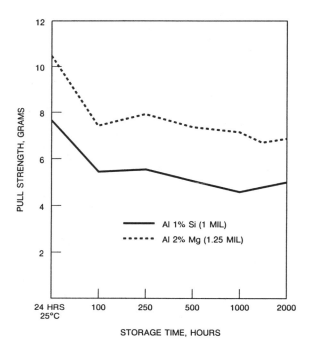

Figure 4.35: Wire bond pull strengths for DuPont 9922 Cu stored at 150°C.[26]

Similar long-term reliability results were obtained by Rockwell when aluminum-to-copper wire bonds were aged for 1000 hrs at 250°C in a nitrogen atmosphere.[27] The resistance of a series of a 222 bond chain was measured and divided by the number of bonds to give the resistance of one link. Under these conditions the results were superior to most of the aluminum-to-gold bonds that were simultaneously tested for comparison (Figure 4.36). However, a different situation arose when these bonds were evaluated in a hermetically sealed hybrid circuit in which components and substrate had been attached with epoxy adhesives. Aging of these circuits at 150°C showed a measurable increase in resistance after 50 hours indicating a degrading effect probably due to adhesive outgassing products.

Though tin-lead solder wetting of copper thick films should theoretically be better than for gold, problems have been encountered because of surface oxide formation and migration of the glassy phase of the copper paste to the surface during firing. Various methods have been used to prepare the surface to obtain better solder wetting, among which are mechanical abrasion, plasma etching, and chemical etching using hydrofluoric acid.[28,29]

Copper, unlike gold, is chemically reactive. It easily forms corrosion products if contaminated with chlorides or other inorganic salts. Copper

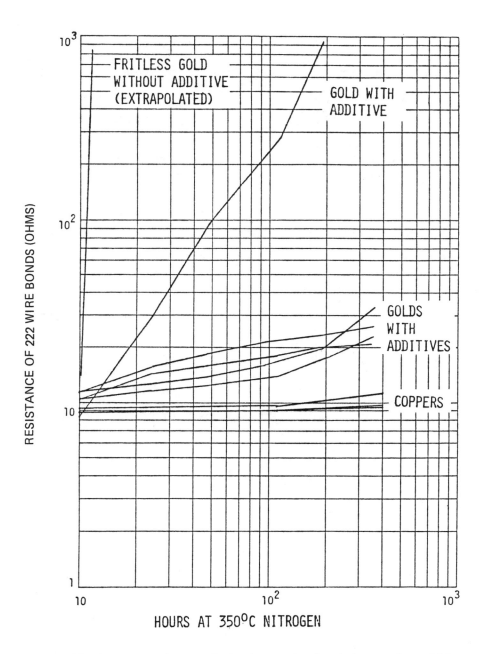

Figure 4.36: Resistance change of aluminum wire bonded to various thick films as a function of time at elevated temperatures.

conductors thus require special cleaning, handling, and storage conditions. Storage in dry nitrogen is recommended. If not hermetically sealed, circuits formed from copper conductors should be protected with a conformal organic coating.[30] Besides its susceptibility to corrosion, copper has been known to migrate between closely spaced conductor lines and through thick film dielectric under conditions of an applied bias, humidity and contaminant ions. Insulation resistance between conductor lines spaced 10 mils apart were 10^{13} to 10^{15} ohms in an initial dry condition, but dropped to 10^6 to 10^7 ohms after a 10 day humidity cycling test.[31] Metal migration that caused interline shorting was also demonstrated in some copper formulations when the water drop test for metal migration was used. Conformal coatings, however, provide effective barriers, generally preventing moisture from condensing on active electronic devices and circuit elements. To be effective the coating must have excellent adhesion to the substrate and must be free of ionic contaminants and pinholes. High-purity silicones have been found to be excellent in protecting copper film circuits.[30]

Processing of Nitrogen-Fired Dielectrics

As with the air-fired dielectric pastes, nitrogen-fired pastes are also screen-printed through mesh that is coarser than that used for the conductors. Typically, a 200-mesh stainless steel screen is used for dielectrics, while 325-400 mesh screens are used for conductors. Subsequent to screen-printing, the dielectric pastes are processed similar to the copper pastes; the same equipment and drying and firing schedules can be used. The furnace temperature profile can be the same as or similar to that employed for the copper conductors (Figure 4.30). To avoid pinholes that might protrude through the entire thickness of the dielectric layer, it is a common practice to screen several layers of dielectric with drying and firing after each screening. This reduces the probability of pin-holes becoming continuous. For air-fired dielectrics, screen-printing two dielectric layers between conductors has been found to be reliable. However, the nitrogen-fired dielectrics have been found to be more porous and metal migration has been reported to occur between conductors in a multilayer system. Thus the application of three and even four dielectric layers between conductors is not uncommon.

The early users of copper thick film systems encountered numerous problems in firing the dielectric. Because of the large amounts of dielectric used in fabricating a multilayer board, complete volatilization and removal of the organic binders in the burn-out zone of the furnace were difficult and resulted in blistering and lifting of the dielectric and splotching of the copper. With air-fired pastes, the organic vehicles are oxidized almost completely to carbon dioxide and water. At the furnace temperatures employed, both these by-products are non-condensable gases easily removed by the air flow and exhaust of the furnace. Nitrogen-fired pastes, however, are formulated using a combination of several organic compounds ranging from low to medium molecular weight polymers. Using very high purity nitrogen these binders are destroyed in the burn-out zone, not by the traditional oxidation mechanism but by pyrolysis (thermal cracking) in which low molecular weight fragments are formed, generally hydrocarbons. These by-products consist of several gases some of which are easily removed while others may condense to liquid droplets in the cooler parts of the furnace, then deposit onto the substrate. As larger multilayer interconnect boards were fabricated on a production scale, the flow rate of nitrogen became critical in removing the high concentrations of organic volatiles. Special modifications had to be made to the furnace to increase the nitrogen flow rate, especially in the burn-out zone. Flow patterns were optimized to sweep out the volatiles before they could condense or carbonize in the hotter furnace zone. Further provisions were made to preheat the nitrogen to avoid "cool" spots in the furnace where gases could condense. The introduction of controlled amounts of oxygen into the burn-out zone was found by others to improve the removal of the organics by allowing both the pyrolysis and oxidation mechanisms to occur simultaneously. With later improvements in the dielectric material and by controlling the amount of oxygen to about 600 ppm in the burn-out zone this problem was obviated. A delicate balance of oxygen was therefore required, high enough to fully remove the organics from the dielectric yet not so high as to oxidize the copper.

As larger multilayer interconnect boards were fabricated still another problem, that of warpage, arose. Applying thick layers of dielectric to one side of a large ceramic board ($>4 \times 4$ inches) induced stresses sufficient to warp the ceramic substrate. Investigations by DuPont into dielectric compositions with expansion coefficients more closely matching those of the ceramic substrates produced new compositions that minimized the warpage. Further, by applying dielectric to both sides of a ceramic board, stress effects could be compensated.

Processing of Nitrogen-Fired Resistors

Nitrogen-fired resistors are screen-printed through 325-mesh stainless steel screens over pre-fired copper terminations, dried at 120°C in air, then fired in the same inert ambient furnace and with the same thermal profile as used for the copper pastes. Only a few companies have developed resistor compositions that are compatible with the nitrogen-fired copper and even these companies have found it difficult to produce a complete range of resistor values from 10 ohms/sq to 1 megohm/sq. that are stable and compatible with the nitrogen-firing process. Generally, stability problems occur at the low and high ends of resistors values, e.g. at 10 ohms/sq and at 100 K to 1 megohm/sq. where very large TCRs and large resistance drifts occur on temperature aging.[32] To obviate the problem of developing stable nitrogen-fired resistor pastes, DuPont introduced the MYDAS system which allows the use of the existing, well-characterized air-fired resistors with copper conductors. According to this approach, the air-fired resistor pastes are screen-printed and fired first at 850°C, then the copper conductors are screened and fired in nitrogen. These copper pastes are fired at 600°C instead of the customary 850-900°C to minimize changes in the resistors.

POLYMER THICK FILMS

Polymer thick films (PTFs), like cermet thick films, are screenable pastes which can form conductor, resistor, and dielectric circuit functions. However, unlike cermet films, PTFs contain polymeric resins which remain an integral part of the final thick-film composition after processing. The key advantage of PTFs is that they can be processed at relatively low temperatures (120-165°C), namely at temperatures required to cure the resin. Cermet thick films, on the other hand, require temperatures of 850-1000°C not only to completely burn off all the organic binders but also to effect melting of the glass frit and sintering of the ceramic particles. A further advantage of PTFs is their low cost for both the material and processing. They have received wide acceptance in commercial products, notably for flexible membrane switches, touch keyboards, automotive parts, and telecommunications. Recently, there has been a resurgence of interest and activity in PTFs because of their application to surface mount devices resulting in replacement of the conventional double-sided plated through-hole boards with single-sided boards having no through-holes. Components may be attached with solder paste, then vapor phase reflow soldered or wave soldered. Thus, an engineer designing printed circuit boards can avoid the use of leaded devices, plated through-holes, and hazardous plating and etching chemicals (Figure 4.37). Because of their low cure temperatures, PTFs may be processed on both ceramic and plastic substrates. The key advantage of using plastic substrates is that thick-film resistors can be batch-screened, avoiding the previous costs in attaching and solder connecting discrete chip resistors. To complement the PTF conductors and resistors, dielectric pastes are also available and are useful in fabricating multilayer substrates.

PTF Conductors

Polymer conductors are formulated with three major ingredients: a polymeric material, a functional material, and a solvent.

The polymeric material is one that can be cured (hardened) at low temperatures. The cured polymer provides integrity to the film and adhesion to the substrate. A variety of polymeric binders may be used. For flexible printed circuits and membrane switches, thermoplastic polymers such as acrylics, polyesters, or vinyl co-polymers are used. For rigid substrates, the thermosetting polymers including epoxy, polyimide, and phenolic may be used.

The functional material is generally a metal that combines high electrical conductivity, good solderability, and low cost. The best compromise is silver. The silver PTFs are solderable using the classical tin-lead or tin-lead-silver solders, but the contact time with the molten solder is critical. Extended contact time causes leaching of the silver and loss of adhesion. Some characteristics of a silver-based PTF are given in Table 4.9.

The solvent should have a high boiling temperature. Its function is to dissolve the resin and control the rheological properties of the paste so that it can be easily screen-printed. The solvent evaporates during the curing cycle.

Shown actual size on the left is a typical 2-sided printed circuit board used in a communication application. Above is the same PC board duplicated as a single-sided polymer circuit. The resulting miniaturization and use of polymer conductors, resistors and dielectrics have reduced its size and weight by 65% with a cost savings of 35%.

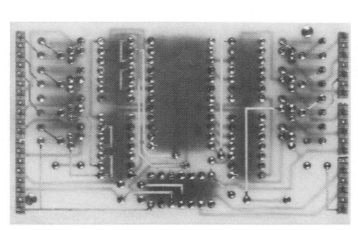

Figure 4.37: Size reduction of printed circuit board using polymer thick films. These PC boards were produced by Modev Developments Inc. using EMCA polymer materials.

Table 4.9: Characteristics of PTF Conductors (Silver Based)*

Electrical conductivity	0.038 ohms/sq/mil
Adhesion	1,500 psi tensile on FR-4 epoxy board
Solderability	>95% coverage with RMA flux and 62/36/2 tin/lead/silver solder
Silver migration	Less than for silver cermet conductors but may require polymer protective coating
Line definition	7–10 mils when screened through a 200 to 250 mesh screen

*EMCA silver PTF cured 30 min at 165°C.

Commercially available conductor pastes are limited to silver compositions, though a few silver-palladium pastes are also available. The amount of metal filler may range from 60 to 80% but 66-77% is typical. Cure temperatures range from 120°C for 1 to 2 hours to 350°C for 30 minutes and may be performed in either convection or infrared furnaces. Curing in air is generally satisfactory except for the high temperature curing compositions where some circuit materials (e.g. copper) may degrade through oxidation. Curing a polymer at temperatures above 250°C for an extended time may also result in decomposition of the polymer and adverse changes in its electrical and physical properties. In such cases, curing may be effected in a nitrogen ambient. PTF conductor pastes are generally screen-printed through 200-250-mesh screens and provide line widths and spacings of approximately 7-10 mils. Cured thicknesses are 25-30 microns.

PTF Resistors

Polymer thick film resistors are formulated similar to conductors except that, instead of metals, resistive fillers are used as the functional component. Resistive fillers consist of carbon or graphite compositions. Resistor values may be varied and controlled by controlling the particle size and concentration of the filler, the type of polymer binder used, and the cure schedule.

Resistor pastes having sheet resistivities ranging from 10 ohms/sq. to 1 gigaohm/sq. are commercially available; however, resistor stabilities are much better for the low sheet resistance pastes. Reported TCRs are −100 to −200 ppm/°C for the 100 ohm to 100K ohm resistor pastes, −500 ppm/°C for the 1 megohm pastes, and −800 ppm/°C for the 10 megohm pastes when measured at 125°C[33] (Table 4.10). Because of their low-temperature curing properties, PTF resistors may be used on a wide variety of substrate materials including low-cost plastic boards such as epoxy, polyimide, polysufone, and phenolic. In contrast, cermet resistors can only be applied to ceramic or other high-temperature-stable substrates because of the high temperatures (>850°C) required to fire them. Resistance values for polymer thick films are much more sensitive to the nature of the

Table 4.10: Characteristics of PTF Resistors*

TCR (25° to 125°C)		±200 ppm/°C for 10; 100; 1K; 10K ohms/sq		
		±300 ppm/°C for 100K ohms/sq		
		±500 ppm/°C for 1 Megohm/sq		
		±800 ppm/°C for 10 Megohm/sq		
ΔR after cure		−5 to −15%		
ΔR after burn-in (85°C/10 hr)		−5 to −10%		
ΔR after burn-in and 85°C storage		1K	10K	100K
24 hr		0	0	0
100 hr		0	0	−0.5%
250 hr		0	0	−0.6%
ΔR after 1,000 hr @ 85°C		<0.5%		
Power rating		5 watts/sq in		

*EMCA Bulletin 515, 4000 Series Polymer Resistor Ink.

substrate than resistance values for cermet resistors. Early applications of PTF resistors were plagued by large changes in resistance that occurred during temperature cycling, elevated temperature burn-in, or aging. It was soon discovered that the glass transition temperature, T_g, of the substrate had a major influence on resistor stability. At or above the T_g the molecular structure of the plastic substrate relaxes and experiences a sharp increase in expansion coefficient. A thermal mismatch then occurs between the resistor and the substrate which degrades the resistors both physically and electrically. It is therefore important to select a substrate that has an expansion coefficient closely matching that of the resistors over the temperature range in which the circuit will be operated and tested. A T_g that is higher than any temperature to which the circuit will be subjected is desirable. Other substrate properties that have been found important in achieving reproducible and stable resistance values include low porosity, high surface smoothness, completeness of cure, and lack of moisture, contaminants or outgassing products.

PTF Dielectrics

Several polymer thick film dielectrics are commercially available and may be used as insulation layers in fabricating multilayer circuits or as overcoatings for resistors and conductors. Vias of 15-20 mils in diameter may be formed during screen-printing. The processing steps and conditions for dielectric pastes are very similar to those for conductors and resistors, for example, screen-printing through a 200 mesh stainless steel screen and curing for 30 min. at 150-175°C or for 10 min. if infrared heating is used. Volume resistivity is greater than 10^{12} ohm-cm and dielectric constants range from 4-8 at 1 KHz, depending on the composition. Breakdown voltages of greater than 1000 volts/mil have been reported.[34]

REFERENCES

1. Franconville, F., Kurzweil, K., and Stalnecker, S.G., "Screen: Essential Tool For Thick-Film Printing," *Solid State Technology*, Oct. 1974.
2. Trease, R.E. and Dietz, R.L., "Rheology of Pastes in Thick Film Printing," *Solid State Technology*, Jan. 1972.
3. Vitriol, W. and Steinberg, J.I., "Development of a Low Temperature Co-fired Multilayer Ceramic Technology," *Proc. ISHM*, 1982.
4. Steinberg, J.I., Horowitz, S.J., and Bacher, R.J., "Low Temperature Co-fired Tape Dielectric Material Systems For Multilayer Interconnection," *Solid State Technology*, Jan. 1986.
5. Smith, R.R. and Dietz, R.L., "An Innovation in Gold Paste," *ISHM Proc.*, 1972.
6. Riemer, D.E., "Optimized Adhesion Mechanism for Thick Film Multilayer Gold," *Electronic Components Conference Proc.*, 1985.
7. Harper, C.A. (Ed.), *Handbook of Thick Film Hybrid Microelectronics*, McGraw-Hill, 1982, "Conductor Materials, Processing, and Controls," by Hicks, W.T.
8. Kohman, G.T., Hermance, H.W., and Downes, G.H., "Silver Migration in Electrical Insulation," *Bell Systems Technical Journal*, Vol. 34, No. 6, Nov. 1955.
9. Short, O.A., "Silver Migration in Electric Circuits," *Tele-Tech & Electronic Industries*, Vol. 15, Feb. 1956.
10. Williams, J.C. and Herrmann, D.B., "Surface Resistivity of Non-porous Ceramic and Organic Insulating Materials at High Humidity with Observations of Associated Silver Migration," *I.R.E. Trans. on Reliability and Quality Control*, PGROC-6, Feb. 1956.
11. Chaikin, S.W., "Study of Effects and Control of Surface Contaminants on Dielectric Materials," *Final Report, U.S. Army Signal Engr. Labs.*, DA-36-039-SC-64454, Stanford Research Institute, Jan. 1958.
12. Chaikin, S.W., Janney, J., Church, F.M., and McClelland, C.W., "Silver Migration and Printed Wiring," *Industrial and Engineering Chemistry*, Vol. 51, No. 3, March 1959.
13. Shumka, A. and Piety, R., "Migrated-Gold Resistive Shorts in Microcircuits," *13th Annual Proc. Reliability Physics Symp.*, 1975.
14. Riemer, D.E., "Material Selection and Design Guidelines For Migration-Resistant Thick Film Circuits With Silver Bearing Conductors," *Proc. 31st Electronic Components Conf.*, 1981.
15. Frankel, H.C., "Water Drop Test For Thick Film Conductor Migration Tendency," U.S. Army Electronics Research and Developent Command, Memorandum File Reports 78-2 and 78-3, 1978.
16. Shah, J.S. and Berrin, L., "Mechanism and Control of Post-trim Drift of Laser Trimmed Thick Film Resistors," *IEEE, CHMT-1*, No. 2, 1978.
17. Coleman, M.V., "The Effect of Nitrogen and Nitrogen-Hydrogen Atmospheres on the Stability of Thick Film Resistors," *Proc. ISHM*, 1979.
18. Agnew, J., *Thick Film Technology*, Hayden Book Co., Rochelle Park, N.J., 1973.
19. Lee, S.M. and Eisenberg, P.H., *Insulation*, Vol. 15, 1969.
20. Wood, D.C., "Water Drop Test For Hermetic Integrity of Dielectrics," *Thick Film Systems*, STP-104, 1978.
21. Soykin, C.A., Perkins, K.L., and Licari, J.J., "Test Methods For Thick Film Dielectrics," *Proc. SAMPE, Natl. Tech. Conf.*, Oct. 1984.
22. VonHippel, A.R., Ed., *Dielectric Materials and Applications*, M.I.T. Press, 1961.
23. Horiike, S., Tanahashi, M., and Aoi, T., "Thick Film Capacitor with Excellent Thermal Characteristics," *ISHM Journal*, Vol. 8, No. 1, March 1985.
24. Ulrich, D.R., "Dielectric Materials, Processing, and Controls," in *Handbook of Thick Film Hybrid Microelectronics*, Harper, C.A., Ed., McGraw-Hill, 1982.

25. Hayduk, E.A. and Taschler, D.R., "Atmosphere Control and the Copper Thick Film Firing Process," *Hybrid Circuit Technology*, Jan. 1986.

26. Pitt, V.A. and Needes, C.R.S., "Ultrasonic Aluminum Wire Bonding to Copper Conductors," Proc. *31st Electronic Components Conf.*, 1981.

27. Kubik, E.C. and Licari, J.J., "An Evaluation of Gold and Aluminum Ultrasonic Bonds To Copper Thick Film Conductors," *ISHM*, Chicago, 1981.

28. Liu, T.S., Pitkanen, D.E., and McIver, C.H., "Surface Treatments and Bondability of Copper Thick Film Circuits," *Proc. 31st Electronic Components Conf.*, 1981.

29. Goldfarb, H., Moudy, L.A., Soykin, C.A., and Licari, J.J., "Surface Preparation of Copper Thick Film Conductors For Improved Solderability," *Proc. NEPCON West*, Anaheim, CA, 1983.

30. Pitkaner, D.E. and Speerschneider, C.J., "Environmental Effects on Thick Films Microcircuits," *IEEE Trans. Components, Hybrids, and Manufacturing Tech.*, CHMT-4, No. 3, Sept. 1981.

31. Licari, J.J., Soykin, C.A., and Watson, J.C., "Stability of Copper Thick Films in Electronic Circuits," *27th Natl. SAMPE Symposium*, May 1982.

32. Watson, J.C., Perkins, K.L., and Licari, J.J., "High Reliability Non-Noble Metal Hybrid Systems," *AFWAL-TR-83-4077*, Sept. 1983.

33. Martin, F.W. and Shahbazi, S., "Polymer Thick Film For Reliability Applications," *Hybrid Circuits Journal*, ISHM-UK, No. 3, Autumn 1983.

34. *EMCA Bulletin 531*, 4552 Polymer Multilayer Dielectric.

5

Resistor Trimming

A major advantage of hybrid thick- or thin-film resistors is their ability to be trimmed to precise values. Trimming is analogous to fine-tuning a component. It consists of the selective and controlled removal of resistive material until a predetermined electrical value is achieved.

Trimming is required because slight variations in hybrid circuit processing make it difficult to guarantee a precise resistor value. Screened and fired thick-film resistors have ±20 percent tolerance as screened, while thin-film resistors, as deposited, have a ±10 percent tolerance. Up-trimming, or cutting into the resistor to increase its resistance value, is the method most often used to reach the selected value. Down-trimming can be done if the layout includes gold pads on the resistor where, by wire bonding, a section of the resistor can be jumpered. Resistors are purposely designed on the low side so that they can be trimmed up to the required value. During trimming, the resistor value is monitored as the material is removed, causing the resistance value to increase until the required value is reached (see Chapter 10 for a description of the mechanism of resistor value increase with removal of material). There are two basic types of trimming: static and dynamic. Static (passive) trimming consists of removal of material to obtain a resistance value within a specified tolerance without power being applied. Dynamic trimming, also referred to as functional trimming, consists of the removal of resistor material while the hybrid circuit is under power. Dynamic trimming is important to obtain specific circuit functions such as a filter characteristic, a comparator trip point, or a voltage regulator correct output voltage.

Thin-film resistors can be trimmed to ±0.1 percent of value and thick-film resistors to ±1.0 percent. Even closer tolerances can be obtained with special design considerations. In trimming resistors that have been screen-printed on thick-film dielectric, extra care and controls are necessary to avoid excessive penetration into the dielectric layer. Trimmed resistors on dielectric are not as stable as trimmed resistors deposited directly onto alumina substrates. However, in either case, resistors (<100 K ohms) drift less than 0.1 percent after 500 hours at 150°C and are satisfactory for most applications.

LASER TRIMMING

Laser trimming of hybrid circuit resistors has been employed since the early sixties. Today it is still the most widely used method for trimming both thin and thick film resistors. The laser system most widely used in the hybrid industry is based on a neodymium-doped yttrium aluminum garnet (YAG) crystal. The YAG laser is better than the CO_2 laser system because its shorter infrared wavelength (1.06 micrometers) permits smaller, narrower cuts and minimizes damage to the peripheral resistive material and to the underlying dielectric. Because the YAG laser beam is invisible, most trim systems also incorporate a helium-neon laser collinear with the YAG. The helium-neon laser emits a visible red beam that facilitates the positioning of the YAG beam. The laser beam is shaped and focused through a series of lenses and mirrors until it ultimately strikes the resistor material whereupon the energy is absorbed and the material heats up. Most of the material vaporizes, some melts and resolidifies, while the remainder is ejected as particles (Figure 5.1).

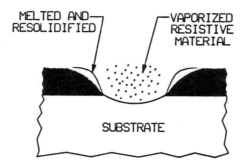

Figure 5.1: Cross section of laser cut (not to scale).

Laser cuts are created by a series of overlapping spots, a typical spot size being 0.001 inch in diameter (Figure 5.2). Controlling the power of the laser beam is important in producing stable resistors. A clean, well-defined kerf must be formed without damaging adjacent resistive material. Residues left in the kerf result in shunting paths for current flow and instability in the resistors. Trimming can also cause thermal shock and microcracking in adjacent resistor material, further degrading resistor stability. Setting the laser parameters to provide short pulses of high peak power minimizes or avoids these problems. Short pulse durations of 50-200 nanoseconds obtained through a Q-switch technique result in pulses of high peak power even though the average power is kept low. These conditions provide rapid vaporization with a minimum of heat flow into the bordering regions. Some common terms associated with laser trimming are defined in Figure 5.3.

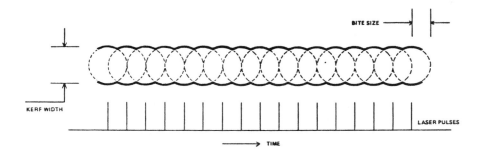

For a fixed hole size, a ragged kerf results if bite size is too large.

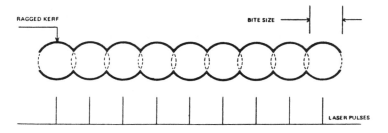

Figure 5.2: Top view of laser trim cut (Courtesy Chicago Laser Systems Inc.).

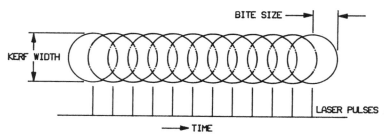

Bite Size—The amount of additional material attacked with each laser pulse.

Kerf Width—The outer width of the cut.

Q-Rate—The number of laser pulses issued per second.

Table Speed, Beam Speed, Cut Speed—The rate of material removal in inches per second.

Hole Size, Spot Size—Diameter of the material removed. The diameter of the hole is related to the optical spot size, but is not necessarily the same. It varies with the material being cut and the power level of the laser pulse.

Figure 5.3: Laser trimming glossary.[1]

A graphic representation of a laser trim system for hybrid microcircuits is shown in Figure 5.4, a block diagram in Figure 5.5, and the mirror system in Figure 5.6. The laser-trim process has been completely automated; the laser beam and the x-y table are controlled by a microcomputer. Most systems are equipped with a TV camera to monitor the trim process. Some commercially available laser trim systems are shown in Figures 5.7-5.10. The ESI trimmer consists of a computer-controlled laser, positioning system, and measurement system. It is capable of producing spot sizes ranging from 0.25 to 2.0 mils diameter over a 3 × 3 inch working area. More advanced laser trimmers that interface with Computer-Aided Design (CAD) equipment are now available. The CAD system assists the design engineer in laying out a resistor so that it can be trimmed faster and more reliably by matching the geometry of the resistor to the tolerance needed. The start and end of the trim area can be derived directly from the CAD plot. Data such as tolerances, trim speeds, and laser rate can be stored in the CAD database at the time of hybrid design. All these advantages, however, are at the expense of having to develop more complex software.

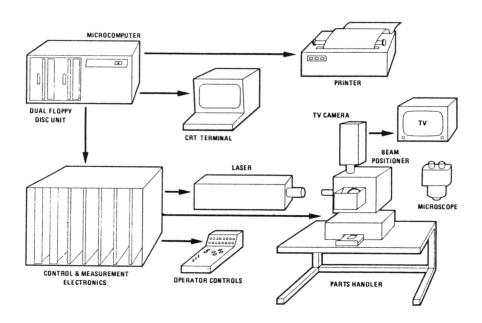

Figure 5.4: Graphic representation of laser trim station.

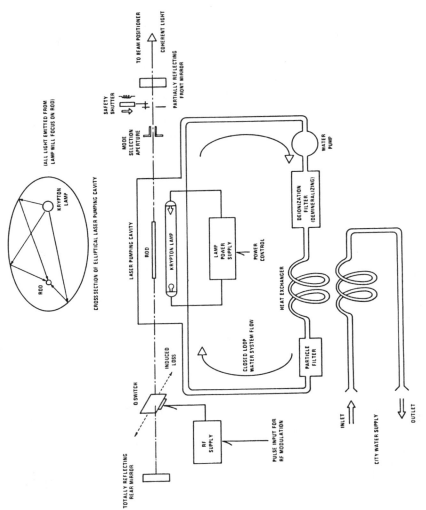

Figure 5.5: Laser system block diagram (Courtesy Chicago Laser Systems).

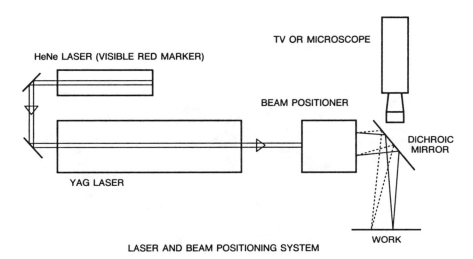

LASER AND BEAM POSITIONING SYSTEM

THE OPTICAL SYSTEM IS FOLDED TO PROVIDE A COMPACT PACKAGE AND THE
WORK IS MONITORED BY A CCTV THROUGH A DICHROIC MIRROR

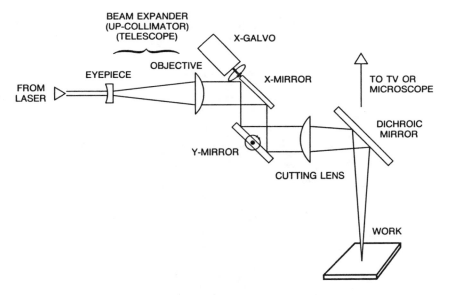

OPTICAL PATH THROUGH BEAM POSITIONER

Figure 5.6: Laser mirror system (Courtesy of Chicago Laser Systems).

Figure 5.7: ESI Model 44 Laser Trimmer (Courtesy ESI).

Figure 5.8: Laser trim station, Chicago Laser Systems 37S (Courtesy Chicago Laser Systems).

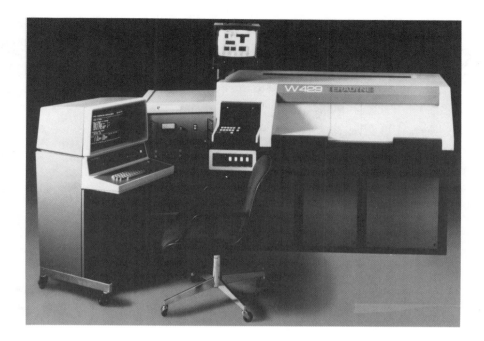

Figure 5.9: Laser trim station, Teradyne W429 (Courtesy Teradyne).

Figure 5.10: Laser optics of Teradyne W429 trim system (Courtesy Teradyne).

Basically, a laser-trim station operates as follows:

1. The resistor is probed.
2. A digital voltmeter measures the resistor value.
3. The laser is positioned at the start point of the trim. This point is programmed into the computer.
4. The laser pulses.
5. The voltmeter takes another reading.
6. The computer compares that reading with the required reading and either shuts the laser off, if the value is within tolerance, or it pulses the laser again.
7. This iterative process continues until the resistor value is within the tolerance specified in the program.

Figure 5.11 is a flow chart of the above description. The trim time may be as short as 0.002 seconds for a wide-tolerance resistor or as long as two seconds for a close-tolerance resistor. If there are many resistors to trim, even these trim times can become a throughput-limited factor. Several options are available to shorten the trim process time even further. Figure 5.12 shows another flow chart for a faster trim technique. The steps for one option are:

1. The resistor is probed.
2. A digital voltmeter reads the resistor value.
3. The computer compares the required value with the measured value.
4. Information previously stored in the computer relates the length of trim with the percentage increase.
5. Based on step 4, the computer calculates the length of trim to bring the resistor within a few percent of the desired value.
6. The computer positions the laser and directs it to trim for the length calculated in step 5.
7. Another measurement is then taken and steps 3 through 7 are repeated.

This technique requires few measurement/laser on-off steps and results in a faster trim sequence. Increased speed is achieved at the expense of longer and more complicated software development but, if a large number of hybrids of the same type are to be produced, the software time and cost becomes small compared with the savings effected by the faster throughput of substrates.

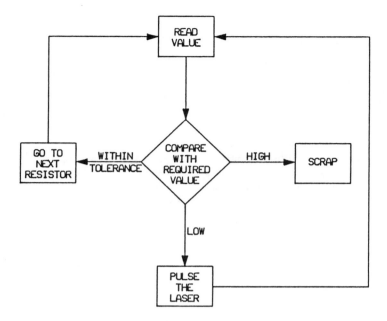

Figure 5.11: Standard laser trim system.

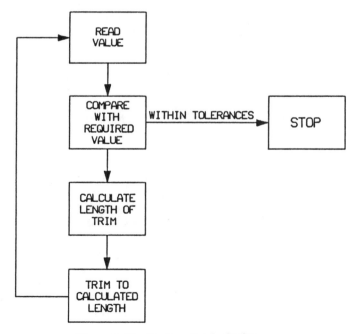

Figure 5.12: Fast method of trim.

ABRASIVE TRIMMING

Abrasive trimming, the first method developed for resistor trimming, is not extensively used today. It is slower than laser trimming, but produces more stable resistors because heat is not involved in the process and therefore no stress or microcracks occur in the material.[1]

In abrasive trimming, fine-grained sand is forced through a small nozzle under high pressure. The sand abrades and removes the unwanted resistor material until the desired resistor value is obtained. Resistor debris, a potential contaminant in a hybrid circuit, is the main drawback in using abrasive trimming. Abrasive trimming can increase the resistor value in one of two ways: (1) By removing a material to form a kerf as with laser trimming. This increases the number of resistor squares, thus increasing the value. (2) By reducing the thickness of the resistor film. Reducing the thickness increases the sheet resistivity, thus changing the value.

Table 5.1 shows the comparison of laser and abrasive trimming.

Table 5.1: Comparison of Trimming Procedures

Advantages	Limitations
. . . . Laser Trimming	
High speed	Intense heat at trim area can
Permits data logging	cause microcracks
Automated	Cracks cause noise in high value
Clean	resistors (over 5 megohms)
Provides self-annealing resistors	Large capital investment
	Requires software development
. . . . Abrasive Trimming	
Ease of setup	Slow process
Low capital cost	Dirty process
Stable resistors	Produces large kerf
Low noise for high value resistors	

RESISTOR PROBING/MEASUREMENT TECHNIQUES

The following discussion deals with probing and measuring resistors that need to be trimmed. To make accurate measurements two criteria must be met. First, the resistor must be accurately probed and then an accurate method of measurement must be applied.

Probe Cards

The first step in setting up an automated trim system is to design a probe card. The cards are typically 4 by 5 inches and contain 30 to 40 probes. The probes must be located so that they do not "shadow" any resistors from the trim medium. If the hybrid circuit contains many resistors, two or three probe cards may be necessary to access all of them. Figure 5.13 shows photos of typical probe cards.

- Commonly Used Probe Cards

- Blade Type Card

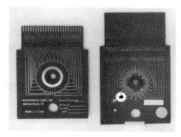

- Epoxy Ring Cards

- C-2200 Round Card and Plug-In
Adaptor Card.

Figure 5.13: Probe card types for resistor trimming (Courtesy Inter-Logic Inc.).

Two probing techniques will now be discussed: two-point and four-point probe techniques.

Two-Point Probing

Two-point probing is the conventional and simplest method for measuring a resistor value.It involves only two probes placed across the resistor being measured (Figure 5.14).

This system is adequate for most resistor measurements. However, errors are introduced by the resistances of the probe, probe contact, and resistor termination pads. Resistance errors can be as much as 0.2 ohms. For a 1000 ohm resistor, the probe introduces only a 0.02 percent error. However, for low-valued resistors, the error is significant; for example, if the resistor value is 10 ohms, the error is 2.0 percent. Therefore, for low-value or high-precision resistors, the two-point probing system is not recommended. Another disadvantage of this system is that the error can vary because of variations in the contact pressure or cleanliness of the probes.

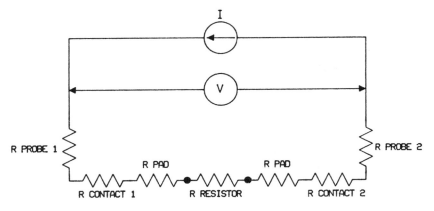

Figure 5.14: Two-point probe system.

Four-Point Probing

This system utilizes four probes (Figure 5.15) and is the method that should be used for low resistor values or for high-precision resistors. Two probes inject the current into the measurement loop and the other two probes are placed close to the resistor termination and connected to the digital voltmeter. In this system the only error term that is introduced is the pad resistance which is less than 0.010 ohms maximum. The probe and probe resistances disappear as error terms since the digital voltmeter is a high-impedance instrument and will not disrupt the current flow in the loop. Even when measuring a 10 ohm resistor, the error term is only 0.1 percent maximum. Since most thick-film gold pads are typically 0.003 ohms, the error in the 10 ohm measurement would only be 0.03 percent. The disadvantage of this method is that it uses twice the number of probes, thus limiting the number of resistors that can be probed per card. Also, since there are four probes per resistor, the hybrid layout must allocate enough space for the extra probes.

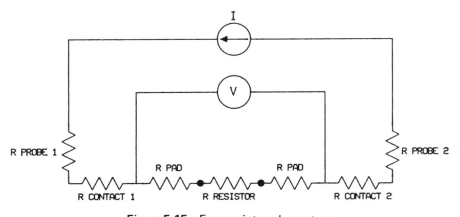

Figure 5.15: Four-point probe system.

Digital Voltmeters (DVM)

The voltmeters used in most trim systems are very accurate. The DVM used in the ESI Model 44 Laser Trim station uses a full Kelvin Wheatstone bridge method to obtain a measurement accuracy of ±0.001 percent.

TYPES OF RESISTOR TRIMS

There are many types of cuts that can be used to trim a resistor but the most popular are: (1) plunge, (2) L-cut, (3) scan, (4) serpentine (Figure 5.16).

Plunge-Cut

The plunge-cut is fast and typically used on resistors of one square or less, and on top-hat resistor designs. This type of trim causes the most disturbance of the current through the resistor. This also causes a hot spot to form at the top of the trim.

Double-Plunge-Cut

The double-plunge, or shadow, cut allows a coarse trim followed by a fine trim in the "shadow" of the first cut. Laser damage is less than in the L-cut since the microcracks do not tend to cause the resistor to open.[2] However, this cut can cause an even bigger "hot-spot" than the single-plunge-cut.[3]

L-Cut

The L-cut provides more accuracy than the plunge-cut. The perpendicular leg provides a coarse trim while the parallel leg provides a fine adjust. The angular and J-cut are more stable with fewer hot spots than the standard L-cut. This is due to the removal of any sharp turns in the trim. As stated previously, laser trimming causes microcracks. Since microcracks are perpendicular to the kerf, in the L-cut trim, they can propagate to the edge of the resistor increasing the value or, worse, causing the resistor to open.

Scan-Cut

The scan-cut is the slowest of those described but has the highest accuracy and provides the most stable resistor because it does not disturb the current flow through the resistor as much as the other trim configurations. Scan-cuts are better for high-frequency operation. They result in low thermal noise because no hot spots are formed in the resistor.

Serpentine-Cut

The serpentine-cut is used when a large resistance change is required. It does this by creating an increase in the current path length. It must be used on large area resistor designs. This is the type that would typically be used during dynamic trims.

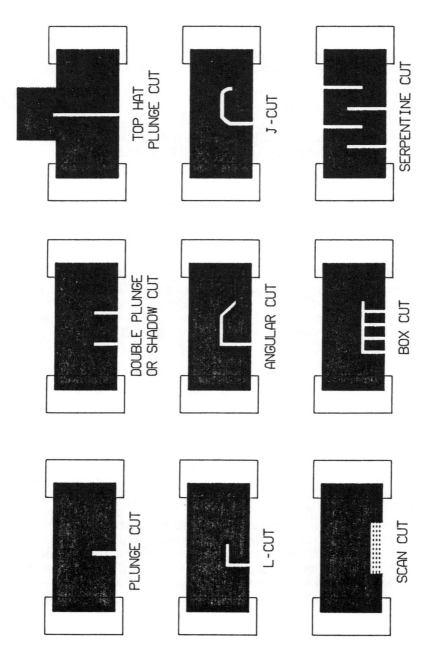

Figure 5.16: Types of laser cuts for resistor trimming.

Digital-Cut

The most accurate and stable resistors are obtained by the digital trim method shown in Figure 5.17. This technique requires a very large area; however, the trimmed resistors are very stable since the trim cuts are not in the current flow.

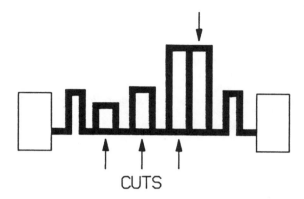

Figure 5.17: Digital trim.

SPECIAL REQUIREMENTS

All telephone equipment, from central switching units to home equipment, require built-in lightning protection. To guard against lightning damage, the voltage to the device is first dropped across a resistor. Due to the high voltage caused by lightning, the customarily used trim patterns such as L-cuts and plunge-cuts that have sharp corners and narrow kerfs cannot be used;[4] the large voltage gradient causes arcing across the narrow kerf.

Two patterns shown in Figure 5.16, the scan-cut and an extension of the J-cut (U-cut, which extends the end of the trim completely back out of the resistor), offer a solution to the lightning problem. Each of these trim patterns eliminates termination points within the resistor and any arcing across the narrow kerf. The drawback of the scan-cut is that it takes a long time, which induces stress in the resistor. The U-cut is best suited for this application; however, it still has sharp corners which induce hot spots, and it is slow since the cut must be made in a start/stop/measure mode.

The Teradyne W429 trim system (Figure 5.9) overcomes the problems associated with the conventional U-cut trims by a technique called algorithmic trimming. In algorithmic trimming, resistor characteristics such as dimensions, nominal value, tolerance, ohms/square, trim speed, and trim direction are programmed into the system. The program then selects an appropriate resistor model from its database along with a corresponding algorithm. It then drives the laser so that a U-cut is performed in real-time at a constant speed and trim energy, instead of the start/stop/measure mode.

As an example of the increased productivity that this technique offers, a 50-ohm resistor can be trimmed to better than 0.1 percent in 300 milliseconds, compared to 3 seconds when standard trim techniques are used.

REFERENCES

1. Spitz, S.L., "Trimmed for Precision," *Electronic Packaging & Production*, October 1985.
2. Chicago Laser Systems, Inc., *An Introduction to Laser Trimming*, 1978.
3. Tektronix, *Thick-Film Hybrid Design Guidelines*, 1985.
4. Lejeune, B., "New U-Cut Algorithmic Trimming for Lightning Protection," *Hybrid Circuits-Journal of the International Society for Hybrid Microelectronics-Europe*, September 1986.

6

Parts Selection

GENERAL CONSIDERATIONS

The selection of component parts to be used in the assembly of hybrid circuits requires different considerations than those used for printed wiring boards. Hybrid parts should be chosen with the following in mind:

1. The type of attachment to be used to mount the device to the substrate. The attachment material (epoxy adhesive, solder, or eutectic) dictates the type of metallization on the backside of the device. For example, a silicon-on-sapphire (SOS) device requires a backside metallization of gold so that it can be alloy attached while a silicon die can be attached directly to a gold pad, since silicon forms a eutectic with gold.

2. The availability of a given device in the die (uncased) form. Almost any device that is capable of being attached to a substrate with epoxy, eutectic, or alloy, and then wire bonded can be used in a hybrid. However, it is to the hybrid manufacturer's advantage to specify parts that are readily available in die form. Producing a hybrid circuit often becomes too costly or suffers a schedule slippage because the hybrid circuit designer selected a part that was difficult to purchase in the uncased form.

3. The electrical characteristics, reliability, environment, and the contract requirements.

4. The utilization of standard parts. The more times a given part is used, the better it is for inventory purposes. Instead of selecting different parts for different hybrids, the system should be designed to use standard parts to the extent possible. Large use of standard parts also drives the price down.

149

The major part types used in the assembly of hybrid circuits are: packages (cases and lids), active devices, and passive devices.

PACKAGES

In selecting parts for hybrid circuits, the first choice to be made is the package. Because all hybrid circuits are fragile and susceptible to atmospheric contaminants, they must be packaged in some manner to protect them. The package provides both mechanical and environmental protection. The die are interconnected with very fine one- or two-mil diameter wires which can be easily damaged in handling. Furthermore, many active devices (even though passivated), aluminum wire bonding pads, aluminum wire, and thin film resistors are susceptible to corrosion from ambient moisture, ionic salts, and other contaminants. Packaging also serves other functions such as providing a standard shape and a defined external lead configuration to facilitate assembly and testing. The protection afforded by the package is not always for the benefit of the hybrid circuit; sometimes it is also for the handler as in the case of high-voltage circuits.

Choosing the right package that meets all the customer's needs is of primary importance to the hybrid manufacturer. Careful consideration must be given to the seal type, plating, package style, and package size before committing a package to production.[1] In selecting the package, an early decision must be made as to the degree of protection that the package is expected to provide; for example, should the package be hermetic or non-hermetic in the sense of meeting definite helium leak rate requirements. Military, space, and some medical applications require hermetic packages while most commercial applications can use non-hermetic packages such as the lower cost epoxy-sealed or plastic-molded enclosures.

After the hermeticity requirements have been established, the next step is to decide on the pin configuration which is determined by the manner in which the hybrid will be assembled at the next level. If the hybrid is to be mounted on a two-sided printed circuit board, the package should be a flatpack type. If the board is single-sided with through-holes, the hybrid should be contained in a plug-in package.

Package Types

Packages may be classified according to the primary material used in their construction—metal, ceramic, or plastic—or their configuration. There are literally hundreds of configurations differing in shape, size, number of leads and manner in which the leads emanate from the package. Generically, however, these configurations fall into one or more of the following categories:

- Plug-in: the output pins are perpendicular to the floor of the package (Figure 6.1)
- Flatpack (also called "butterfly"): the pins emanate axially from the sides of the package (Figure 6.2)

- Integral-lead: the output leads are integral with the package (no glass-to-metal seals) (Figure 6.3)

- Cavity packages: the package provides a cavity in which the hybrid circuit is contained whose ambient can be controlled

- Epoxy-sealed: a cavity package, either metal or ceramic, in which the lid is attached with epoxy adhesive

- Plastic-encapsulated: a solid plastic body package in which the circuit has been transfer-molded or cast with epoxy, silicone, or other plastic

Top and Side View

Example of Plug-in Package

Figure 6.1: Plug-in package (Courtesy Augat, Isotronics PI-4947S; 88 pins).

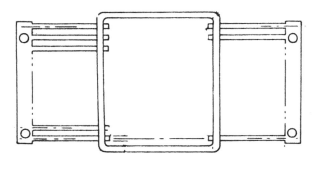

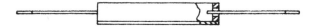

Top and Side Views

Example of Flatpack Package

Figure 6.2: Flatpack package (Courtesy Augat, Isotronics, IP 1113, 55 pins).

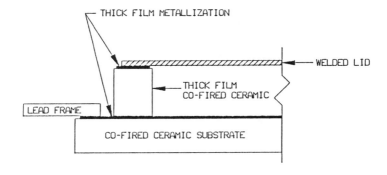

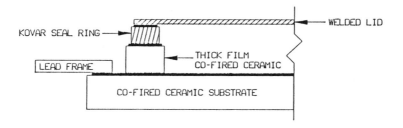

Figure 6.3: Integral-lead ceramic configurations.

The packages most widely used for hybrids are cavity types constructed of metal or ceramic. Epoxy-sealed ceramic packages are used for some commercial applications while plastic molded hybrid circuits are used for the simpler commercial hybrids that contain but a few devices. Besides the standard packages mentioned above, package manufacturers can produce almost any custom package that can be designed. Examples are:

Microwave modules with RF feedthroughs

Power packages constructed of copper or molybdenum. These packages can have thick leads of 20 mils in diameter to carry increased current.

Fiber optics packages with both flat leads and round leads

Machined packages with individual compartments

Packages with mounting ears on the sides

Flatpacks with double rows of leads

Metal Packages. The material most widely used for the construction of metal package cases and covers is Kovar, an alloy of 53% iron, 29% nickel, and 18% cobalt. Kovar is used with either a gold or nickel plating to prevent its corrosion. For military and high reliability applications the plating requirements are specified in MIL-G-45204, Type I, Class 1. The finish must be a high-temperature-resistant 24-karat gold plated to a thickness of 100 microinches or greater. The underplating should be an electroless nickel plating containing 8 to 12% phosphorus plated to a thickness of 100 to 200 microinches. Cases and lids may be completely nickel-plated or tin-plated as low cost alternatives to gold; however, there are some potential problems associated with nickel and tin (see Chapter 12).

The output leads are sealed into the Kovar package sidewalls or floor through glass-to-metal seals. A borosilicate glass such as Corning 7052 is used because its thermal expansion coefficient (about 5 ppm/C) closely matches that of Kovar.[1] The seal is formed by first oxidizing the metal surrounding the apertures in the package walls, then placing the Kovar leads through glass beads and inserting the combination in the package apertures. The assembly is fixtured and heated to the melting point of the glass (about 500°C) whereupon the metal expands slightly more than the glass and, upon cooling, holds the glass in compression. Metal packages consist of two general configurations: plug-in or flatpack types.

Plug-in Packages. The plug-in package uses either straight pins or so-called "nail-head" pins that are sealed to the floor of the package through glass beads and are perpendicular to the bottom plane. The pins are generally 18 mils in diameter on spacings of 100 mils. Packages with pin counts as high as 170 are available. To maximize the number of pins in a given area, the pins may be set in rows as shown in Figure 6.4. Plug-in packages are available in three styles—platform, modular sidewall, and uniwall.

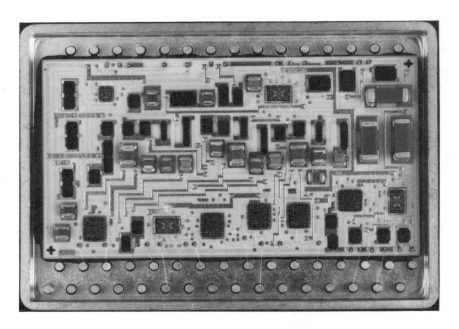

Figure 6.4: Hybrid mounted in a multi-row plug-in package (Courtesy Rockwell International).

The platform plug-in package (Figure 6.5) consists of a flat metal base (up to 60 mils thick) with no sidewalls and perpendicular leads around the perimeter that are sealed through glass beads. After the substrate has been attached to the base and devices have been assembled, it is sealed by soldering or welding on a dome-shaped lid. Platform packages are more difficult to handle than packages with sidewalls, and greater care must be taken to prevent damaging wire bonds and die. However, wire bonding and inspection are quite easy.[2]

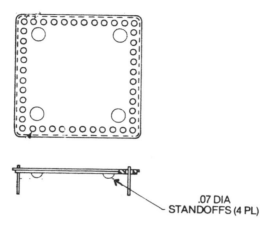

Top and side view configuration

Examples of platform packages

Figure 6.5: Platform plug-in package (Courtesy Airpax).

Flatpack ("Butterfly") Packages. The flatpack, also called the "butterfly" package, has the external leads emanating from one or more of the sidewalls. The wall thickness is about 40 mils and the leads consist of 10-mil-thick and 15-mil-wide Kovar ribbons spaced on 50- or 100-mil centers. Depending on the outside dimensions of the package, lead counts can reach 100 or more. Compared with the plug-in style, the flatpack can accommodate a larger substrate with a higher lead density in a given outside dimension. As with the plug-in packages, the sidewalls of flatpacks may be either modular or uniwall.

Sidewall Construction. Plug-in packages and flatpacks may be either of the modular or uniwall construction. Modular plug-in packages are fabricated from a 20-mil-thick base that has either a metal ring or individual sides brazed to it. These sidewalls are about 40 mils thick. Holes are stamped into the base, then the glass seals and leads are added. The leads are typically on 50 mil centers. Modular packages are more expensive than uniwall types and sometimes present problems in obtaining a hermetic seal due to the surface not being flat at the edges.

In the uniwall package (the entire package is formed from one piece of metal) the sidewalls are not separately brazed to a base plate. The construction begins with a platform of Kovar about 40 mils thick which is then formed into the package by deforming it around a set of dies. Diagrams comparing the uniwall with the modular construction for both plug-in and flatpacks are shown in Figure 6.6.

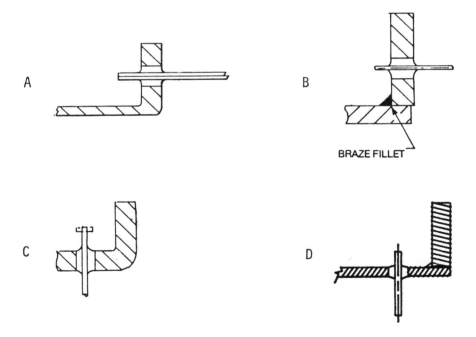

Figure 6.6: Package sidewall constructions—A. uniwall flatpack; B. modular flatpack; C. uniwall plug-in; D. modular plug-in.

Ceramic Packages. A second widely used package material, also providing hermetic qualities, is alumina ceramic. Ceramic packages are manufactured by the co-fired tape process and have an advantage over metal packages in avoiding the expensive and fragile glass-to-metal seals. Ceramic packages can be designed and built as integral-lead packages, leadless chip carriers, and leaded chip carriers. For the integral-lead package, the external metal leads are brazed or soldered onto thick-film conductor traces that emanate from the ceramic. These leads are formed as an integral part of the package by co-firing and sealing them between ceramic layers. Integral-lead packages are normally produced by the co-fired tape process though they may also be produced using the thick-film, screen-printing and firing process. Designs can be tailored for custom circuits. The substrate (usually alumina, though it can also be beryllia) is integral with the base of the package. The interconnect circuitry is printed on the base in multilayer fashion. A ceramic ring is next attached to the base making the cavity hermetic. The top surface of the ceramic ring is metallized so that a metal lid can be welded or soldered to it. The input/output metal traces that run under the ceramic ring can be attached to a metal lead frame or wire bonded. An advantage of the integral lead package is that the input/output leads can be spaced very closely (about 7 mils on center). Ceramic packages may also be constructed with Kovar seal rings (also known as "window frames") so that they may be sealed with metal lids by the conventional seam welding or belt-furnace sealing processes. Otherwise, ceramic lids may be attached with a low-temperature-melting glass preform or epoxy adhesive. Still another variation involves metallizing the sealing surfaces of both the ceramic case and the lid and joining them with a solder preform.

Epoxy-Sealed Packages. Both metal and ceramic packages can be sealed by attaching the lid with epoxy adhesive and curing the adhesive. The adhesive may be used either as a paste or preform. Epoxy-sealed hybrid circuits are widely used in commercial applications because of their low cost. Though quite reliable for most applications, epoxy-sealing cannot be considered hermetic in the sense that metallurgically sealed packages are. In a high humidity environment, moisture eventually permeates the width of the adhesive bond line and enters the package cavity. Some would like to believe that plastic sealed packages are hermetic because they initially pass the helium leak test specified in MIL-STD-883; however, the penetration of moisture into such packages is only a matter of time (see Chapter 7).

Plastic Encapsulated Packages. Hybrid circuits may be encapsulated with plastic resins either by transfer molding or by casting. Though plastic packaging is a widely used low-cost process for single-chip devices (ICs, transistors, resistors) it is seldom used for hybrid circuits, especially for the high-density complex types where the probability of the plastic stressing the wire bonds or supplying ionic impurities that can chemically and electrically degrade the devices is very high. In those cases where plastics are used, the high purity epoxy or silicone resins having low thermal-expansion coefficients are recommended. Generally, the high pressures that are used in transfer molding can distort wire and damage wire bonds. Casting is much more benign; the liquid resin is poured onto the circuit under normal pressure. However, even here, stresses due to shrinkage of the plastic on curing and differences in expansion between the plastic and hybrid components must be dealt with.

Package Testing

Most package manufacturers perform the following quality conformance inspection and screen tests as defined in MIL-STD-883 in the following sequence:

Test	MIL-STD-883, Method Number
Visual Inspection	2009
Seal (Leak Test)	1014
Thermal Shock	1001
Seal (Leak Test)	1014
Heat Resistance for Gold	460°C/5 min (no discoloration)
Solderability	2003
High Temperature Storage	1008
Solderability	2003
Lead Integrity	2004 B2
Seal (Leak Test)	1014
Lead Integrity	2004 A
Seal (Leak Test)	1014

In addition, manufacturers may also perform qualification tests that a customer may require. The following is a sequence used for a typical qualification.

Moisture Resistance	1004
Salt Atmosphere Exposure	1009
Temperature Cycling	1010
Seal (Leak Test)	1014
High Temperature Storage	1008
Seal (Leak Test)	1014
Constant Acceleration	2001
Seal (Leak Test)	1014
Shock	2002
Seal (Leak Test)	1014
Vibration Fatigue	2005
Seal (Leak Test)	1014
Vibration (Variable)	2007
Seal (Leak Test)	1014

ACTIVE DEVICES

Active devices that are used in hybrids are typically uncased, bare silicon chips or die and consist of transistors, diodes, and integrated circuits. As previously stated, any type of device can be used in a hybrid. It is possible to install leaded components when they are not available in the uncased form. These leaded devices are usually diodes or transistors that, due to their special electrical requirements, have been packaged and fully tested. When a leaded part is used in a hybrid, the leads must be formed in a planar fashion so they can be surface mounted to the substrate. These leads can be soldered or microgap welded to the substrate. It is advantageous to avoid using solder in a hybrid package; therefore, whenever possible, welding or wire bonding should be used.

Active die are DC electrically tested in the wafer form and the rejects are marked with colored ink so they may be discarded after separation from the wafer. Since these devices must be probed to be tested, AC and temperature testing are not possible. This may cause a yield problem at the hybrid level when the devices are tested at high and low temperature. One of the keys to high yields at the hybrid level is high-quality incoming inspection and testing of devices. This should include visual inspection to MIL-STD-883 Method 2017 and 2010, and testing of sample packaged parts at high and low temperature. An additional test that should be performed is material bondability. This entails bonding wires to the pads, then testing them for bond strength. If the sample passes, the lot may be assumed to be good and released to assembly.

Passivation

The semiconductor chip devices used in hybrid assembly are purchased with a passivation layer of either silicon nitride or silicon dioxide. These coatings are applied by the manufacturer at the wafer stage as one of the last steps in the fabrication of devices. They are applied by evaporation, sputtering, or chemical vapor deposition to the entire surface of the die, except for the wire bond sites. Though the passivation layer protects the active surface of the die from particles, moisture, ionic residues, and general handling damage, it is thin (2000-6000 angstroms) and sometimes porous. Therefore it should not be considered a hermetic seal; however, it does lower the leakage current on devices. Silicon nitride is better than silicon oxide passivation.

Metallization

Almost all of semiconductor and integrated circuit devices that are used in hybrid microcircuits have aluminum as the top conductor metallization. The aluminum is typically 8,000 angstroms thick. Thickness is a critical parameter for a reliable wire bonding process. If the bond pad metallization is too thin, "punch through" can occur during the bonding operation causing damage to the underlying silicon. In a few cases, the device bonding pads consist of gold. This is excellent if gold wire is used for bonding because it provides a completely monometallic interface at both the die and substrate levels.

Transistors

Nearly all transistor types are available as chip components, including saturated switches, general-purpose amplifiers, RF amplifiers, and power switches. One of the drawbacks of using uncased transistor die is the inability to fully test and grade them. As an example, the 2N2222 transistor chip comes from a large family of 2N types, all employing the same basic fabrication process. These are screened and graded into 2N types ranging from a commercial-grade 2N2218 to a military-grade JANTXV2N2222A part. In the hybrid uncased version, the hybrid manufacturer must rely on the room temperature DC wafer probe tests to grade the part, and this is typically not adequate to ensure electrical performance. This grading procedure applies to the entire spectrum of semiconductor devices and must be considered when selecting and specifying chip components for hybrid circuits.

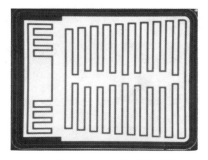

Figure 6.7: Power transistor die.

Diodes

Signal diodes, rectifier diodes, and regulator diodes present no special problems in chip form. However, temperature-compensated very stable reference diodes pose a problem. Even if the temperature and voltage characteristics of a reference diode are known prior to installation in a hybrid, the assembly and bonding processes may change these characteristics drastically. If this type of stability is needed, it is recommended that the device be used in its cased form and that it be mounted external to the hybrid.

Linear Integrated Circuits

These chips perform an analog function and are comprised of amplifiers, comparators, regulators, line drivers, line receivers, mixers, timers, and a variety of other functions. As previously stated, any device that is available in the packaged form can be purchased in the die form. Some devices may have to be handled differently. For instance, there are some FET input operational amplifiers that actually contain three chips in the package: two input FETs and an amplifier. Even this "die" can be purchased in the three pieces and used in a hybrid.

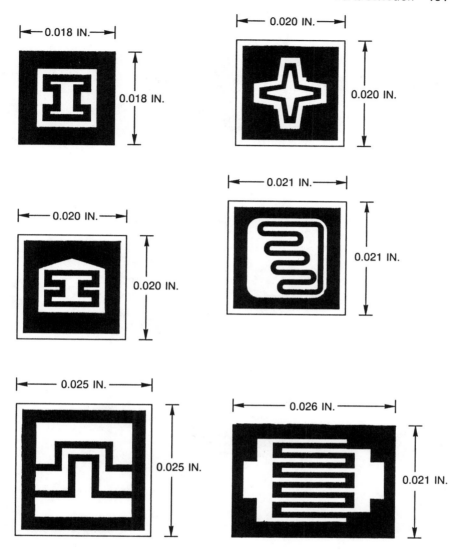

Figure 6.8: Topologies for small-signal transistor die.

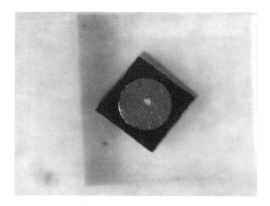

Figure 6.9:   Diode die.

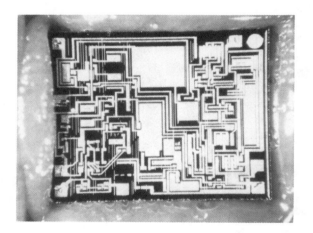

Figure 6.10:   Linear operational amplifier die.

Digital Integrated Circuits

Digital integrated circuits are the most frequently used semiconductor die in hybrid microcircuits. They consist of gates, flip-flops, counters, memory devices (random access memories and read-only memories), shift registers, multiplexers, gate arrays, and combinations of other digital devices. The major types are bipolar and metal oxide semiconductors.

Bipolar devices are fabricated with NPN and PNP transistors including: standard TTL (transistor-transistor logic), high-speed TTL, low-power Schottkey TTL, emitter-coupled logic (ECL), small-scale integration (SSI), medium-scale integration (MSI), large-scale integration (LSI), and very-high-speed integrated circuit (VHSIC).

Metal oxide semiconductors (MOS) are fabricated by p-channel and n-channel techniques. The two techniques are combined in Complementary MOS (CMOS).

Figure 6.11: Digital gate die.

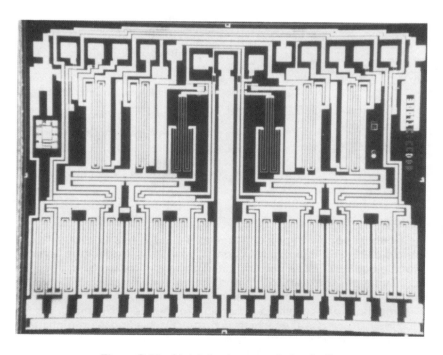

Figure 6.12: Multiplex integrated circuit die.

PASSIVE DEVICES

Passive devices consist of capacitors, resistors, RC arrays, and inductors (Table 6.1). These devices may be either batch-fabricated on the hybrid substrate or may be purchased separately in chip form and then assembled on a hybrid substrate. The decision as to which to use depends on the availability of substrate area and the stability and tolerances expected. Capacitors and inductors that are screen-printed onto hybrid substrates take up a considerable area, and the capacitors are limited to small values (0.18 to 100 pf).[3] Thus the use of chip capacitors and inductors is more prevalent.

Table 6.1: Passive Devices Used in Hybrid Microcircuits

Device Type	Description	
Discrete Resistors	Substrate	- thin-film - thick-film
	Add on	- thin-film on silicon, sapphire, or glass - thick-film on ceramic
Resistor Arrays	Substrate	- thin-film - thick-film
	Add on	- thin-film on silicon, sapphire, or glass - thick-film on ceramic
Discrete Capacitors	Substrate	- thin-film - thick-film
	Add on	- thin-film on silicon, sapphire, or glass - ceramic chip - solid tantalum chip
Capacitor Arrays	Substrate	- thin-film - thick-film
	Add on	- thin-film on silicon, sapphire, or glass
R-C Arrays	Substrate	- thin-film - thick-film
	Add on	- thin-film on silicon, sapphire, or glass
Discrete Inductors	Substrate	- thin-film - thick-film
	Add on	- wire-wound toroid - thin-film chip

Capacitors

The most popular types of capacitors used in hybrid circuits are the ceramic chips and thin-film MOS on silicon chips. Capacitors based on other dielectrics such as porcelain and glass are available but not widely used. Tantalum chip capacitors find special applications where their inherently high capacitance values are required (Table 6.2).

Capacitors are classified according to the dielectric materials from which they are constructed which, in turn, are classified according to their temperature coefficients as defined by either EIA or MIL-C-55681A (Table 6.3). Chip capacitors are available in three basic dielectric types. In order of decreasing temperature coefficient, these are BX, X7R, and NPO and

Table 6.2: Chip Capacitor Characteristics

Value Range	Dielectric Material	Tolerance	Temperature Coefficient	Size Range	Advantages/ Disadvantages
0.5 pf to 0.027 μf	Ceramic NPO	To ±5% or ±0.5 pf, whichever is larger	30 ppm/°C	0.040 x 0.030 x 0.040 inch to 0.225 x 0.210 x 0.065 inch	Good stability High cost Large case size vs capacitance value
100 pf to 1.5 μf	Ceramic K1200	To ±5%	±15%	0.040 x 0.030 x 0.040 inch to 0.225 x 0.210 x 0.070 inch	Large capacitance vs case size Large capacitance change with temp
0.1 μf to 10 μf	Solid Tantalum	-20 + 40%	±15%	0.050 x 0.100 to 0.150 x 0.285	Large capacitance vs case size Low voltage
0.5 pf to 300 pf	MOS Chips	±10%	+35 ppm/°C	0.020 x 0.020 to 0.060 x 0.060	Low leakage Smaller area Can have multiple values

Table 6.3: Summary of Capacitor Characteristics of Materials Currently Available

Type	Dielectric Constant	Temperature Coefficient	Operating Temperature Range	Dissipation Factor	Typical Cap Δ With DC Voltage Bias	Typical Cap Δ With AC Voltage Bias
COG/NPO	80	0 ±30ppm/°C	−55°C to +150°C	.05%	0%	0%
COG/NPO	80	0 ±10ppm/°C	−55°C to +150°C	.05%	0%	0%
W5R/BX	1600	±15% Max	−55°C to +125°C	2.0%	−10%	+20%
K2000	2000	±15% Max ·	−55°C to +125°C	2.0%	−20%	+20%
K2500	2500	±15% Max	−55°C to +125°C	2.5%	−25%	+20%
K5000	5000	+20% −50%	+10°C to + 85°C	2.5%	−40%	+30%
K5000	5000	+20% −85%	−55°C to +125°C	2.5%	−40%	+30%

classified as either Class I or II. Class I dielectrics are based on low K dielectrics (<150) and are composed primarily of titanium dioxide or titanates admixed with other oxides. Class I capacitors are used where high stability and low loss (high Q) over a full temperature range are required. NPO capacitors are examples of the Class I type. Class II capacitors are based on high dielectric constant materials such as barium titanate (K > 500). They exhibit non-linear characteristics and are less stable than the Class I types.[4]

Special considerations in selecting and specifying capacitors include:

1. NPO capacitors should be specified only when necessary to avoid their higher cost and higher case-size versus capacitance values.

2. Capacitor manufacturers customarily supply capacitors of 200 pf and lower with NPO dielectric, regardless of what was specified.

3. Capacitor data sheets indicate tolerances as low as 1 percent. Specifying low tolerances, if not needed, should be avoided because it results in longer delivery times and higher cost. Instead, if a tight tolerance is required in an RC time constant, one should take advantage of the trim capability of the substrate resistors to achieve the required time constant.

Figure 6.14 shows a silicon chip capacitor array which has trim capability. The die is 0.020 × 0.030 inches. Figure 6.15 is a schematic representation of the array. Trimming is effected by wire bonding to the various taps, thus paralleling the various sections to achieve the necessary value.

Resistors

Normally resistors are deposited or screen-printed onto the substrate during the fabrication process. However, chip resistors are used for a variety of reasons.

1. There may be only one or two resistors required for a particular design. In this case it is more economical to use

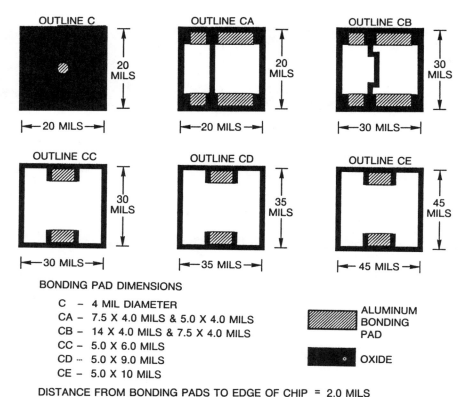

BONDING PAD DIMENSIONS

    C – 4 MIL DIAMETER
    CA – 7.5 X 4.0 MILS & 5.0 X 4.0 MILS
    CB – 14 X 4.0 MILS & 7.5 X 4.0 MILS
    CC – 5.0 X 6.0 MILS
    CD – 5.0 X 9.0 MILS
    CE – 5.0 X 10 MILS

ALUMINUM BONDING PAD

OXIDE

DISTANCE FROM BONDING PADS TO EDGE OF CHIP = 2.0 MILS
CHIP THICKNESS = 6.0 MILS ± 1 MIL

Figure 6.13: Configuration and size of various silicon chip capacitors.

Figure 6.14: Chip capacitor array.

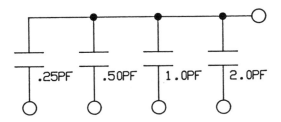

Figure 6.15: Schematic of chip capacitor array.

chip resistors instead of depositing or screen-printing resistors, avoiding the expense of added process steps.

2. On a thick-film substrate, there may be a few resistors that are significantly different in value from the others such that it would require processing another sheet-resistance ink. In this case it is more economical to use chip resistors for the few that are different in value.

3. In a thin-film circuit where all the resistors are formed from a low sheet resistance material, there may be a need for a few high-value resistors. Again the practical solution is to use chip resistors for the few high values.

4. Thick-film resistors have high temperature coefficients of resistance. If the circuit design requires thick film to achieve a multilayer structure but also needs precision resistors, then thin-film chip resistors should be used.

5. Thick-film resistors have a larger noise index than thin-film. Thin-film chip resistors should therefore be used in circuits that require low noise.

Thin-film chip resistors are commercially available as 30-mil-square chips with either nichrome or tantalum nitride as the deposited resistive material. They are produced in various configurations including single-value resistors, center-tapped resistors (Figure 6.16), and multiple resistors in custom-designed arrays. Figure 6.17 shows an example of a resistor array that contains seven resistors; Figure 6.18 is the schematic for this array. Chip resistors are available with all the bonding pads on the top or with one bonding pad on top and the second connection on the bottom. The standard substrate material for thin-film resistors is silicon, though quartz or polished alumina may also be used.

Thick-film chip resistors are also commercially available and are used when very small or large resistance values are required. Power chip resistors may also be trimmed to 0.005 percent and dissipate 3 watts in free air. The standard substrate material for thick-film resistors is alumina; however, for power applications, beryllia is preferable.

Figure 6.16: Center tapped resistor die.

Figure 6.17: Resistor array die.

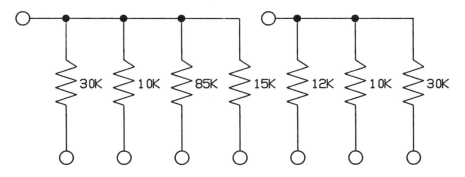

Figure 6.18: Schematic of chip resistor array.

Table 6.4 compares the three types of chip resistors and provides standard parameters. However, any of the parameters may be custom adjusted by the vendor or user. For example, if very stable resistors are required, vendors may produce precision thin-film resistors that have a 0.02-percent tolerance (user trimmable to 0.0005 percent), a temperature coefficient of 0.6 ppm per deg C, and stability to within 0.05 percent after 2,000 hours at 70°C.

The military specification that controls chip resistors is MIL-R-55342C.

Inductors

Inductors because of their large size are generally not used inside of hybrids; instead they are assembled external to the hybrid on a printed wiring board. However, three types of inductors can be assembled in hybrids: Thick-film screen printed or thin-film deposited inductors, chip inductors, and wire-wound inductors.

Film Inductors. Thick- or thin-film inductors can be fabricated on the substrate (Figure 6.19). They range in value from 10 to 100 nh but occupy considerable substrate area.[5]

SQUARE SPIRAL

ROUND SPIRAL

Figure 6.19: Typical configurations of thick film spiral inductors.

Table 6.4: Chip Resistor Characteristics

Resistive Material	Resistance* Range	TCR Range	TCR Tracking	Typical Tolerance	Power Dissipation	Typical Chip Size (inch)	Advantages/ Disadvantages
Nickel- Chromium (NiCr)	10Ω to 1 meg	0 to +50 ppm/°C	5 ppm/°C	±5% standard, ±0.01% special	250 mw at 70 C	0.030 x 0.030	Good temperature characteristics
Tantalum Nitride	10 to 1 MΩ	+50 to -150 ppm/°C	5 ppm/°C	±5% Standard, ±0.1% special	250 mw at 70 C	0.030 x 0.030	Stable-moisture resistant
Thick Film	0.10 to 50 MΩ	±50 to ±350 ppm/°C	N/A	±10% Standard to ±1%	30 watts/sq inch	0.050 x 0.050	Least expensive - Highest noise - Highest resis- tance values available

*Larger values are available on larger chip sizes (consult supplier data sheets).

Two equations that have proved to be in close agreement with measured values for round and square inductors fabricated on a substrate are as follows:[6]

Round Inductor.

$$L \text{ in nH} = (0.8a^2n^2)/(6a + 10c)$$

Where: $a = (d_o + d_i)/4$
$c = (d_o - d_i)/2$
$d_o = $ Outside diameter in mils
$d_i = $ Inside diameter in mils
$n = $ Number of turns

Square Inductor.

$$L \text{ in nH} = (0.0216S^{0.5}n^{1.67})$$

Where: $S = $ Surface area of coil in square mils
$n = $ Number of turns

Chip Inductors. Chip inductors are also available with values ranging from 0.1 to 1000 μH.

Wire-Wound Inductors. The wire-wound inductors can be attached to the substrate with epoxy and then the leads welded or wire bonded to the conductor metallization (Figure 6.20). Values are available up to 1000 μH.

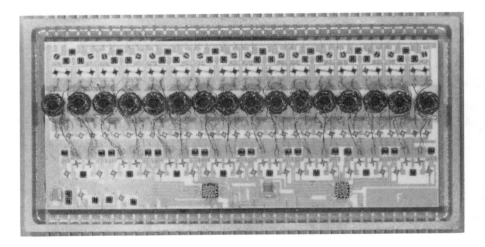

Figure 6.20: Hybrid using wire wound inductors (Courtesy Rockwell International).

Procurement

Devices should be purchased to Source Controlled Drawings (SCDs) to ensure the reliability and configuration of the chips. A typical SCD should include the following:

1. A pictorial of the chip topology, including pad locations and outline dimensions.

2. DC electrical test requirements for 100-percent testing of chips at room temperature.

3. Lot-sample testing (active devices) for AC and DC parameters at temperature extremes. These tests are performed on cased samples from the lot.

4. Qualification Tests

MIL-STD-883 Method 5008 defines the requirements for device Source Control Documents (SCD). To meet this specification SCDs must be prepared for all devices.

REFERENCES

1. Pokrzyk, J., Glass-To-Metal Packaging For Hybrid Circuits, *Hybrid Circuit Technology*, April 1985.
2. Lockheed Missiles & Space Company, Inc., *Thick-Film Hybrid Microcircuits Notebook*, 1979.
3. Tektronix, *Thick-Film Hybrid Design Guidelines*, 1985.
4. *Understanding Chip Capacitors*, Johanson Dielectrics Inc., Burbank, CA, 1974.
5. Jones, R.D., *Hybrid Circuit Design and Manufacture*, Marcel Dekker, Inc., 1982.
6. Harper, C.A., *Handbook of Thick Film Hybrid Microelectronics*, McGraw-Hill, 1982.

7

Assembly Processes

INTRODUCTION

The processes used to assemble hybrid circuits are now fairly standard in industry. They differ primarily in the choice of attachment materials (epoxy or metallurgical attachment), interconnection processes (ultrasonic, thermosonic, or thermocompression wire bonding), and sealing methods (seam welding or belt furnace sealing). The assembly processes are generally common to both thin and thick film circuits.

There are several assembly sequences depending on whether the substrate is alloy-attached or epoxy-attached to the inside of the package. If alloy-attached, the substrate must be attached in the package first since this is a high temperature operation (about 310-320°C using gold-tin solder preform). The subsequent die and component attachments are performed in the order of decreasing processing temperatures. Thus, if some components are to be attached with soft solder (tin-lead or indium alloys) and others with epoxy, the higher temperature solder attachments should be conducted first. It is, of course, desirable and less costly to employ only one attachment process, preferably epoxy bonding because it is the least expensive and easiest for rework.

When epoxy is used as the attachment material, there are two sequence options. The substrate may be attached to the inside of the case first, then the die attached to the substrate—or all the die may be attached to the substrate, then the substrate attached to the inside of the package. With multiple-up substrates (laser scribed on the back side for subsequent separation) it is economical to batch-fabricate the conductor/resistor/dielectric patterns, batch-apply the epoxy adhesive by screen-printing or automatic dispensing, automatically picking and placing the die, and curing the adhesive—all prior to separating the individual substrates and inserting them into cases.

After die attachment, the die are interconnected by bonding gold or aluminum wire from the device pads to metal bonding sites on the substrate.

174

Numerous cleaning steps are required throughout the assembly sequence. Cleaning must be thorough enough to remove all contaminants, yet not so severe that it will degrade the die or the wire bonds. No ideal cleaning solvent or process yet exists so, at best, the chosen procedure is a compromise.

Upon final cleaning, the hybrid circuits are vacuum-baked to remove traces of moisture and other adsorbed volatiles, then sealed in dry nitrogen. In some cases the circuits are coated with a thin polymeric coating prior to vacuum-baking and sealing to immobilize any loose particles that may remain on the circuit after cleaning. The coating also prevents particles from breaking loose after sealing and during actual use. Parylene (Union Carbide Corp.) is the most commonly used coating for this purpose. A flow-diagram showing the major assembly steps is given in Figure 7.1. Materials and process details for these assembly steps are discussed in the following sections.

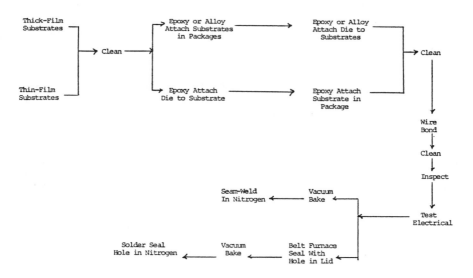

Figure 7.1: Flow diagram for hybrid assembly processes.

DIE AND SUBSTRATE ATTACHMENT

Types and Functions

The most widely used method for attaching die to the interconnect substrate or for attaching substrates to the inside of packages is epoxy adhesive bonding. It is estimated that over ninety percent of all hybrid microcircuits produced employ epoxy adhesives for both device and substrate attachment, largely because of their low cost and ease of rework. A second attachment method is metallurgical joining which can be effected either by eutectic scrubbing or by using an alloy preform. Metallurgical attachment is generally used for selected hybrid circuit applications, notably for high power circuits which require efficient thermal dissipation or for circuits that must meet a very low moisture content. Still a third, more recently introduced method, is silver-glass adhesive bonding, though this method has found more extensive applications in single-chip package assembly than in hybrid assembly.

Attachment materials serve three functions: mechanical, electrical, and thermal. The foremost consideration for all attachment materials is that they have sufficient bond strength to assure that components stay in place, not only for the life of the circuit but also during the accelerated mechanical, thermal and chemical stresses that may be imposed on the circuit during processing and screen testing. For example, the attached die and substrates must not detach during the constant acceleration screen test required by MIL-STD-883. Generally, epoxy-attached die and substrates will pass 5000g in the Y(1) axis, but become marginal at 7,500 to 10,000g, whereas metallurgically attached components can withstand 10,000g or greater. Further, the components must not detach during temperature cycling or elevated temperature exposure as encountered in burn-in and life testing.

In some cases, the attachment material must function as an ohmic contact providing an electrical conduction path from the device to the substrate circuit pads as with chip capacitors, chip transistors, and some resistors. In these cases electrically conductive adhesives (silver or gold filled epoxies) or low-melting solders are used.

Lastly, the attachment material may have to function as an efficient thermal transfer medium, conducting heat from the die to the substrate and from the substrate to the metal or ceramic package. Solder or metal alloys are best for this purpose, but often can't be used because they are also electrically conductive, render rework difficult or impossible, or result in wire bond or device degradation due to the high temperatures required in melting the solder. In such cases, epoxies filled with metals or with thermally conductive oxides (alumina, beryllia) are used as a compromise.

Adhesive Attachment

Epoxy Adhesives. Two types of adhesives, electrically conductive and electrically insulative (nonconductive), are popular for hybrid assembly (Figure 7.2). Conductive epoxies may either be silver-filled or gold-filled, though the silver-filled versions dominate the market. Integrated circuit die that require no ohmic contacts are generally attached with non-conductive epoxies as are ceramic substrates that are attached to the inside floors of metal packages. Silver-filled epoxies are used to attach devices that require electrical contacts such as transistors, capacitors, and diodes. However, if carefully applied, the same conductive epoxy may be used to attach all devices. This is desirable from a production standpoint because one can screen-print or dispense a single epoxy instead of several epoxies. Metal-filled adhesives have thermal conductivities higher than their insulative counterparts. This is an added advantage in using silver-filled adhesives, even for those parts that do not require ohmic contact.

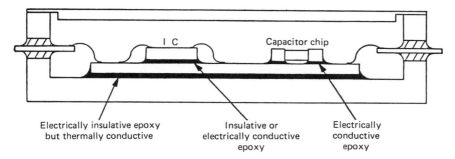

Figure 7.2: Cross-section of a hybrid circuit showing applications of epoxy adhesives.

Though epoxy adhesives, or as a matter of fact, any organic materials, were initially looked askance for use in high reliability microcircuits, considerable work has been done during the past ten years in evaluating, optimizing, establishing requirements and specifications, and qualifying adhesives so that today hybrid microcircuits assembled with adhesives are considered highly reliable.[1-6] NASA, foreseeing the need to control the quality of adhesives, under several contracts with Rockwell International, prepared and issued in 1982 the first specification MSFC-SPEC-592 "Specifications For the Selection and Use of Organic Adhesives in Hybrid Microcircuits." This specification, modified to some extent, served as the basis for Method 5011 of MIL-STD-883.

Two key concerns that led NASA to generate an adhesives specification were the integrity of the bonded devices during temperature cycling and the amounts and effects of outgassing products from the adhesive on the circuit performance.[7]

Solder, previously used to attach capacitors, had been shown to crack and detach during temperature cycling. Some early epoxy adhesive formulations also were reported to crack during extensive temperature cycling (500 cycles from −65°C to 150°C).[8] Bond shear strength requirements were therefore established for various mechanical and thermal stress conditions. These should however be augmented with electrical test data in the case of electrically conductive adhesives since small microcracks or detachments at the bond line may result in high-resistance joints yet pass the strength requirement.

The initial selection of the epoxy material is a key factor in assuring reliability. Literally thousands of epoxy formulations are available. They differ widely in properties such as thermal stability, amounts and types of outgassing products, and purity. Specification requirements were therefore set for the total weight loss as a function of temperature, amount of water evolved, amounts of chloride, sodium and potassium ions, and total ionic contents.[1]

Other specification requirements included electrical resistivity under various stress conditions, electrical properties (dielectric constant, dissipation factor), and thermal properties (thermal conductivity, coefficient of linear thermal expansion).

In addition to the selection of an adhesive based on its inherent materials properties, the manner in which the adhesive is processed is equally important in assuring reliability. For example, in dispensing a silver-filled epoxy, care should be taken to prevent flow or spreading over or between circuit lines which risks electrical shorts, metal migration, or particle detachment. "Bleed-out" can also occur. This is a separation and migration of one of the constituents of the epoxy during cure. Either the resin or the hardener/catalyst component of the epoxy may creep along a ceramic or gold-plated surface. The degree of bleed-out differs with different adhesives, their degree of cure, the type of surface used, and variations in the cleanliness and surface conditions of the substrate. Though not well understood, bleed-out contaminates wire bond pad sites and can affect both the initial bondability and long-term integrity of wire bonds. Cleaning in an oxygen or oxygen-argon plasma has been found effective in removing the bleed-out material.[9]

Adhesive Forms. Electrically conductive and insulative epoxy adhesives are available in two forms: *paste* or *film* (also referred to as tape adhesive). Paste adhesives are formulated to have optimum flow properties so they can be dispensed either by automated epoxy dispensing machines or by screen printing. Paste adhesives are best purchased as pre-mixed, degassed, and frozen in small tubes. Degassing, mixing, and packaging by experienced adhesives manufacturers avoids human errors that can be made if the user processes the adhesive. For example, mixing a two-component epoxy introduces air, and if not adequately degassed, the air becomes entrapped in the cured adhesive creating small voids in the bond line. These voids decrease both the electrical and thermal conductivity and may even degrade the bond strength. Frozen adhesives have a fairly long shelf life when stored at $-40°C$. They are convenient to use since only the number of tubes that are to be used in a given time need to be thawed.

Film or tape adhesives are partially cured, B-staged, epoxies similar to those used in printed circuit fabrication. The film, cut to the exact size, is placed between the two mating surfaces, held with pressure, and cured by heating. Film adhesives have several production advantages over paste adhesives, especially in controlling the amount of adhesive that is applied and assuring complete coverage. These factors are important in attaching large area substrates or large die. The film adhesive can be purchased pre-cut to the sizes required or may be purchased in large sheets and cut by the user. With paste adhesives, errors often occur in dispensing too much or too little adhesive. With small amounts, complete continuous coverage may not be attained which reduces the bond strength; excessive adhesive flows up and around the sides of the substrate or die and can contaminate the top circuit. Excessive flow from silver-filled adhesives can short the circuit immediately or in time as a result of silver migration. Film adhesives have not been too popular in attaching small die because of the logistics in handling large numbers of very small sizes. Also with the advent of automated pick-and-place machines and automated dispensers, the paste adhesives are more suited to the attachment of die.

Polyimide Adhesives. Polyimide adhesives for die attachment were first introduced in 1978 to meet the need for hybrid circuits and single chip devices that would encounter temperatures higher than 200°C in actual use or in accelerated temperature testing. Both electrically conductive (silver filled) and electrically insulative versions were introduced. Polyimides are superior to epoxies in their thermal stabilities at temperatures of 250°C or higher. The isothermal weight loss curves at 250°C for a typical polyimide compared with one of the better high-temperature-stable expoxies (an anhydride cured epoxy) are shown in Figure 7.3. The continued decomposition of the epoxy is evident from its linear weight loss as a function of time. Because there are very few applications in which hybrid circuits may be exposed to such high temperatures, polyimides are not extensively used. One exception is in the oil/gas drilling industry where hybrid circuits must withstand the high temperatures and harsh environment of the deep well. For such applications, hybrids assembled with polyimide adhesive are used for digital/analog processing in sensing and measuring the deep well environment.[10]

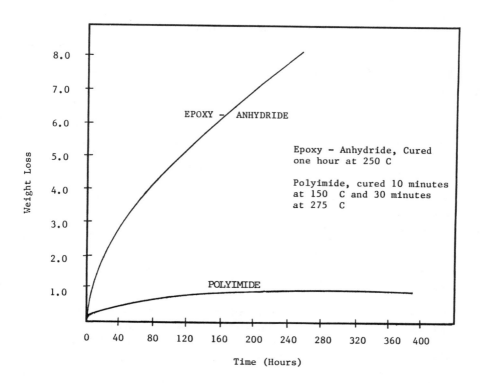

Figure 7.3: Comparison of polyimide and epoxy weight loss at 250°C in nitrogen.

The use of polyimides in hybrid assemblies is limited because of two problems: the high temperatures that must be used to cure the adhesive (generally a step cure culminating at 275°C) and the potential for entrapment of solvents or water that is released during cure. Unlike epoxies, where compositions of 100% solids (no solvents) can be formulated, polyimide adhesives require highly polar solvents (such as N-methyl pyrrolidone) to keep them in solution. This has presented a problem in the entrapment of solvent beneath large die and substrates even after curing. Further, many formulations are condensation polymer types, that is, they polymerize by the elimination of water between molecules. This can result in the entrapment of water and formation of voids within the bond line, a reduction in thermal dissipation, and degradation of the adhesive bond strength. More recent polyimide formulations are based on addition polymerization which obviates the problem of water evolution, though the solvent entrapment problem remains since the formulations still require organic solvents. Controlling the bond line to a very thin layer ameliorates both bubble entrapment situations.[11,12]

Adhesive Bond Strength. A prime requirement for both adhesively attached devices and substrates is that they have sufficient strength to last the life of the circuit and withstand the thermal, mechanical, and chemical exposures that they will be subjected to during processing, screen testing, and accelerated life testing. Generally strength is no problem since epoxies are among the most adherent materials available and shear strengths of over 3,000 psi are easily attainable. However, several materials and process parameters must be controlled to obtain reliable and reproducible results. Surfaces to be bonded must be well cleaned to remove both organic and inorganic (ionic) residues. Often both polar and non-polar solvents may be required to adequately remove all contaminants. In some difficult-to-bond situations, abrasion (scuffing of the surface) or plasma cleaning must be used. Cleaning using an oxygen, argon, or oxygen-argon plasma has become popular as an efficient method for removing traces of oils and other organic residues; though a pure oxygen plasma must be used with caution since it may oxidize metals. Silver-filled epoxies are known to darken and even blacken when exposed to the oxygen plasma due to oxidation of the silver. Adhesion to gold-plated Kovar cases may also be marginal because of the inherent inertness of gold. Some hybrid manufacturers abrade or scuff the gold on the inside floor of the package to expose some Kovar while others employ a primer to enhance adhesion. Plated gold varies in texture, surface finish, and porosity all of which may adversely affect adhesion. Plating bath ionic residues that have not been completely removed or oily residues that have become entrapped in a porous gold plate will also degrade the adhesive strength. To assure complete removal of these contaminants the user may have to further clean the packages with deionized water to remove ionic residues and plasma clean to remove tenacious oily residues.

In selecting and qualifying an adhesive it is a good practice to assemble prototype circuits and subject them to various cleaning solvents, processing temperatures, and mechanical screen tests. Bond shear strength requirements for adhesively attached die, substrates, and capacitors are given in MSFC-SPEC-592 for conditions of room temperature, 150°C, after solvent immersion, after temperature cycling, and after 1000 hrs aging at 150°C. For the last three conditions a 70-80% retention of the initial bond strength is required. The constant acceleration (centrifuge test) in the Y(1) axis is an excellent test to assure that all devices and the substrate will not detach, since the force in the Y(1) axis is pulling away from the bonded part. For military/space programs hybrid circuits are tested on a 100% basis at 5,000 g. Most paste adhesives become marginal at 7,500g and will detach at 10,000g. By controlling the uniformity and thickness of the adhesive, for example, by using a preform or applying the paste by screen-printing, centrifuge forces of 15,000g can be withstood. However, to consistently meet requirements exceeding 10,000g, eutectic or alloy attachments are recommended.

Electrical Conductivity. The electrical conductivity and stability of electrical parameters at elevated temperatures, after aging at elevated temperature and under power, are basic considerations in the selection of die attach adhesives.

The electrical conductivity of metal-filled adhesives is a function of the extent to which the metal particles contact each other; the higher the filler content the greater the probability for metal-to-metal contact and the higher the conductivity. There is a limit, however, to the amount of filler that can be blended in an epoxy resin without affecting its flow properties or its ability to be dispensed. Even more significant than the weight percent of filler is its volume fraction. For example, though the weight percent of silver in an epoxy paste adhesive may be high (70-80%), the volume percent in the cured adhesive can be as low as 21%. Thus to maximize conductivity, adhesives should be selected that have as high a volume percent of filler as practical without affecting their processing properties.[13] Other factors important in improving electrical conductivity are the shape and size of the metal particles. Silver is normally used in the form of flakes or combinations of flake and powder. The flake form provides a higher number of contact points than either spherical or cubic particles.

The conductivities of the best silver-filled epoxies are of the order of magnitude of 10^{-4} to 10^{-5} ohm-cm. Though still several orders of magnitude worse than pure silver metal, the electrical conductivities of silver or gold-filled epoxy adhesives are adequate for most circuit applications. Electrical conductivities of silver-filled epoxies are quite stable, decreasing only slightly at elevated temperaturies (Figure 7.4), while aging for 1,000 hours at 150°C generally improves the conductivity (Table 7.1). In spite of these excellent characteristics, there have been instances where the device and circuit parameters have shifted due to the formation of a resistive path at the bond-line interface between the die and adhesive. This resistive layer sometimes occurs initially during the cure of the adhesive or may develop later as a function of time, temperature, and current density. In such instances, several techniques have been successful in assuring a reliable and stable electrical path.

1.   Separately scrubbing the back-side of the silicon die to a gold-plated surface to form a gold-silicon eutectic layer.

This requires an extra step but removes oxides on the silicon and assures a good ohmic contact.

2. Separately scrubbing the die onto gold-plated Kovar or molybdenum tabs, then adhesively attaching the tabs to the circuit. In addition to assuring a reliable electrical interface with the device this method has the added advantage of spreading the heat dissipated by the device over a wider area.

3. Sintering die that have backside gold metallization, which results in interdiffusion and again in an improved electrical path. The diffusion should be controlled to prevent driving most of the gold into the silicon thus resulting in a silicon-rich surface or high resistance contact.

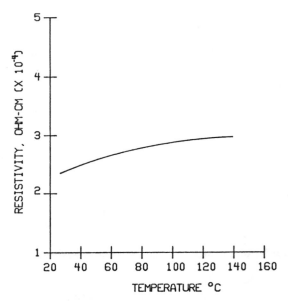

Figure 7.4: Volume resistivity vs temperature for a silver-filled epoxy paste adhesive.

Table 7.1: Volume Resistivities for Silver-Filled Epoxy Adhesives (ohm-cm)

Adhesive	At 25 C	At 60 C	At 150 C	At 25 C after 1000 hrs at 150 C
Ablebond 36-2[a]	2.4×10^{-4}	2.6×10^{-4}	2.9×10^{-4}	1.4×10^{-4}
Ablebond 84-1-LMI[a]	7.0×10^{-5}	8.0×10^{-5}	9.0×10^{-5}	5.0×10^{-5}
Amicon C868-1[b]	3.4×10^{-4}	3.8×10^{-4}	4.8×10^{-4}	2.1×10^{-4}
Epi-Bond 7002[c]	8.0×10^{-5}	4.8×10^{-4}	5.0×10^{-4}	5.5×10^{-4}
Epo-Tek H35-175M[d]	2.0×10^{-4}	2.5×10^{-4}	2.5×10^{-4} (at 125 C)	2.7×10^{-4}

(a) Ablestik Laboratories   (b) Amicon Corp   (c) Furane Products Company   (d) Epoxy Technology, Inc

Weight Loss. A convenient way of assessing and comparing epoxy adhesives is to measure the total weight loss as a function of time and temperature. The temperature used is 150°C because this is the maximum temperature that hybrid circuits generally encounter in either screen testing or actual operation. A plot of the isothermal weight loss as a function of time up to 1,000 hours gives a good picture of an adhesive's thermal stability and the amount of volatiles that can be expected to outgas. Figure 7.5 shows weight loss curves for two silver filled epoxies, one a 100% solid type (no solvent in the formulation) and the second, a solvent based epoxy. The much higher weight loss for the solvent based system may be due to a combination of entrapped solvent being released and gradual decomposition of the epoxy.

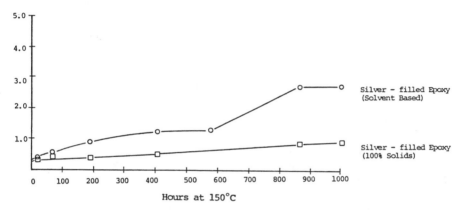

Figure 7.5: Weight loss of conductive adhesives vs time at 150°C in nitrogen.

Because of the long duration of this test a more expedient test, thermogravimetric analysis (TGA), is often used. In TGA a small sample, 10-30 milligrams, of the cured adhesive is heated at a constant temperature rise (10°C/minute is common) on a very sensitive electrobalance. The weight loss is dynamically recorded as a function of the temperature increase. TGA is an accurate method that is sensitive to very small changes in weight. TGA curves may also be used to establish an optimum cure schedule to minimize outgassing. Figure 7.6 compares TGA curves for two cure schedules for a commonly used silver-epoxy adhesive, the extended cure time providing a lower weight loss. TGA curves provide a quick picture of the temperature at which the material begins to decompose and the temperature at which volatiles are released, but it is not an isothermal weight loss as is the first test. A complete thermal assessment should therefore employ both tests and probably also a DSC (differential scanning calorimetry) analysis. The latter provides heat change data, exothermic and endothermic reactions that occur as a function of temperature and reflect changes of state such as polymerization, melting, oxidation, decomposition, and transition from a glassy to an amorphous state (glass transition).[14-16]

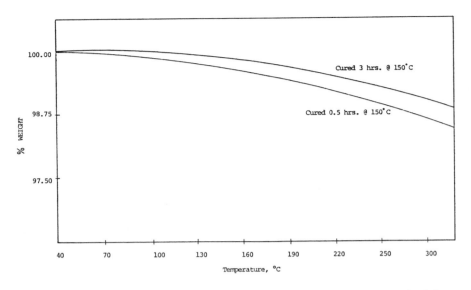

Figure 7.6: Thermogravimetric curves for a silver-epoxy at two cure schedules.

Outgassing of Adhesives. A major concern in using organic adhesives to assemble hybrid microcircuits has been the effect that outgassing products have on the electrical performance of the devices and circuits. The main concern has been water; however, other constituents, though released in smaller amounts, may be even more deleterious than water. Though many studies have been performed on the effects of moisture on chip devices and microcircuits the damaging effects are due to a combination of factors including the sensivitity of the devices and circuit, the nature and integrity of the device passivation, the amount of moisture, the time and temperature of exposure, gases other than water, and ionic contaminants on the surface in conjunction with moisure. In view of the numerous parameters and the difficulty in controlling all of them, industry and government agencies have settled on specifying the amount of moisture that is allowable in a hermetically sealed circuit package. The moisture requirement for Class B circuits has been set at 5,000 ppm(v) maximum and for Class S circuits at 3,000 ppm(v). Although hybrid circuits can meet these requirements it is still not a guarantee that the circuit will be reliable. Other constituents that may be evolved such as ammonia, amines, ketones, alcohols, chlorinated hydrocarbons, hydrogen chloride, and boron trifluoride must be reported, if detected, but quantitative requirements for these gases have not been established.

Moisture in itself, if pure, is probably not deleterious. However, in a microcircuit it is highly unlikely that water would be pure or would remain pure for long. First, there are other constituents from the epoxy that could contaminate the circuit including chlorides, metal ions, and amines. Secondly, there are always trace ionic residues on the circuit that cannot be completely removed even with the best cleaning solvents and methods.

Moisture then acts as a medium mobilizing and transferring these ions and contaminants to other portions of the circuit. Specific effects may be: corrosion of aluminum metallization causing electrical opens, wire bond deterioration with increases in resistance or bond lifts, device leakage currents, metal migration and electrical shorting, and electrochemical corrosion (for example of thin film nichrome resistors). To avoid these failure mechanisms, other tests in addition to moisture outgassing have been specified in MSFC-SPEC-592. These include weight loss at elevated temperature, corrosivity, total ionic impurities, and chloride, sodium, and potassium ion concentrations.

In recent years, adhesives manufacturers have made significant improvements in reducing the amounts and types of outgassing products and impurity ions in their formulations. For example, both epoxy resins and hardeners have been purified through distillation or extraction with solvents to remove chloride and other ionic contamination. Ammonia and amine generating hardeners have been avoided or removed. The hybrid circuit manufacturers also have made extensive improvements in reducing outgassing by optimizing the vacuum bake schedule used prior to hermetically sealing the circuits. Vacuum baking at 150°C for 16 to 96 hrs has been found effective in removing most of the moisture and other volatiles from the adhesive and other surfaces of the circuit and package.

Corrosivity. Many accelerated tests exist for measuring the corrosion potential (corrosivity) of adhesives.[17] In generating MSFC-SPEC-592, a rather simple, inexpensive test was adopted—one that qualitatively determines whether an adhesive is corrosive. According to this test, small patches of the uncured adhesive (mixed, if a two-component type) are placed on the aluminum side of aluminized Mylar film, then allowed to stand in room ambient for 48 hrs. The adhesive is then removed by dissolving in acetone. If the adhesive is corrosive it will etch the thin film of aluminum with which it was in contact. Any etching is evident by holding the Mylar film to light and observing light transmission. Thin film aluminum is ideal for this test because it simulates actual device and hybrid circuit conductor line and bonding pad metallization. Thin film aluminum is also very susceptible to corrosion in the presence of small amounts of ionic impurities (especially chloride ions) and moisture.

Ionic Contents. Ionic impurities in an adhesive are a key factor in causing electrolytic corrosion. The first generation of epoxies used in electronic assemblies as adhesives and molding compounds were commercial grades. These formulations inherently contained sodium and chloride ions which are by-products of the synthesis of epoxy resins from bisphenol A and epichlorohydrin.[18] These ions, especially chloride were responsible for many of the early device failures because of chloride induced chemical corrosion of aluminum. Epoxies that are produced today for semiconductor and hybrid applications are purified to reduce the ionic content or are synthesized by procedures that avoid the sodium chloride by-product. Quantitative analysis of the individual ions in an adhesive may be performed by atomic absorption spectrophotometry or ion chromatography. A list of the concentration requirements for various ions and some typical values are given in Table 7.2.

Table 7.2: Ionic Content of Die Attach Adhesives (ppm)

	Cl⁻	Na⁺	K⁺	NH₄⁺
NASA/MSFC 592 requirement	<300	<50	<5	NR*
Proposed MIL-A-87172 requirement	<300	<50	<50	<100
1st generation silver epoxies	200	30	10	0–150
New purified silver epoxies	10–20	<10	<10	<10
1st generation insulative epoxies	150–600	50	<1 (ND)	0–150
New insulative epoxies	<10	<10	<1 (ND)	<1 (ND)
Silver polyimide	<5–10	<5–10	<5–10	<5–10

*NR, no requirement; ND, none detected.

A measure of the total ionic content may be obtained by measuring the electrical resistivity of a water extract of the adhesive since the ions are readily soluble in water. The water extract resistivity is therefore an indication of the purity of the adhesive. The test basically involves pulverizing a weighed amount of the adhesive, digesting it in 100 ml of deionized water of measured resistivity, and measuring the decrease in resistivity after a specified period of time. The resistivity of a blank water sample must also be measured and subtracted from that of the adhesive sample. The total ion content is then calculated according to Method 7071 of FED. STD 406 and reported as parts per million of sodium chloride.

Metallurgical Attachment

There are two metallurgical methods for the attachment of die and substrates: *alloy attachment* with a binary or ternary alloy preform at its melt temperature or *direct eutectic attachment* in which two metal surfaces are heated at or above the eutectic temperature to form a bond. In the phase diagram for a binary alloy, the eutectic temperature is the temperature at which the liquid and solid phases are in equilibrium. Using either of the alloy constituents alone as an attachment material would be impractical because of their very high melting temperatures. However, at a specific composition of the two (the eutectic), the melting point is significantly depressed. The eutectic is a unique composition that occurs in some phase diagrams where the molten alloy transitions from a liquid to a solid state without going through a "mushy" phase. In examining the phase diagram for gold/silicon (Figure 7.7), it is seen that both gold and silicon have high melting points, 1063°C and 1414°C, respectively. The eutectic composition of these metals, however, has a practical melt temperature of 363°C. Alloys of gold with tin, germanium, or silicon are commercially available in the form of wire, ribbon, or preforms. Generally, the gold content of these alloys is greater than 70%. Many other binary and ternary alloys of gold, indium, tin, lead, germanium, silicon, and silver are available. They are used not only to attach die and substrates but also to attach lids to packages to effect the final hermetic seal. A list of the more commonly used eutectic alloys is given in Table 7.3.

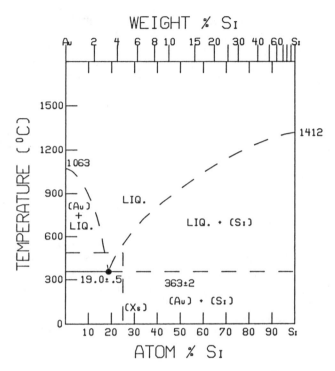

Figure 7.7: Gold-silicon phase diagram.[19]

Table 7.3: Typical Eutectic Alloys and Their Melt Temperatures*

Alloy	Eutectic Temperature (°C)
97 In, 3 Ag	143
62.5 Sn, 36.1 Pb, 1.4 Ag	179
63 Sn, 37 Pb	183
96.5 Sn, 3.5 Ag	221
80 Au, 20 Sn	280
97.5 Pb, 2.5 Ag	303
88 Au, 12 Ge	356
96.76 Au, 3.24 Si	363

*Abstracted from a list of 223 metals, alloys, and eutectic alloys with melt temperatures ranging from 10.7° to 1063°C from the Indium Corporation of America, Utica, NY.

Eutectic attachment can also be accomplished directly without a preform, for example, by scrubbing the backside of a silicon die to a gold conductor pad at the melt temperature. The circuit substrate is placed on a hot stage and heated. The die is then picked up with a heated colet and rubbed back and forth (scrubbed) onto the pad generally above the eutectic temperature (400-500°C) whereupon the two surfaces are bonded. The mechanical scrubbing action is necessary to displace oxides of silicon that invariably are present on the surface. Further, the scrubbing action removes oxides of the molten alloys from the die-to-substrate interface, minimizing voids that interfere with thermal conduction. The presence of surface oxides have the damaging effect of lowering the surface energy and reducing the wettability of the solder. A theoretical treatment and review of surface solder wetting and adhesion has been given in the literature.[19] The direct eutectic attachment method is the standard used in the semiconductor industry for the attachment of single chip silicon devices in individual packages. The parts must be bonded in an inert atmosphere, such as forming gas, to avoid oxidation of one or both of the metals which can occur rapidly in air at the melt temperatures.

Though metallurgical attachment is widely used in the assembly of single die in packages, several guidelines must be followed when eutectically attaching die in hybrid circuits. For example, if both epoxy adhesive and alloy eutectics are used in the same circuit, the higher temperature operation (alloy attachment) must be performed first; otherwise the epoxy will decompose at the high temperatures. Similarly, in reworking a circuit, removing and replacing the epoxy presents no problem because of the low temperatures involved (150°C or less), whereas reworking the alloy attached devices requires re-exposing the entire circuit including the epoxy to very high temperatures. Depending on the number of rework cycles and the total high temperature exposure, aluminum-to-gold wire bonds may also suffer degradation. A general comparison of eutectic/alloy attachment with epoxy adhesive attachment is given in Table 7.4.

Silver-Glass Adhesives

Silver-glass pastes, introduced and marketed by Johnson-Matthey Inc., are primarily used as alternatives to eutectic or alloy attachment of ICs and other semiconductor die. The pastes consist of silver, glass, an organic binder, and solvent. Step-temperature processing culminating at 400-450°C is required. At 100-150°C, the solvents are removed by evaporation, then at 300°C the organic binders are removed, and finally at 400-450°C the glass melts and joins the die to the substrate.[20] This stepped temperature profile is shown in Figure 7.8.

The drying step is especially important. The time and temperature that should be used is a function of the thickness of the paste and the size of the die. Large die require lower temperatures and longer times to avoid solvent entrapment and the generation of voids during subsequent high temperature steps. Processing schedules for various die sizes are given by the manufacturer or may be worked out experimentally by the user.[21] An air atmosphere is required for the burn-out step since the organic binder

Table 7.4: Comparison of Alloy and Epoxy Attachment Methods

Eutectic/Alloy	Epoxy
Provides electrically conductive path	Can be electrically conductive or insulative depending on the epoxy used
High thermal conductivity	Low thermal conductivity for insulative epoxies; better for metal-filled epoxies
High material cost	Low material cost
Difficult to rework	Easy to rework
High temperature process (to 500 C depending on alloy)	Low temperature process(< 150 C)
No outgassing	Outgassing of water,hydrocarbons,and other species; requires vacuum bake-out and special controls
May require fluxing and extra cleaning steps to remove flux	No fluxes required
Can be rigid and brittle causing cracking of large die	Inherently flexible, providing stress relief
No bleed-out	Epoxies may produce bleed-out of resin during cure

must be oxidized and decomposed to volatile by-products. The remaining silver-glass matrix, whose composition is about 80% silver and 20% glass, after melting, joins the die to the substrate. Its properties are given in Table 7.5. Among the key advantages of silver-glass over eutectic or alloy materials are:

Low material cost (eutectic alloys often contain gold as a constituent)

Amenable to automated dispensing: high speed stamping or dot transfer

Uniform void-free coverage, especially for large silicon die

Low thermal stress on silicon die

High thermal dissipation

To date, the major use for silver-glass adhesive has been in the semiconductor industry where it is used as a low cost alternate to gold eutectic or alloy attachment for single die in single packages.[22] There have been few applications of silver-glass to hybrid circuits, primarily because of the high processing temperatures required. Repeated exposure of the entire circuit to temperatures of 400°C, for example in reworking hybrid circuits, can degrade wire bonds and other passive or active die.

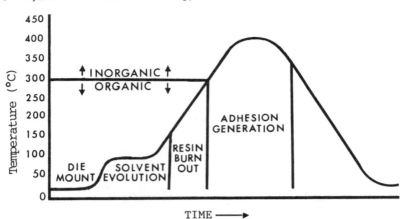

Figure 7.8: General processing temperature profile for silver-glass (Courtesy Johnson-Matthey Inc.).

Table 7.5: Properties of Silver-Glass Die Attach Adhesive*

Viscosity(Brookfield RVT,T-C Spindle,25 C)	19-27 Kcps at 10 rpm 13-21 Kcps at 20 rpm
Shelf Life	Six months at room temperature

Electrical
  Sheet Resistance
  Volume Resistivity
	8.1 x 10(-3) ohms/sq(500 uinch)
	5 x 10(-6) ohm-cm

Thermal Dissipation	θ_{JA} lower than for gold-silicon eutectic
Thermal Conductivity(Bulk)	0.187 cal/sec cm C
Thermal Expansion Coefficient	17 +-2 ppm/C
Adhesion	Typically 3 times Mil-Std-883 requirement.Unaffected by surface roughness or degree of oxidation of silicon backside
Residual Gas(Moisture)	Equal to Au/Si eutectic
Ionic Contamination	Na and K: less than 30 ppm Cl less than 10 ppm
Thermal Cycling	No failures or significant decrease in adhesion after 2000 cycles(-65 to +150 C)
Thermal Aging	No degradation in conductivity or adhesion on storage for 1000 hrs at 400 C
Reliability	No failures in operating life 2000 hrs at 125 C or in thermal or mechanical shock tests(Mil-Std-883)

*Compiled from Johnson-Matthey Inc. Product Bulletin for JMI4613 and reference 22.

WIRE BONDING

Processes

After the c'   'avices have been attached to the substrate, they must be electrically ι. ¬rconnected so they can function as a circuit. Interconnecting the chip circuitry to the substrate circuitry constitutes one of the more critical steps in the manufacture of hybrid circuits.[23] Since it is estimated that twenty-five percent of hybrid circuit failures are due to faulty wire bonds (see Chapter 12), it is clear that wire bond interconnections must be well understood and carefully produced, tested, and inspected.

Interconnections are generally made with very fine gold or aluminum wire (0.7 to 2 mils in diameter). The top surface metallization of almost all semiconductor and integrated circuit devices is thin film aluminum while the substrate metallization is gold (either thin film or thick film). Thus, depending on the wires used, a bi-metallic joint is formed either at the device level (if gold wire is used) or at the substrate level (if aluminum wire is used). From a theoretical standpoint it is desirable to have a totally mono-metallic system; in practice this is difficult and expensive to achieve. Indirect methods such as first depositing a barrier metal then thin-film aluminum onto the thick film gold conductor pads of the substrate have been used for some specific reliability applications. For the majority of applications bi-metallic bonds are reliable, provided that prolonged high temperature exposures, contaminants, and moisture are avoided.

Many wire bonding processes can be used to interconnect hybrid circuits. (For simplicity the term wire bonding refers to both wire and ribbon bonding processes.) The major types are:

Thermocompression

Ultrasonic

Thermosonic

Microgap

Beam lead

Tape carrier

Of these, thermocompression, ultrasonic, and thermosonic bonding are most widely used for fine wire interconnection of hybrid circuits.

Thermocompression Bonding. Many metals can be joined to themselves or to other metals at a temperature lower than the melting temperature of either of the metals if pressure is applied.[24] The temperature can be elevated by using either pulse or steady state tool heating methods. Thermocompression (TC) wire bonding consists of joining two metals by diffusion at elevated temperature and pressure. The elevated temperatures used (generally 250 to 300°C) maintain the metals in an annealed state while a molecular bond is formed.[25] The softer metals (gold, platinum, aluminum, copper, silver) are more readily diffusion bonded than harder metals such as nickel. There are two variations of thermocompression bonding: wedge bonding and ball bonding.

Wedge Bonding. Wedge bonding, the first type of thermocompression bonding developed, commonly uses tools like those shown in Figure 7.9.

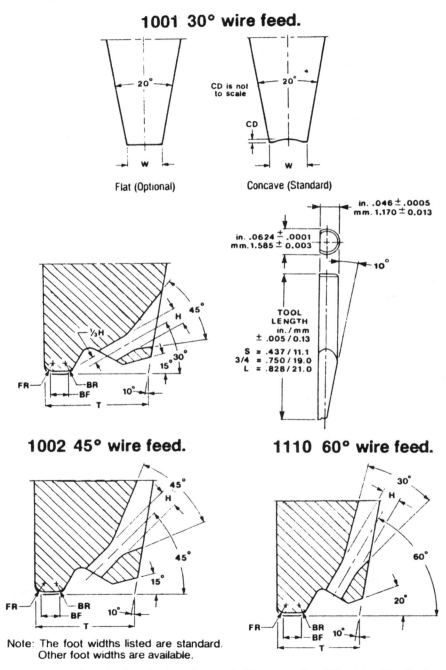

Figure 7.9: Standard wedge bonding tools (Courtesy Small Precision Tooling).

Different wire-feed angles (30°, 45°, and 60°) are used when increasing wire clamp clearance is required behind the tool to clear nearby components or the package wall. A diagram of the tool and wire clamp arrangement is shown in Figure 7.10. In wedge bonding, the first and second bonds are aligned in a straight line parallel to the direction of the wire from the supply spool to the tool. The wire extends under the bonding tool foot. The tool is aligned for the first bond where it is lowered and the bond is formed using a predetermined dwell time and force. The bonding tool is raised and positioned on the second bonding pad directly behind the first where the second bond is formed. The tool is raised and the clamp carefully tugs off the wire from the second bond and then shoves the wire back under the bonding foot, readying it for the next bond. This cycle is essentially the same for ultrasonic, thermocompression, or thermosonic wedge bonding.

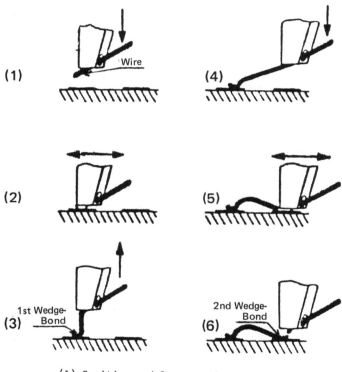

(1) Position and lower tool to the first bonding position.

(2) Bond wire (with defined pressure).

(3) Raise tool.

(4) Form loop.

(5) Form second bond.

(6) Break wire and raise tool to Position 1.

Figure 7.10: Steps for wedge-wedge bonding.

Ball Bonding. Ball bonding is similar to wedge bonding except that gold wire is used and one end of the wire is formed into a ball which is bonded into a "nailhead" configuration (Figure 7.11). The wire is fed through a wear resistant (tungsten carbide, titanium carbide, alumina, beryllia) bonding tool then "flamed-off", that is, cut with an electronic flame-off (EFO) so that a ball forms at the end. In ball bonding, gold wire is usually used since it melts and readily forms a ball even in an air ambient. Aluminum wire cannot readily form a ball and quickly oxidizes unless the area is flushed with an inert gas. A key advantage of ball bonding over wedge bonding is the freedom of movement of the bonding tool; it can be moved in any direction after the first bond has been formed. The capillary bonding tool is positioned and lowered over the intended bonding site (usually the semiconductor die). The bonding force plastically deforms the ball into a "nailhead" configuration producing the first bond. The tool is then raised, moved, and positioned over the second bond site. Since at this point no ball can be formed on the wire, a wedge bond is produced with the contoured circumference of the capillary. After the second bond has been completed, the capillary is raised a short distance, the wire is clamped then pulled as with the wedge bond. The wire is then prepared for the next operation by again "flaming-off" the end to form a new ball. This cycle is repeated many times throughout the circuit to completely interconnect all the devices.

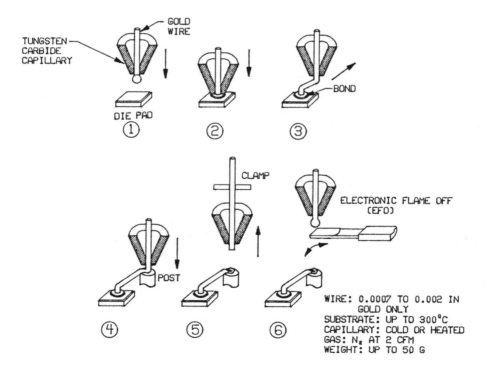

Figure 7.11: Steps for ball bonding.

Of the two methods for producing thermocompression bonds, the ball-wedge combination is preferred especially for automated bonding because the tool can be moved in any direction after the first bond has been made.

The main limitation of TC bonding is the high temperature required to produce reliable bonds. Even pulse heating the tool can damage some heat-sensitive devices. Prolonged steady-state stage heating can degrade the substrate elements (resistors, conductors) and the chip devices. If the devices are attached with epoxy adhesive, heating above 150°C can soften the epoxy and degrade its adhesive strength. This problem can be alleviated by using a pulse heated TC bonder which usually does not require heating the substrate. A current pulse is sent through the bonding tool which heats the tip to produce the bond, thus localizing the heat to the area of the bond pad. Pulse TC bonding is used extensively for thermal-sensitive devices and for die that are attached with epoxy to avoid thermal degradation of the epoxy.

A further limitation of TC bonding is its sensitivity to trace organic contaminants on the bonding pads. Organic volatiles that condense on the bonding pads during the curing of the epoxy die attach adhesive have been shown to degrade wire bondability and bond strengths. Mechanical burnishing or plasma cleaning after die attachment have been found effective in removing the residues and improving bondability. The efficiency of plasma cleaning was demonstrated by Auger analysis of the bonding pads before and after plasma cleaning. The carbon content found before cleaning disappeared after cleaning.[26]

During TC bonding care must be taken to prevent the capillary tip from becoming clogged and preventing proper wire-feed. This can occur when foreign material enters or adheres to the inside bore of the capillary. In addition, if too small a ball is produced it will stick in the capillary orifice. This situation can occur when there is relative movement between the tool and substrate during bonding.[27] The tip may be cleaned by forcing a small diameter tungsten wire through the capillary or by ultrasonically cleaning the tool in acetone or other solvent.

Ultrasonic Bonding. Ultrasonic bonding differs from TC bonding in that ultrasonic energy instead of heat is used to form the bond. The ultrasonic energy is coupled through a transducer to the bonding wedge tool similar in configuration to TC wedge bonding (Figure 7.12). The oscillator is tuned to the resonant frequency of the transducer and tool. In other ultrasonic power supplies the oscillator is designed to sweep over a range of frequencies to ensure that the members being joined are subjected to the resonant frequency one or more times to ensure a quality bond. Ultrasonic bonding employs ultrasonic energy to scrub the wire onto the bonding pad resulting in a microfriction bond mechanism. This scrubbing action, coupled with pressure, is the means of forming the molecular bond. The metals to be joined are softened by mechanical motion and localized heat due to friction instead of by externally applied heat.

The bonding sequence is the same as for TC wedge bonding, and, as with wedge bonding, the second bond must be made directly behind the first (Figure 7.13). The wire is fed through a clamp, then through the bonding wedge. After the first bond has been formed, the substrate is moved to the second position and the second bond is formed. The wire, while held in the clamp, is then pulled and severed at the edge of the second bond. The clamp then shoves the proper length of wire under the tool for the next bond.

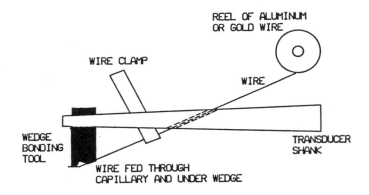

Figure 7.12: Side view ultrasonic bonder.

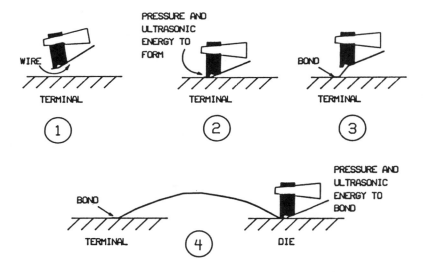

Figure 7.13: Ultrasonic wedge bonding sequence.

Ultrasonic bonding is essentially a "cold" process and can be used to join a wide variety of metals, thus rendering it a popular bonding technique. Both gold and aluminum wire can be used and wire sizes, like those of TC bonding, are typically 0.7 to 2 mils in diameter. For power devices larger diameter wires ranging from 5 to 20 mils in diameter are used to obtain higher current carrying capacity. Typical pull strength values for 1-mil aluminum wire bonded to thick film gold metallization range from 8 to 10 grams.

The key advantages of ultrasonic over TC bonding are:

1. There is no chance for degrading active or passive devices due to heat. Ultrasonic bonding is a room temperature process; no external heating is necessary.

2. Ultrasonic bonding is more forgiving of surface contaminants than TC bonding. The ultrasonic scrubbing action removes oxides and organic contaminants from the bonding site. However, some residual oxide actually assists in the localized heating process by increasing friction.

3. Void-free interfaces are produced which result in high quality, low-resistance bonds. SEM photographs of ultrasonic bonds are shown in Figure 7.14.[28]

4. Both gold and aluminum wire can be used.

5. It is faster than TC bonding

Among the disadvantages are:

1. Ultrasonic energy can cause mechanical damage

2. The acoustical properties of the members being joined can change causing variations in strength

3. Careful control of the ultrasonic energy is required

To achieve high production rates the bonding machine must perform reliably for extended periods of time without cleaning, service, or replacement of the bonding tools. Probably the most critical component of a bonding machine is the bonding tool. The tool shown in Figure 7.15 combines a tungsten carbide shank with a high density osmium alloy bonding tip. Osmium alloy has been shown to be an excellent transmitter of ultrasonic energy and highly wear resistant thus minimizing tool problems.[29] Other bonding tools are produced from titanium carbide, alumina, or beryllia.

Thermosonic Bonding. The third main wire bonding process combines the best features of thermocompression and ultrasonic bonding. Pressure, temperature, and ultrasonic energy are combined to produce the bond. The bonding sequence is usually the same as that for the TC ball-wedge process—the first bond is a ball bond while the second is a wedge bond. Thermosonic bonding is rapidly supplanting other bonding processes largely because of its automation capability. Automated thermosonic bonders are now commercially available that can be programmed to produce several hundred wire bonds in several minutes which manually would take hours. The key advantages of thermosonic bonding over TC bonding are that it employs lower temperatures (about 150°C) than TC

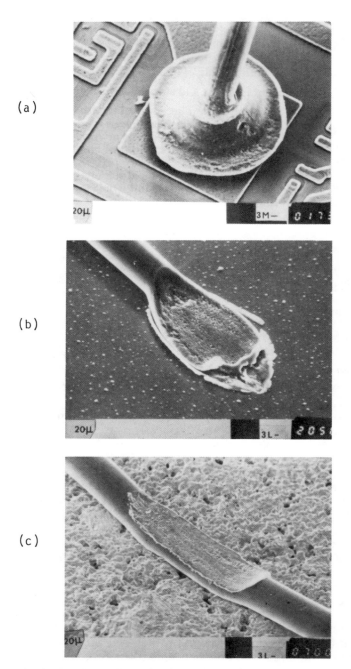

Figure 7.14: Bond configurations produced by ultrasonic wire bonding:[28] (a) ball; (b) wedge or second bond by capillary tool (ball/wedge); (c) wedge by wedge/wedge tool.

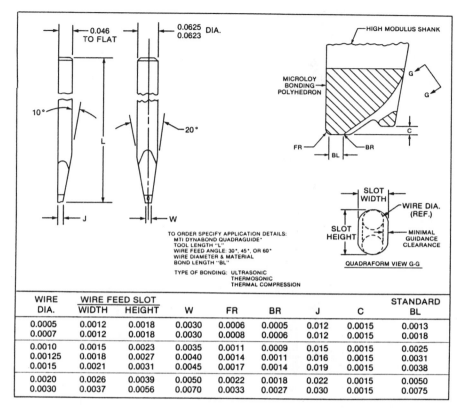

TO ORDER SPECIFY APPLICATION DETAILS:
MTI DYNABOND QUADRAGUIDE*
TOOL LENGTH "L"
WIRE FEED ANGLE: 30°, 45°, OR 60°
WIRE DIAMETER & MATERIAL
BOND LENGTH "BL"

TYPE OF BONDING: ULTRASONIC
THERMOSONIC
THERMAL COMPRESSION

WIRE DIA.	WIRE FEED SLOT		W	FR	BR	J	C	STANDARD BL
	WIDTH	HEIGHT						
0.0005	0.0012	0.0018	0.0030	0.0006	0.0005	0.012	0.0015	0.0013
0.0007	0.0012	0.0018	0.0030	0.0008	0.0006	0.012	0.0015	0.0018
0.0010	0.0015	0.0023	0.0035	0.0011	0.0009	0.015	0.0015	0.0025
0.00125	0.0018	0.0027	0.0040	0.0014	0.0011	0.016	0.0015	0.0031
0.0015	0.0021	0.0031	0.0045	0.0017	0.0014	0.019	0.0015	0.0038
0.0020	0.0026	0.0039	0.0050	0.0022	0.0018	0.022	0.0015	0.0050
0.0030	0.0037	0.0056	0.0070	0.0033	0.0027	0.030	0.0015	0.0075

Figure 7.15: Ultrasonic bonding tool specifications (Courtesy, Microminiature Technology Inc).

bonding (225 to 270°C), is omnidirectional, less sensitive to surface contaminants, and is a very rapid process when automated. Among its limitations—it requires larger bonding pads than those used for ultrasonic wedge bonding.

The development of a ball bonding process for aluminum wire would be highly desirable. Besides the obvious lower cost of aluminum wire, aluminum provides a mono-metallic interconnection at the device level, thus avoiding any potential for intermetallic formation. Extensive studies have been performed on many aluminum wire compositions in an attempt to develop a process in which aluminum can be flamed-off to form a ball. Aluminum wire containing 2% magnesium proved most successful, but the process is still in a pre-production state.[30]

Microgap Bonding. Microgap bonding is a diffusion process in which a split-tip bonding tool is used (Figure 7.16). The tool is lowered onto the wire and current is passed from one electrode half to the other through the upper member being joined while force is simultaneously applied. Individual

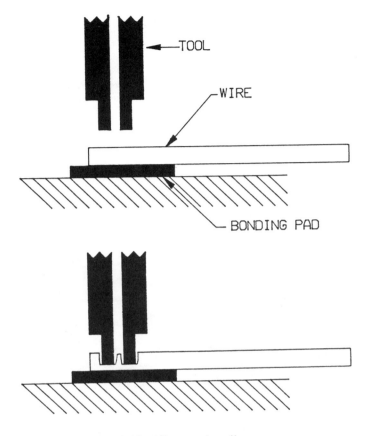

Figure 7.16: Microgap bonding process.

bonds are formed under each half of the electrode. Microgap bonding is used where large diameter wires, ribbon wire, or more difficult to bond material such as copper are required as interconnects for high power circuits. It is among the more reliable bonding processes provided that certain precautions are taken in using the electrodes among which are:

Electrodes must be clamped securely

Electrode area should be large enough to prevent excess heating

Electrodes should be burnished, using an alumina substrate, after 5-10 bonds have been made and dressed after 20-40 bonds

Sample bonds should be made after dressing to remove any residues

Bonding schedules should be developed that prevent electrodes from glowing or sparking

Beam-lead Bonding. Beam-lead bonding offers an alternate to chip-and-wire bonding. Die are purchased from the semiconductor device manufacturer with cantilevered gold beams instead of the usual bonding pads. Hence there is no need to apply wire or other leads and half of the interconnections have already been made and tested by the manufacturer. The beams are about 0.5 mils thick, 3 mils wide and 10 mils long and are isolated from the side of a semiconductor die by a passivation layer that extends onto the beam. Because of the larger contact area the interconnections have higher strength than wire bonds. The die are assembled by positioning them face down on the substrate so that the beams mate with corresponding bonding pads. They are then thermocompression bonded in one of three ways:

> All beams are bonded simultaneously by a tool that contacts and applies pressure and heat—a process called gang bonding
>
> Beams are bonded simultaneously a row at a time by tilting the bonding tool and rotating it around the device—a process called wobble bonding
>
> Each beam is bonded separately with a TC bonding tool

If the parallelism of the beams, tool, and substrate is maintained, gang bonding is the fastest process. In practice, this parallelism is difficult to attain, so the second process, wobble bonding, is most often used.

Some ten years ago, beam-lead bonding held the promise of replacing chip-and-wire bonding for all hybrid microcircuits but this has not materialized for several reasons:

1. The unavailability of many devices in the beam lead configuration. Hybrid manufacturers found themselves mixing chip-and-wire and beam-lead processes in the same hybrid circuit, which increased costs.

2. The cost of many beam-leaded devices was higher than estimated.

3. The die had to be bonded face down so inspection could not be performed after assembly.

4. Contaminants became entrapped beneath the die and were difficult to remove.

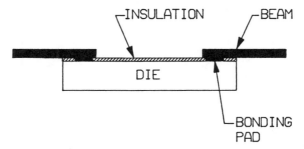

Figure 7.17: Beam-lead structure.

Unique Bonding Processes. Besides the previously discussed more conventional bonding processes, there are many unique variations. Because of the wide variety of hybrid circuits and their often unique types, wire bonding improvisations have had to be made. For example, electrical connections must sometimes be made with wire that resists conventional bonding. Some wires are not readily bondable by either thermocompression or ultrasonic techniques, for example, the low-temperature-coefficient wire LTC-65 (65% gold, 35% nichrome) which is used to minimize thermal drain of a component or to provide a heat or photon source. For this application, ball-entrapment TC bonding may be used. According to this method, the wire on the pad is positioned and temporarily attached with a TC ball bond. The excess wire is then removed at the nailhead and the bonding completed by using a microgap bonder to permanently attach the ball and the wire to the pad (Figure 7.18). The resulting bond, which provides a stress-relieving feature, has been shown to be highly reliable.[31]

Another unique bonding process is required where devices are so sensitive to pressure that the normal force used in conventional TC bonding is too high and causes damage to the device. One method of bonding wire to these sensitive pads is to first attach a small diameter indium sphere to the bonding pad at a low temperature (about 125°C), using a conical cavity bonding tool. Then the gold wire is positioned over the indium and a low-force TC wedge bonding tool is used to diffusion bond the wire to the indium projection (Figure 7.19).[31]

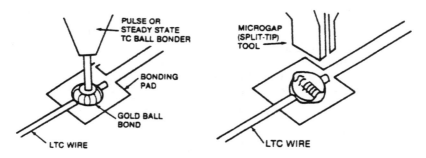

Figure 7.18: Ball entrapment TC bonding method.

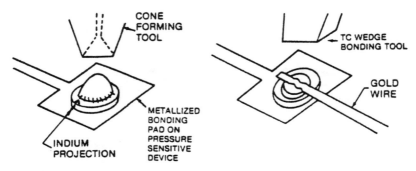

Figure 7.19: Low stress wire bonding method.

Automated Bonding

There are three widely used automatic bonding methods: tape automated bonding (TAB), automatic wedge bonding, and automatic ball bonding.

Tape Automated Bonding. TAB is an established interconnection process primarily used for bonding single die in individual packages. However, TAB can also be used in hybrid circuit assembly. In the TAB process, semiconductor die with bumped interconnect pads are placed in apertures of a tape (a dielectric film, generally, polyimide Kapton), then gang-bonded to photoetched metallized pads on the tape. The tape (film), with sprocket holes along its sides is similar to movie film; it may be 16 mm, 35 mm, or 70 mm. Like movie film, it is handled on reels. The bumps on the die pads are an essential part of the process. They are required to provide effective contact between the die pads and the corresponding pads on the tape and also to raise the tape slightly above the die thus preventing shorting at the edges (Figure 7.20). Bumps are formed on the die at the wafer stage as a batch process. An alternate approach is to form the bumps on the tape instead of on the die. To form the bumps, the classical tri-metal system, titanium/palladium/gold is vapor deposited or sputtered over the aluminum metallized pads of the wafer and their thickness is increased by electroplating gold. An alternate approach is to form the bumps on the tape interconnect pads (Figure 7.21). This is referred to as bumped tape automated bonding or BTAB. The BTAB process is particularly useful to a user to configure devices in TAB form when they are not available with bumps from the semiconductor manufacturer. Typical TAB patterns have etched 5-mil wide conductor lines and spacings, though 2-mil lines and spacings have also been produced.[23] Four different TAB patterns are shown in Figure 7.22.

Once the tape has been fabricated, the die are aligned under the cantilevered beams that extend from the tape over the apertures. The tape is held by the sprocket holes and the die are positioned with the aid of a microscope or TV monitor. After the chips have been aligned, interconnections are made simultaneously by thermocompression bonding—a process called *inner-lead bonding* (Figure 7.23). At this stage the individual chips can be burned-in and electrically tested. This pre-testing of chip devices is a major benefit of the TAB process for hybrid circuits; it increases the first-time-through yield and avoids costly rework at the assembled hybrid stage.

The last step in the TAB process consists in gang-bonding the outer leads of the tape to an interconnect substrate such as the thin- or thick-film ceramic substrates used for hybrid circuits. This involves severing the chip and leads from the tape, forming the leads, then TC bonding, a process called *outer-lead bonding*. Equipment for inner and outer lead bonding is commercially available, two models of which are shown in Figures 7.24 and 7.25.

In summary, the main advantages of TAB are that the die can be pre-tested at the tape stage and burned-in prior to assembly, thus eliminating device infant mortality, the process is highly automated, and the interconnections are very strong and rugged. The average pull-strength of TAB leads is about 50 grams compared to 5 to 9 grams for 1-mil-diameter wire

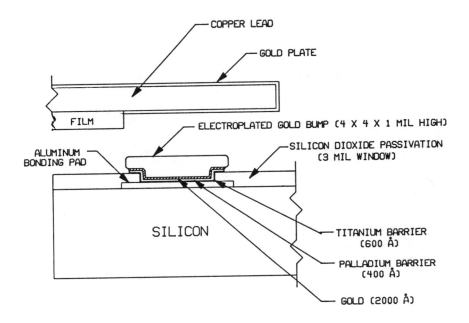

Figure 7.20: Gold-bumped device for TAB interconnection.

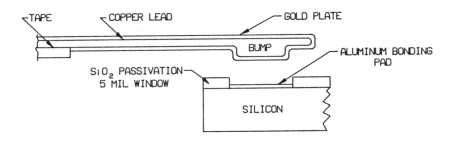

Figure 7.21: Gold bumps on tape (BTAB interconnection).

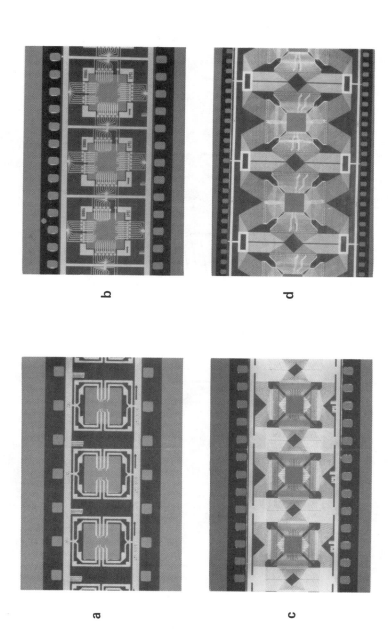

Figure 7.22: Examples of TAB interconnect patterns. (a) 16mm polyimide tape, 8 leads 3.0 mils wide on 8.0 mil centers, gold plated. (b) 35mm polyimide tape, 42 leads 4.0 mils wide on 8.0 mil centers, gold plated. (c) 35mm polyimide tape, 164 leads 2.0 mils wide on 4.0 mil centers, tin plated. (d) 70mm polyimide tape, 224 leads 3.0 mils wide on 6.0 mil centers, gold plated. (Courtesy of International Micro Industries, Inc.).

ALIGN

BOND

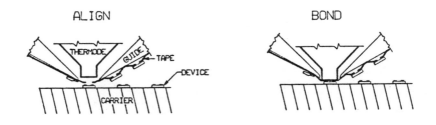

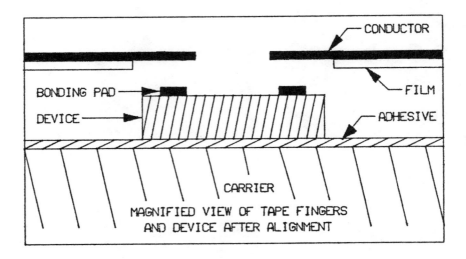

MAGNIFIED VIEW OF TAPE FINGERS
AND DEVICE AFTER ALIGNMENT

PICK

FEED

Figure 7.23: Inner-lead bonding sequence.

Figure 7.24: Inner lead bonding equipment (Courtesy of International Micro Industries, Inc.).

Figure 7.25: Outer-lead bonder, Model 4835 (Courtesy Jade, subsidiary of Kidde Inc.).

bonds. The TAB process has been useful in interconnecting and packaging custom devices with as few as 16 I/Os to as many as 300 and is finding many applications for VHSICs (Very High Speed Integrated Circuits).[32] On the negative side, there is an initially high investment in equipment and tooling—therefore not practical for the production of small quantities of many part types—and many die are not available in bumped (TAB) format.

Automatic Wire Bonding. The introduction of automatic wire bonding machines has not only reduced the labor time and cost of producing hybrid circuits but has also done much to reduce wire bond failures due to human errors. Once a machine is programmed, it consistently and reproducibly performs the desired steps. The computer can also sense variations and adjust the bonding parameters accordingly. Automatic thermosonic ball-bonding using gold wire has become the dominant chip-and-wire bonding process. A hybrid containing 50 chip devices and 500 wires that required 40-60 minutes to bond manually now takes less than 5 minutes to bond automatically.

An automatic bonding machine consists of a computer-controlled bonding head, a closed-circuit TV camera, an x-y table, and computer programming capability. The Hughes Model 2460 (Figure 7.26) and the Kulicke and Soffa Model 1419/DAWN are two of the most widely used automatic wire bonders. Software (for these machines) allows high-speed bonding in a wide variety of complex hybrids. The software takes into account variations in chip heights, wire lengths, and chips procured from different manufacturers. It can also control the size of the wire ball, vary the bonding schedules, and control the shape of the wire loop.

Automatic wire bonders are generally equipped with pattern recognition features so that parts can be referenced prior to bonding. Parts can be referenced by pattern recognition even if they are displaced as much as 30 mils from their originally programmed positions or rotated up to 15 degrees. Both the bond forces and bond times are computer-controlled. Another powerful feature of automatic bonding machines is the precise control of the shape of the wire loops that can be achieved by controlling the trajectory of the capillary after the ball bond has been placed. This now makes stitch bonding close to the base of a silicon chip possible without the risk of shorting to the chip's edge. Long wire (up to 130 mils in length) can be placed with reliable loop shapes. The algorithm controlling loop shape is such that it is automatically revised whenever the bond height or wire length changes, thus maintaining a relatively constant loop height and shape regardless of variations from wire to wire.

Besides automatic gold-ball wire bonders, automatic aluminum wire bonders are also available and used in hybrid circuit assembly. The Hughes Model 2470 and Kulicke and Soffa Model 1470 are two examples. The machines have specifications similar to the thermosonic ball bonders but employ ultrasonic wedge bonding and can use either aluminum or gold wire in sizes from 0.7 mil to 2 mils diameter.

Quality and Reliability

Because of the high incidence of failures due to wire bonds and the large numbers of variables that can affect wire bond reliability, thorough

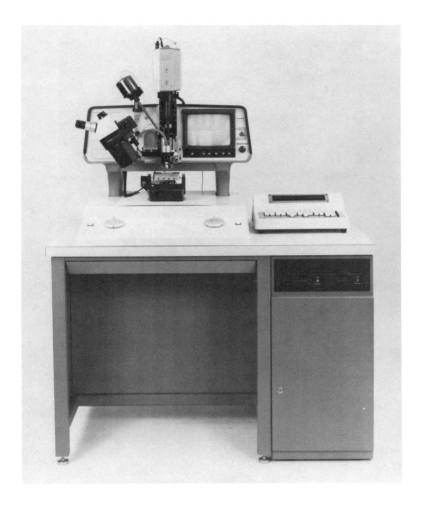

Figure 7.26: Wire bonder for automatic thermosonic gold wire bonding (Courtesy Hughes Aircraft, Industrial Products Division).

inspection and testing are important before sealing and delivering a hybrid circuit. Visual inspection criteria are specified in MIL-STD-883, Method 2017 which defines and pictorially depicts acceptance and rejection criteria for wire bonds. These criteria are based largely on geometries, locations, and appearance of the bonds. Thus an overbond condition, caused by excessive force or temperature, is rejected because of its highly deformed shape and thinned out bond edge (Figure 7.27). An underbond condition shows no deformation (Figure 7.28) while an acceptable bond is intermediate (Figure 7.29).

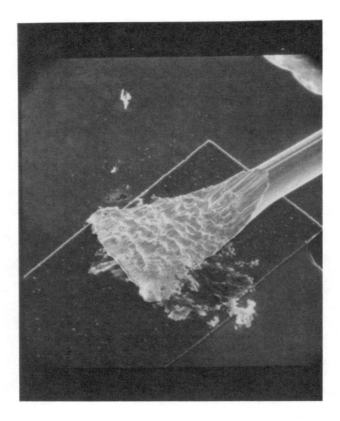

Figure 7.27: SEM photograph of 1-mil gold wire bond to thin-film aluminum for overbond condition (magnification 530X).

Though visual inspection is helpful in detecting and removing marginal bonds, it is not foolproof. Bonds that pass visual may still be weak or may have latent defects (chemical or physical) that can cause them to fail later. For this reason many manufacturers, especially those producing hybrids for military, space, or medical applications, augment visual inspection with mechanical tests, both non-destructive and destructive. In non-destructive pull testing each wire is pulled to a specified force that will break marginal/weak bonds but assure that the surviving bonds are not degraded. The The force that is applied depends on the wire size and type of wire bond. For example, a 2-gram force is used to test 1-mil-diameter aluminum wire bonded to gold. Other values are specified in MIL-STD-883. To perform this test, non-destructive wire pull testers with digital readouts are commercially available.

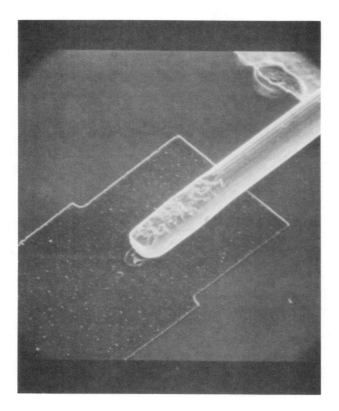

Figure 7.28: SEM photograph of 1-mil gold wire bond to thin-film aluminum for underbond condition(magnification 530X).

Destructive pull tests are performed on a lot-sampling basis. A minimum of ten bonds are pulled to destruction and the pull strengths and types of failure are recorded. An identically prepared second set is aged at elevated temperature (example 225°C for 2 hrs) and also pulled to failure. Average values and histograms for each set are then compared to established minimum values.

Extensive qualification tests may also be performed in which a statistical sampling of wire bonds are aged at three elevated temperatures for various periods of time up to 1,000 hours and measurements made of electrical resistance changes and wire pull strengths. These data can then be plotted and used to calculate the activation energy and, by using the Arrhenius equation, the life expectancy at lower temperatures. These tests and the short term accelerated test described above are especially valuable

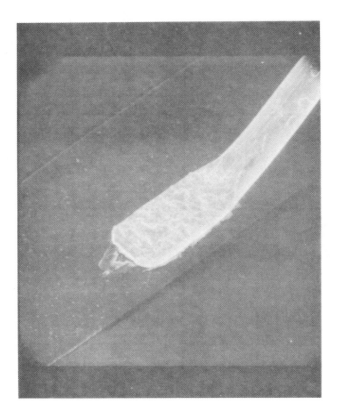

Figure 7.29: SEM photograph for a high-reliability production, 1-mil gold wire bond to thin-film aluminum (magnification 530X).

in disclosing any propensity for intermetallic (Au-Al) formation that can occur at lower temperatures over a longer period of time.

Intermetallic compounds formed in aluminum-to-gold bonds have been extensively studied. Bonds that had been exposed to elevated temperature, >200°C, developed a purple coloration called the "purple plague". These bonds were mechanically weak and high in electrical resistance—sometimes even open. At first it was thought that only one intermetallic ($AuAl_2$), the purple version was formed, that it was brittle and nonconductive, and the main contributor to failures. Later studies showed that at least five intermetallics could form ($AuAl_2$, $AuAl$, Au_2Al, Au_5Al_2, Au_4Al) and that the white-tan form Au_2Al was responsible for bond degradation.[33-36] This white intermetallic compound was shown to be brittle and to have a high electrical resistance, while the purple intermetallic compound proved to be of high strength and electrically conductive.

Besides high temperatures, it is thought that many impurities in or on the metallization catalyze and accelerate intermetallic growth. Among these are trace amounts of nickel in gold plating, impurities in thick-film gold pastes, fluxes, traces of tin and lead from solder splatter, and silicon. By understanding and controlling these variables, reliable bimetallic wire bonds can be produced.

CLEANING

Cleaning may be considered the "unsung" process of hybrid circuit manufacture. Often taken for granted, it has never been glamorous enough to receive the full attention of engineers and chemists. Yet cleanliness, at every step of the assembly process, is essential in assuring reliability and high yields. Some hybrid circuit problems that have been the result of contaminants and inadequate cleaning include: high electrical leakage currents, electrical shorts or opens, corrosion, PIND test failures, poor wire bondability, wire bond degradation, difficulty in lid sealing, and metal migration (whisker or dendritic growth). There are two aspects in maintaining circuits free of contaminants. The first is to prevent contaminants from getting into a circuit during its manufacture. Use of finger cots, vacuum tweezers for handling, storage of parts in dry nitrogen, assembly in laminar flow stations and clean rooms, and use of hair caps, smocks, and shoe brushes minimize contaminants and permit less severe cleaning processes. The second aspect is the selection of efficient solvents and processes for cleaning the hybrid at various stages, during and after assembly.

Contaminants and Their Sources

Contaminants found in hybrid circuits may be classified as particulates, ionic residues, inorganic residues, or organic residues. A major source of contaminants is through handling. Human skin continuously sheds particles and transfers salts (sodium chloride), ammonium compounds (through perspiration), and natural oils. Other contaminants derive from cosmetics and clothing worn by workers and transferred to the circuits during assembly and handling. Among these are hair-sprays, hand lotions, facial creams, deodorants, and synthetic and natural fibers. A compilation of potential contaminants and their sources is given in Table 7.6.[37,38]

Solvents

The solvents most commonly used for hybrid circuit cleaning are: deionized water, Freon TF, azeotropes of Freon TF, and isopropanol. Azeotropes are constant boiling mixtures of fixed compositions of two or more solvents. Their key feature is that they distill at a boiling point lower than either of the constituents without fractionating or changing in composition. Azeotropes, because they combine both polar and non-polar solvents, are particularly efficient solvents for vapor degreasing and removing a wide range of contaminants in one step. Solvents may be classified as either hydrophilic or hydrophobic.

Table 7.6: Typical Contaminants Found In/On Hybrid Circuits and Their Sources

CONTAMINATION TYPE	POTENTIAL SOURCES
FIBERS (NYLON, CELLULOSE)	Clothing, paper towels, tissues, and other paper products.
SILICATES	Rocks, sand, soil, fly ash
OXIDES AND SCALE	Oxidation products from some metals
OILS AND GREASE	Oils from machining, vacuum pumps, fingerprints, body greases, hair sprays, tonics, lotions, and ointments.
SILICONES	Hair sprays, shaving creams, aftershave lotions, hand lotions, soap.
METALS	Slivers and powders from grinding, machining, and fabricating metal parts, pieces of gold or aluminum wire from bonding operations, eutectic alloys and silver particles from attachment processes, silicon chips from die, nickel plating, silver and tin dendrites from metal migration.
IONIC RESIDUES	Ammonium compounds from perspiration, sodium chloride from fingerprints, residues from cleaning solutions containing ionic detergents, certain fluxes such as the glutamic acid-hydrochloride types, residues from previous chemical steps such as etching or plating.
NONIONIC RESIDUES	Rosin fluxes, nonionic detergents, organic processing materials.
SOLVENT RESIDUES	Cleaning solvents and solutions.
ORGANIC RESIDUES	Epoxy adhesives, photoresists
CERAMIC	Chips of alumina or beryllia from substrates, thick-film resistors, and thick-film dielectrics.

Hydrophilic Solvents. These are highly polar molecules that have a strong affinity for water and for water-like solvents. Structurally, they contain groups that have a high polarity such as OH (hydroxyl), NH_2 (amino), C=O (keto), CHO (aldehyde), or COOH (carboxyl). They are generally completely miscible in water. Examples include: water, alcohols (methanol, ethanol, or isopropanol), formaldehyde, acetic acid, acetone, and methyl ethyl ketone.

Hydrophobic Solvents. These can be either polar or non-polar compounds and have an affinity for oils and greases, which they readily dissolve. They are miscible with oils and grease but, immiscible in water. Examples include: Freon TF (DuPont), trichloroethylene, methylene chloride, toluene, and xylene. The principle of the ancient chemists: "similia similibus solvuntur" or "like likes like" still applies. Thus, structurally similar compounds have a strong affinity for each other and will be miscible. Freon TF and all Freon azeotropes are extensively used in the electronics industry because they are excellent solvents for greases and oils and are inert to the circuit components. The following are Freon cleaning solvents listed in order of increasing solvent strength (all Freons are products of DuPont):

Freon TF, trichlorotrifluoroethane

Freon PCA, a high purity form of Freon TF

Freon TMC, an azeotrope of 50% Freon TF and 50% methylene chloride

Freon TE, an azeotrope of Freon TF and 4% ethyl alcohol

Freon TES, Freon TE with a stabilizer

Freon TMS, an azeotrope of Freon TF and 5.7% methanol and stabilizer

Freon TA, an azeotrope of Freon TF and 11% acetone

Freon T-WD, an emulsion of Freon TF with 6% water and 2.5% surfactant (detergent)

Table 7.7 lists a few of the common defluxing solvents. High Kauri-Butanol values indicate a high degree of solubility for organic materials. This solubility parameter is used to select the solvent that most closely matches the contaminant to be removed.

Table 7.7: Azeotropic Solvent Compositions[39]

Azeotropic Composition Wt. %		Trade Name	Boiling Range, °F	Toxicity[1] TWX	Kauri-Butanol Value	Solubility Parameter
Methylene Chloride	92.7	M-Clene-S[2]	101	100	>136	9.9
Methanol	7.3					
Fluorocarbon 113	90.3	Freon TA[3]	112	1000	46	7.4
Acetone	9.4	Genesolv DA[4]				
Fluorocarbon 113	95.5	Freon TE[3]	112	900	40	7.4
Ethanol	4.5	Genesolv DE[4]				
Fluorocarbon 113	92	Freon TMS[3]	108	510	48	7.4
Aliphatic Alcohols	8	Genesolv DMS[4]				
1,1,1-Trichloroethane	96	Alpha 565[6]	165	345	124	8.6
1-Propanol	4					
1,1,1-Trichloroethane	93	Prelete[6]	164	510	124	8.6
Aliphatic Alcohols	7					

1. ACG1H, 1981
2. Diamond Shamrock Corp., Cleveland, Ohio
3. E.I. duPont De Nemours & Co. (Inc.), Wilmington, Del.
4. Allied Chemical Corp., Morristown, N.J.
5. Alpha Metals, Jersey City, N.J.
6. Dow Chemical Co., Midland, Mich.

In order for a solvent to effectively clean a surface it must "wet" that surface. Wetting is the ability of the solvent to flow in an unbroken pattern across a substrate or printed circuit board. Efficient wetting results in a continuous sheet of solvent, while poor wetting causes the solvent to "bead" and roll off the substrate, much like water on a freshly waxed car.

The theory of wetting is well established; the more acute the contact angle of solvent to substrate, the better the wetting (Figure 7.30). As the angle between the surface and the fluid approaches zero, the solvent spreads across the surface more quickly. As the angle increases, the solvent will flow more slowly until it forms beads.

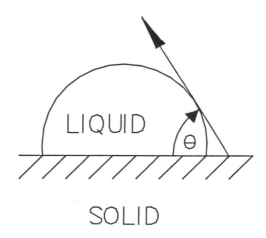

Figure 7.30: Contact angle of solvent to surface.

The wetting characteristics of a solvent vary depending on the type of substrate that is being cleaned. For example, the wetting characteristics of an epoxy-glass printed circuit board differ considerably from those of a ceramic substrate. In some cases the solvent does not contact the substrate, but only the film of contaminant on its surface. Efficient solvents will quickly wet a variety of different surfaces and have low surface tension to these surfaces.

Hybrids that use solder-attached devices must be thoroughly cleaned of all flux residues. Flux residues are particularly detrimental to wire bonds. Figures 7.31 and 7.32 show the damage that can be caused by the entrapment of solder and flux in a hybrid circuit package. The selection of a flux-removing solvent should not be based entirely on the solubility of the flux in the solvent because, in practice, flux when exposed to soldering conditions changes in chemical composition and generally becomes less soluble.

It has been reported that, in relation to wetting, capillary action and mass transport, chlorocarbon blend solvents wet and penetrate tight spaces (less than 10 mils) better than fluorocarbon blends under normal operating conditions.[39] Aqueous or detergent systems reach a practical limit of 10 mil spaces; for tighter areas, organic solvents should be used.

The solvent's ability to penetrate tight places is primarily a function of its surface tension, density, and viscosity. High density solvents with low surface tension and low viscosity provide the best wetting characteristics.

Figure 7.31: SEM micrograph of failed wire bond due to chloride contamination.

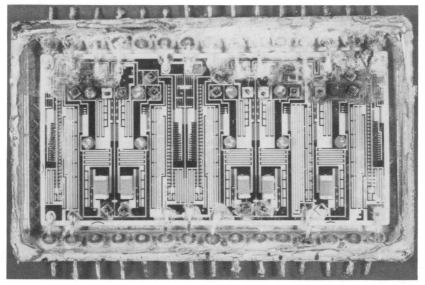

Figure 7.32: Corrosion of hybrid circuit due to solder/flux contamination during seal.

The relationship frequently used to express wetting as a function of these three parameters is:

Wetting Index = (Density × 1,000)/(Surface Tension × Viscosity)

Cleaning Processes

Basically there are three generic methods for cleaning: manual, batch, and plasma.

Manual (Hand) Cleaning. This is a non-conveyorized procedure for cleaning package exteriors, specific areas inside the package, and for flushing large particulates from the package. The following are examples of several manual cleaning procedures for removing various contaminants:

> Flux Removal—Immerse in Freon TMC, agitate for 2 minutes, immerse in fresh Freon TMC and again agitate for 2 minutes, rinse with clean Freon TMC, then rinse with Freon TF. If flux remains on the exterior of a sealed package, it may be scrubbed off with a cotton swab dampened with isopropyl alcohol. If this procedure is ineffective, immerse the parts in water-white-rosin flux at 55 to 65°C and agitate for 2 minutes, immerse 2 minutes in isopropyl alcohol, and flush with a spray of Freon TF.

> Device (Die) Cleaning—Place uncased die in a holding fixture and immerse in Freon TF, move up and down for 2 minutes, transfer to fresh Freon TF, immerse for 1 minute, then dry with nitrogen at 18 psi.

> Cases and Covers—These may be cleaned by immersing in isopropyl alcohol, flushing with fresh alcohol, and drying with nitrogen. However, a better cleaning method is to place the cases and covers in a beaker of deionized water and ultrasonic agitate for 10 minutes, spray rinse with deionized water, spray with alcohol, dry with nitrogen, then vacuum bake at 150°C for 2 hours. This method is especially effective in removing plating salts that may not have been cleaned off by the vendor.

> Substrates—Prior to metal vapor deposition or thick-film screen-printing, substrates should be prepared in a Class 1000 environment or better as defined in FED-STD-209. Substrates are immersed for 30 minutes in a solution of hydrogen peroxide, water, and ammonium hydroxide then transferred to heated high-purity water for 3 minutes. They are then spray-rinsed with deionized water and immersed in water until the alkalinity (pH) of the water returns to neutral. The substrates are then immersed in isopropyl alcohol and blow-dried with nitrogen. Batch methods using detergent (discussed later), may also be used.

For best results manual cleaning should be followed by conveyorized ing in that it involves a conveyorized process.

Batch Cleaning. Batch cleaning is distinguished from manual cleaning in that it involves a conveyerized process.

In a typical conveyorized process (Figure 7.33) the hybrids are loaded into baskets then sequentially subjected to immersion, (with or without ultrasonics), solvent vapor phase (vapor degreasing), room temperature spray, and drying. Each zone of the equipment is programmed and under computer control.

Figure 7.33: Conveyorized cleaning console (Courtesy Delta Sonics).

For vapor degreasing, the hybrid circuits are suspended in the vapor of boiling solvent. The hot vapors condense on the cool surfaces of the hybrids, bathing all exposed surfaces with liquid solvent. After the parts reach the temperature of the solvent vapor, condensation ceases and cleaning stops. To repeat the process, the hybrids are immersed or sprayed with fresh solvent at room temperature, cooling the surfaces and allowing vapor phase condensation to reoccur. Solvent cleaning may sometimes be undesirable because of flammability, toxicity, or the high cost of some solvents. In such cases cleaning in aqueous detergent solutions may provide an alternate approach.

Substrate Cleaning. During the fabrication and assembly of hybrid circuits, little or nor particulate matter is generated from the ceramic substrate. Therefore if adequate handling procedures are followed after the substrates are received, no cleaning, other than blowing with dry nitrogen, is required. However, since meticulous handling is difficult and cleanliness can not be assured, it is recommended that manual or batch cleaning be used prior to processing the ceramic substrates. Alumina ceramic is a highly sorptive material; ceramic exposed to the atmosphere for any period of time will absorb/adsorb a surface film of organic vapors from the atmosphere. Even if substrates are stored in clean-room environ-

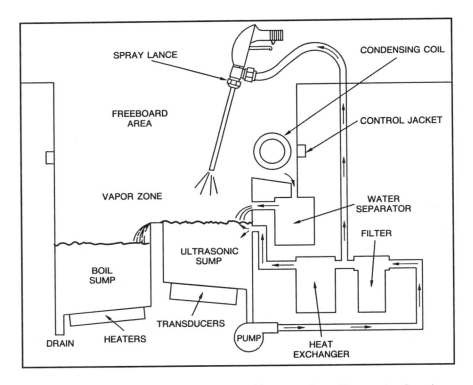

Figure 7.34: Ultrasonic vapor defluxer (Courtesy Crest Ultrasonics Corp.).

ments, it is impossible to filter out organic and inorganic gaseous materials, such as automobile exhaust gases, hydrocarbons, and sulfur compounds. Adsorption or absorption of these gases renders the ceramic surface hydrophobic (non water absorbing). The removal of adsorbed organic materials can be accomplished by cleaning in strong alkaline detergent solutions (discussed under Hand Cleaning), by plasma cleaning, or fire-cleaning. The latter consists in conveyorizing the substrates through a belt furnace at 850 to 1000°C in air to burn off the organics. Cleaning with organic solvents, though effective in removing gross amounts of oils and greases, is not as effective in removing absorbed organic films. A popular test to detect the presence of these organic films is the "Water Break-Free Test". According to this test, a surface is considered clean when it can maintain an unbroken layer of water for at least one minute when positioned vertically. Sometimes this test gives the user a false sense of cleanliness because some unrinsed detergents result in a zero contact angle. Thus any detergent cleaning of a substrate should be followed by a dilute acid and water rinse to neutralize and remove the detergent residues.

Cleaning any surface should involve some "scrubbing" action to be efficient. Ultrasonic cleaning provides the best scrubbing action but is not recommended for assembled hybrid circuits because ultrasonic energy can weaken or destroy fine wire bonds. Ultrasonic cleaning is best used to clean empty cases, lids, and substrates. An example of a conveyorized method for cleaning substrates, cases, or covers[40] is as follows:

Clean the parts ultrasonically in an aqueous detergent solution having a pH between 10 and 13 at a temperature of 60-70°C for 2 minutes. (In the complete cleaning cycle the fluids are filtered at a rate of one-half the tank volume per minute).

Spray rinse with deionized water or filtered softened water, at ambient temperature to remove gross amounts of the detergent solution.

Neutralize by immersion in ultrasonically agitated dilute acid solution at 20-30°C for 2 minutes.

Spray rinse with deionized water at ambient temperature.

Rinse in an overflow of deionized water with ultrasonics for 2 minutes.

Spray-rinse with deionized water.

Dry the parts by water displacement and vapor rinsing with an organic solvent.

Figure 7.35 depicts graphically a seven-stage ultrasonic detergent cleaning and drying system.

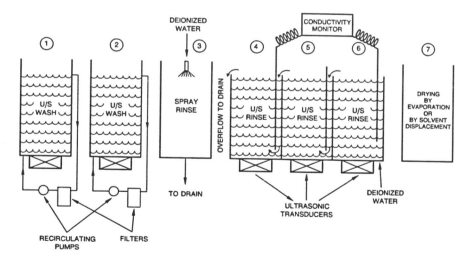

Figure 7.35: Seven-stage ultrasonic detergent cleaning and drying system.

Hybrid Assembly Cleaning. The following is an example of a sequence of steps used in cleaning hybrid assemblies without ultrasonics:

Spray-rinse with Freon TF.

Transfer hybrids to the vapor phase of boiling Freon TF (117°C).

Spray-rinse with Freon TF while in the vapor zone. (Hybrids should remain in the vapor zone for about 3 minutes.)

Transfer parts from the vapor zone and allow to dry while in a vertical position.

If any gross particles remain on the hybrid, they can be flushed off with a stream of isopropyl alcohol, followed by rinsing with Freon TF, and drying in a stream of nitrogen.

Solvent Safety. Precautions as specified by the supplier or manufacturer should be followed in using any chemicals or organic solvents. The toxicity and flammability of each material must be known and understood before implementing it in production. Exhaust hoods and scrubbers should be used to remove toxic or flammable fumes, and rubber gloves used for handling chemicals and solvents. With proper precautions and adhering to OSHA requirements, a wide variety of chemicals and solvents can be used safely. Data on the physical, chemical, toxicity, and flammability properties of industrial solvents and chemicals may be found in several handbooks.[41-43]

Plasma Cleaning. When a low-pressure gas is subjected to a high-energy input such as radio-frequency or microwave energy, the gas dissociates and ionizes through collisions with high energy electrons. The resulting mixture of electrons, unreacted gas, neutral and ionic species is referred to as a plasma and is highly effective in physically and chemically removing contaminants from a surface. Even an inert gas such as argon becomes highly reactive after being subjected to RF energy. Activated argon atoms and ions bombard the surfaces and physically dislodge contaminant residues. A pure oxygen or oxygen-argon plasma is much more effective than argon alone in removing organic residues because, in addition to the physical mechanism, activated oxygen reacts chemically with organic residues converting them to carbon dioxide and water—both volatile gases. However, a pure oxygen plasma can oxidize some thin film metals changing their electrical or physical properties. A plasma cleaner consists of a chamber equipped with an RF power supply and filled with a gas such as argon or oxygen (Figure 7.36).

Plasma was first used in 1967 to strip photoresist; in 1970 it was used in wafer processing, and finally in 1980 it was applied to cleaning hybrid circuits. Plasma can be used to:

- Remove traces of residual photoresist from device surfaces.

- Clean wire bonding pads to remove trace organic residues resulting from epoxy outgassing, epoxy smear, or epoxy bleed-out.

- Clean a surface to improve its wettability and adhesion of organic coatings, adhesives, or solder.

The most widely used application of plasma cleaning is to prepare hybrid circuits prior to wire bonding. Plasma cleaning of wire bonding pads greatly improves bondability and bond strength of thermocompression bonds.[26] After device attachment, circuits are placed in a plasma chamber and evacuated to 0.05 Torr, then backfilled with argon of 0.175 Torr. The gas is then subjected to 50 to 75 watts of RF energy for 3 to 5 minutes. Table 7.8 summarizes the parameters for plasma cleaning in oxygen and argon. The effectiveness of plasma cleaning is demonstrated by Figure 7.37 in which trace residues on integrated circuit bonding pads were completely removed.

Figure 7.36: Plasma cleaner (Courtesy Yield Engineering).

Table 7.8: Plasma Cleaning Parameters

	Oxygen Plasma	Argon Plasma
Cleaning Process	Chemical	Physical
Temperature	150 C	70-100 C
Etch Rate	800-1000 A/min	500 A/min
Pressure	1 Torr	0.2 Torr
Power	75 watts	75 watts
Gas Flow	2-5 cc/min	2-5 cc/min
Time	2-5 min.	2-5 min.

(B)

(A)

Figure 7.37: Plasma cleaning of bonding pads (Courtesy of March Instruments, Inc.): (A) photograph of bonding pads prior to cleaning with argon plasma; (B) photograph of bonding pads after 3 minutes of cleaning with argon plasma.

Epoxy bleed-out, another contaminant frequently encountered in hybrid circuits, can also be removed by plasma cleaning. "Bleed-out" results from resin separating from the filler and hardener during the cure of epoxy die-attach adhesives. The resin penetrates small crevices in the substrate and migrates along the surface, often contaminating wire bond pads. The thickness of this bleed-out may range from 10 to 2,500 Angstroms. After the epoxy has cured, the bleed-out is barely perceptible even under high magnification. The efficiency of plasma in removing epoxy bleed-out on a gold-plated surface around a transistor can be seen in Figure 7.38.

PARTICLE IMMOBILIZING COATINGS

Though organic coatings are not generally used in hermetically sealed hybrid circuits, some NASA and military programs have required their use to assure that any particles that remain on the circuit after cleaning or that slough from the circuit after sealing do not migrate and cause electrical shorts. The function of the coating is to freeze in place all loose particles and to protect the circuit from any particles that may subsequently detach during vibration, constant acceleration testing, or actual use. Most hybrid manufacturers use the PIND (Particle Impact Noise Detection) test on a 100% basis to assure the absence of loose particles. According to this screen test (defined in MIL-STD-883, Method 2020) the sealed packages are placed on a shake table and acoustically coupled to a transducer that detects and amplifies any sound due to particles that move within the package while the package is vibrated or shocked. The sound produced by the vibrating particles is picked up visually on an oscilloscope or audibly by a speaker. Though the PIND test is useful for the majority of applications, there is no guarantee, if a package has passed, that particles won't subsequently detach. Thus some programs require that the devices or hybrid circuits be coated prior to sealing.

Parylene Coatings

Parylene coatings have received the greatest attention as particle immobilizing coatings for hybrid circuits. The Parylenes, a tradename of Union Carbide for highly purified p-polyxylylene polymers, were introduced for commercial applications in 1965.[44] They have the distinction of being among the very few polymers that can be vapor deposited and the only polymers that can be vapor deposited on a commercial scale by the thermal decomposition of a solid dimer. On heating the solid dimer to 550-600°C, it decomposes, forming gaseous diradicals; these being unstable, quickly combine on a substrate to form high molecular weight ultra thin coatings (see Figure 7.39). Three Parylene coatings differing in the number of chlorine atoms that each phenyl group contains, were introduced by Union Carbide. Parylene N contains no chlorine groups; Parylene C contains one chlorine atom per phenyl group; and Parylene D contains two

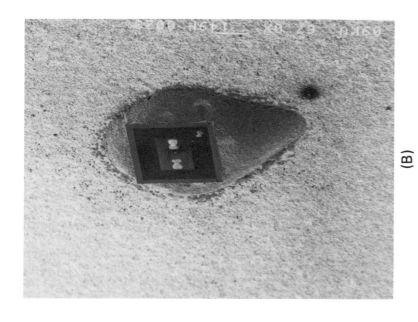

(A)

(B)

Figure 7.38: Removal of epoxy bleed-out by plasma cleaning (Courtesy of March Instruments Inc.): (A) SEM photo of a transistor epoxy bonded to a plated gold surface showing the epoxy bleedout; (B) the same transistor after 3 minutes of cleaning with argon plasma.

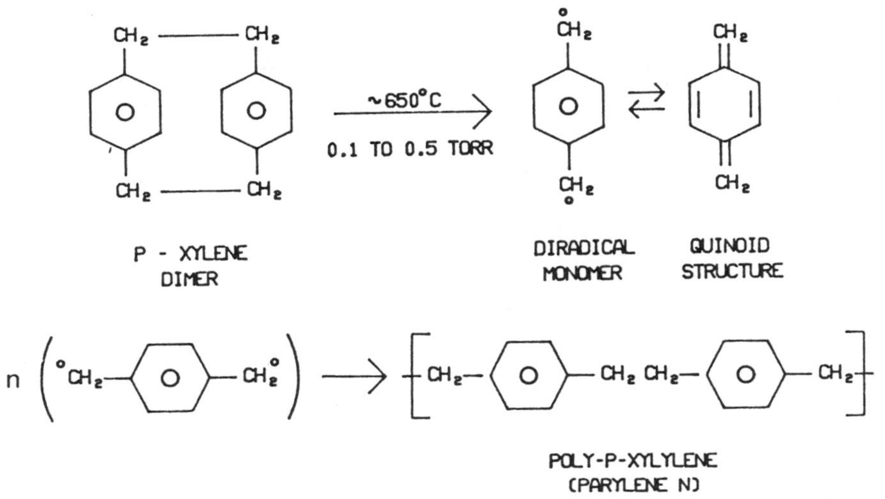

Figure 7.39: Parylene synthesis reactions.

chlorine atoms.[37] The importance of the Parylene coatings in microelectronic applications was first recognized by Licari and Lee[45] who conducted studies on the effects of Parylene coatings on the electrical parameters of CMOS/SOS devices through a study contract with NASA. The high purity of these coatings (no ionic contaminants) and their ability to be deposited as ultra-thin, porous-free coatings with efficient coverage provide excellent protection for semiconductor devices and integrated circuits. Further, the Parylenes not only are benign to small diameter (1 mil) wire bonds but actually reinforce the bonds mechanically and provide a barrier from the corrosive effects of moisture and contaminants. This is in contrast with other organic coatings such as epoxies or thick silicones that can stress and break wire bonds. Characteristic properties of Parylenes of importance in electronic applications are given in Tables 7.9, 7.10, and 7.11. A high capacity Parylene generator having a vertically mounted 24 inch by 18 inch diameter deposition chamber is shown in Figure 7.40.

Parylene would be the ideal coating for electronics were it not for some inherent limitations in its application, inspection, and removal. In applying Parylene, masking of the sealing surfaces and the external pins of the hybrid package is difficult and costly. Because Parylene is vapor deposited it penetrates and coats areas that would otherwise be inaccessible. Thus Parylene's key feature of thoroughly and evenly coating an entire circuit becomes a problem in preventing it from penetrating beneath the maskant and coating the seal and pin areas. Pressure-sensitive tape has been used as a maskant but considerable time is necessary to firmly apply the tape without damaging the leads. Subsequent to coating, the removal of the tape requires fastidiousness and care to avoid distorting the leads. Furthermore, depending on the tape used, adhesive residues may be difficult to

**Table 7.9: Thermal and Mechanical Properties of Parylene Coatings
(Courtesy, Union Carbide Corporation)**

	Parylene N	Parylene C	Parylene D
Tensile Strength, psi.	6,500	10,000	11,000
Yield Strength, psi.	6,100	8,000	9,000
Elongation to Break, %	30	200	10
Yield Elongation, %	2.5	2.9	3
Density, g./cm.^{3}	1.11	1.289	1.418
Coefficient of Friction			
Static	0.25	0.29	0.33
Dynamic	0.25	0.29	0.31
Water Absorption,			
24 hours	0.06 (0.029″)	0.01 (0.019″)	—
Index of Refraction,			
$n_D 23°C$.	1.661	1.639	1.669
Melting or Heat Distortion			
Temperature, °C.	405	280	>350
Linear Coefficient of			
Expansion, (10^{-5}/ °C.)	6.9	3.5	—
Thermal Conductivity,			
(10^{-4} cal./sec./			
cm.^{2}– °C./cm.)	∿3	—	—
Data recorded following			
appropriate ASTM method.			

**Table 7.10: Electrical Properties of Parylene Coatings
(Courtesy, Union Carbide Corporation)**

	Parylene N	Parylene C	Parylene D
Dielectric strength, short			
time, volts/mil at 1 mil	7,000	5,600	5,500
Volume resistivity,			
23°C, 50% RH, ohm-cm	1×10^{17}	6×10^{16}	2×10^{16}
Surface resistivity			
23°C, 50% RH, ohms	10^{13}	10^{14}	5×10^{16}
Dielectric constant			
60 Hz	2.65	3.15	2.84
10^3 Hz	2.65	3.10	2.82
10^6 Hz	2.65	2.95	2.80
Dissipation factor			
60 Hz	0.0002	0.020	0.004
10^3 Hz	0.0002	0.019	0.003
10^6 Hz	0.0006	0.013	0.002

Note: Data recorded following appropriate ASTM method.

Table 7.11: Barrier Properties of Parylene Coatings
(Courtesy, Union Carbide Corporation)

| Parylene | Gas Permeability .....cc-mil/100 in² –24 hr–atm (23°C) ..... | | | | | | Moisture Vapor Transmission Rate g-mil/100 in² 24 hr (37°C, 90% RH) |
	N_2	O_2	CO_2	H_2S	SO_2	Cl_2	
N	7.7	39.2	214	795	1,890	74	1.6
C	1.0	7.2	7.7	13	11	0.35	0.5
D	4.5	32	13	1.45	4.75	0.55	0.25

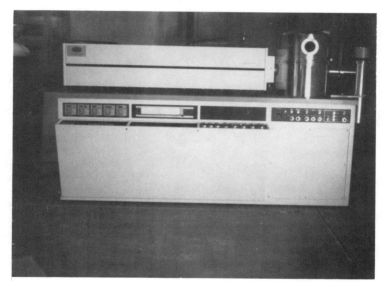

Figure 7.40: Parylene deposition console, Model 1040 (Courtesy of Union Carbide Electronics Division).

remove and may require special cleaning methods. A second masking method employing a strippable latex maskant is similarly limited. The strippable maskant is difficult to apply in controlled amounts to selected areas and its removal is tedious in that the Parylene overcoating makes it difficult to grip and remove the maskant. The best masking technique employs special tooling designed so that a rubber or foam gasket is pressed tightly against the seal area during the deposition. One such tool consists of two metal plates: a bottom one which may be flat or contain cavities into which the hybrid circuits are placed and an upper metal plate with apertures corresponding to the cavities in the bottom plate. Two thin sheets of foam or elastomer, also with apertures, are inserted between the plates such that the hybrid packages are sandwiched between the two rubber sheets and plates and the seal areas and pins are protected (Figure 7.41).

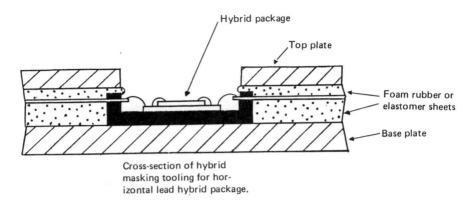

Cross-section of hybrid masking tooling for horizontal lead hybrid package.

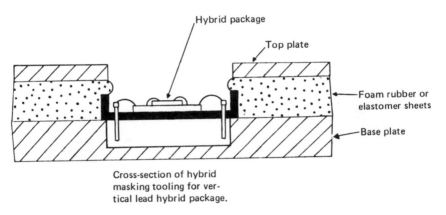

Cross-section of hybrid masking tooling for vertical lead hybrid package.

Figure 7.41: Masking tools for Parylene coating.

A unique method for depositing Parylene on hybrid circuits that have already been sealed consists in vapor depositing the gaseous monomer through a small hole in the lid. The hole, drilled or punched in the lid, permits the dimer free radicals to enter as vapor, then polymerize and condense on the inside surfaces of the package. The hole is then sealed using a tin-lead solder. This method was developed as an expedient to salvaging already assembled and tested hybrids where it was shown that there was a risk of particle entrapment. However, drilling or punching the hole requires great care to avoid even more metal particles being generated and becoming entrapped in the hybrid package.

In reworking a Parylene-coated circuit, the Parylene must first be removed. This is relatively difficult and costly. Parylene is extremely insoluble in the normal organic solvents and, though classified as a thermoplastic, is of such high molecular weight that it will not soften or melt at temperatures compatible with the hybrid circuit elements. The best removal method is plasma etching using oxygen. This method is not selective; thus all of the Parylene must be removed, then reapplied after the rework. In general, 0.001 to 0.002 inches of Parylene can be completely removed in 20-30 minutes in the oxygen plasma. Any exposed silver-filled epoxy tarnishes under these conditions, but it has been found that plasma cleaning with argon as the last step will remove the tarnish. The effects of oxygen plasma on other sensitive elements of the circuits, for example, on some semiconductor devices and nichrome resistors, should be evaluated before using this process.

Lastly, Parylene applied in very thin films is transparent, making visual inspection to assure complete coverage difficult. This problem has been largely resolved by adding a fluorescent compound to the solid dimer and co-depositing the two. Anthracene added to the dimer in small quantities (0.75-1%) has been found effective. It imparts a blue fluorescence to the coating when inspected under ultraviolet light.

Solvent-Soluble Coatings

High-purity coatings have been investigated as alternates to Parylene because of the difficulties that have been encountered in removing Parylene from circuits that need to be reworked. The most useful of these have been the block of copolymers of silicones and styrenes. These styrene-modified silicone coatings are readily and completely soluble in toluene, xylene, the Freons, and chlorinated solvents. The coatings can be removed quickly and easily from the entire circuit by vapor degreasing with Freon TF solvent. In the original studies performed by Licari and Weigand, Dow Corning's DCX9-6326, a block copolymyer of dimethylsiloxane and alpha-methyl styrene, was characterized for electrical, physical, and chemical properties, then applied to both hybrid microcircuits and printed wiring boards.[46-48] These studies were later corroborated and expanded by David.[49] In the studies, extensive long term reliability data were collected on the effects of the coating on sensitive chip devices and on gold and aluminum wire bonds to thick-film gold conductor pads. Wire bond pull strengths and electrical resistance of a series of wire bonds were measured after 1,000 hrs at 150°C and after 1,000 temperature cycles from −55 to +125°C. In all cases the performance of the coated wire bonds was almost identical to the uncoated bonds and no electrical failures of devices were noted. Though these coatings are more economical to process from the standpoint of ease of removal for rework, ease of masking, and the lower costs of spray application and equipment, they are still experimental and limited in availability.

Particle Getters

Plastic resins that cure to a tacky state may be used as alternates to coatings in immobilizing particles. These particle getters are soft silicone gels and are applied to the center of the inside of the lid, prior to sealing. The effectiveness of silicone getters has been demonstrated by several companies through controlled experiments in which known quantities of metal particles of various sizes were purposely added (seeded) to a circuit containing the getter, then vibrated and evaluated for loose particles by PIND, X-ray, and electrical testing.[50] The types of particles used were representative of those that might be found in a hybrid package. They included gold wire, aluminum wire, solder balls, and pieces of eutectic alloy. It is important to apply the getter to the center of the lid so that the seal area does not become contaminated with organic "bleed-out" from the silicone which could degrade the seal. A distance of 0.1 inch minimum from the seal perimeter area should be kept free of all organic contaminants. Another consideration in the use of silicone getters is the amount of outgassing. As with other organic materials, the extent of outgassing is a function of the material used, its degree of cure, and the vacuum bake schedule. Silicone gels are available from both General Electric and Dow Corning.

VACUUM-BAKING AND SEALING

Vacuum-Baking

The last assembly operations are vacuum-baking and sealing. Both steps are extremely important in maintaining the reliability of the circuit for long periods of time under any adverse environments that the hybrid might encounter. Vacuum-baking removes adsorbed and absorbed moisture, but is also effective in outgassing other volatile materials from the circuit, especially organic volatiles from epoxy adhesives. If not removed, these constituents become entrapped in the package and can later cause corrosion or otherwise degrade the electrical functioning of the circuit. Though vacuum-baking is considered a critical step, it has been difficult to standardize on an optimum vacuum-bake cycle applicable to all hybrid types, because of differences in materials and devices, types and amounts of epoxy, and processing conditions used by the manufacturers. Thus manufacturers have empirically established vacuum-bake schedules that are best suited to their hybrids and often use several schedules depending on requirements. The primary criterion for the selection of the vacuum-bake cycle is the moisture requirement that the sealed package must meet. For example, vacuum-baking at high temperatures for extended periods of time may be necessary to meet the 3,000 ppm (maximum) moisture requirement for Class S military and space-graded circuits, while less severe schedules might suffice for commercial hybrids. A survey showed that there were almost as many vacuum-bake schedules used as there were hybrid manufacturers. Some of the more widely used schedules are:

24 hours at 150°C in nitrogen, followed by 1 hour at 150°C in vacuum.

16-24 hours at 150°C in vacuum

4 hours at 135°C in vacuum

10 hours at 150°C in vacuum

3 hour nitrogen bake at 100°C

72 hours at 150°C in nitrogen, followed by 16 hours in vacuum at 150°C

The oven used for vacuum-baking should be capable of being maintained to at least 150°C and a vacuum of 50 microns of mercury. The oven is connected to a dry-box so that the lid sealing operation can be performed in a nitrogen ambient. The sealing chamber contains two double-pass doors: one from the vacuum-bake chamber to the dry-box, the other from the dry-box to the outside ambient. After the specified vacuum-bake period, dry nitrogen or a mixture of 20-30% helium in nitrogen is introduced in the chamber and the parts are transferred to the dry-box for sealing. Once in the dry-box the parts should be sealed within 24 hours, preferably within several hours. If the hybrids remain in the dry-box for any extended time they can reabsorb moisture, small as it is, from the dry-box atmosphere. Since the vacuum-bake operation has a higher throughput than the sealing operation, the circuits often remain in the dry box ambient for extended periods of time before being sealed. The nitrogen ambient must be extremely dry (less than 50 ppm water) but even under these conditions epoxies and other hygroscopic surfaces of the hybrid will getter moisture from the nitrogen. Since there is a continuous flow of nitrogen, the amount of water absorbed by the hybrid continues to increase with time. Therefore, if the hybrid circuit packages cannot be sealed within a short period of time after vacuum-baking, the vacuum-bake cycle should be repeated.

Several factors influence the amount of moisture in the dry box. Among these are:

- The moisture content of the nitrogen entering the dry box. This should be controlled to less than 5 ppm.

- The moisture content of the nitrogen gas in the dry box. This should be controlled to less than 50 ppm, preferably less than 20 ppm.

- The seal integrity of the dry box. The box must register a positive pressure to prevent air from entering through the seals.

- Human intervention. Both doors of the interconnecting pass-through should not be opened simultaneusly which would allow air and moisture to enter.

- Introduction of tools or other materials into the dry box without prior vacuum bake-out. Back-filling an interlock that contains tools is not as efficient in removing moisture as vacuum-baking. Figure 7.42 shows that evacuating for 10 minutes is as effective as flushing with dry nitrogen for 2 hours. Not only does vacuum accelerate the drying time, but it also assures that oxygen and moisture are desorbed from the surfaces.[51]

- The flow rate of the incoming nitrogen. Figure 7.43 relates moisture (in ppm) to time at different flow rates. A flow rate of 10 cubic ft/hr should be used for a 1.5 cu ft enclosure to reach 50 ppm moisture within one hour.

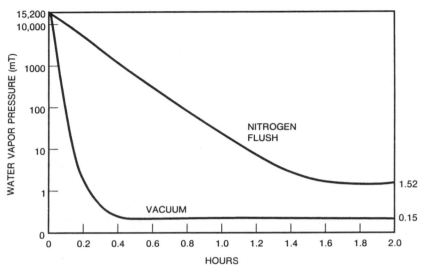

Figure 7.42: Drying times—vacuum vs nitrogen (Courtesy Benchmark Industries).[51]

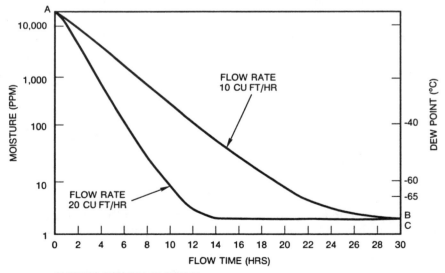

INTERNAL VOLUME: 24.5 CU FT
INITIAL HUMIDITY: 20,000 PPM
GAS HUMIDITY: 2 PPM

Figure 7.43: Drying times vs nitrogen flow rates (Courtesy Benchmark Industries).[51]

Sealing

Throughout any discussion of package sealing the word hermetic is used to describe the effectiveness of the sealing process and the integrity of the sealed package. It is important to note that hermetic is not used in the absolute sense of being completely air-tight. All hybrid and device packages, even those that are metallurgically welded, have finite leak rates. The term hermetic is therefore applied to packages that have very low leak rates, generally less than 1×10^{-7} atm-cc/sec of helium, while packages that have leak rates of 10^{-6} atm-cc/sec or higher are considered non-hermetic. Hybrid circuits must be sealed to prevent moisture, oxygen, and other ambient contaminants from entering the package, thus assuring long-term reliability. Secondly, sealing prevents mechanical damage due to handling. Hybrid circuits contain uncased die, thin-film metallization, and very fine, fragile wire and wire bonds. Though the die are passivated, there are differences in the type, quality, and integrity of the passivation layers used by different die manufacturers and there may be variations from lot to lot from the same manufacturer. Thus one cannot rely on the passivation layers for full hermeticity and long term protection. Furthermore, the wire bond sites, the wire itself, and, generally, thiln-film resistors are not passivated. There are three general methods for sealing hybrid circuits: metallurgical, glass bonding, and epoxy bonding. Metallurgical sealing may consist of soldering or welding.

Soldering. Packages may be solder-sealed by inserting a solder preform between the lid and package seal area and heating to a temperature that will melt the solder.

There are four solder sealing processes depending on the method by which the heat is applied. These are: hand soldering, belt-furnace sealing, seam soldering, and platen soldering. The solder may be one of two types: solf solder—a low melting alloy such as 63% Sn/37% Pb, m.p. 183°C or hard solder—a higher melting alloy such as 80% Au/20% Sn, m.p.280°C.

Hand-Soldering. Hand soldering is the simplest and oldest type of sealing. It is a manual process in which a soft solder such as tin-lead is melted using a heated solder-iron. Hand-soldering generally requires a flux to remove surface metal oxides, reduce surface tension, and improve wetting and adhesion. Soft solder sealing of hybrid circuits carries the risk of both flux and solder splatter becoming entrapped in the package and degrading the reliability of the circuit. A key advantage of hand-soldering is the low temperatures that can be employed which may be necessary for some heat sensitive circuits or devices. Indium alloys having melting temperatures even lower than the 63 Sn-37 Pb are available. (see also this chapter, Die and Substrate Attachment).

Belt-furnace Sealing. Belt-furnace sealing is a conveyorized process in which a solder preform is inserted between the lid and seal surface of the package, held together with a clamping fixture, and heated in a furnace to the melting temperature of the solder. Belt-furnace soldering uses a hard solder such as Au-Sn or Au-Ge. The furnace is continuously flushed with dry nitrogen so the parts are sealed in an inert atmosphere. Belt-furnace sealing is a popular, rather economical process, widely used for sealing single chip devices in small packages. In belt-furnace sealing hybrid circuits having many epoxy-attached devices, a breather hole of 20-40 mils diameter is drilled or punched in the lid before sealing so that outgassing products from the epoxies can escape. The packages are then vacuum-baked and the hole is closed with soft solder in a nitrogen dry-box. This process entails several extra steps but, without the breather hole, gases evolved from the epoxy expand and cause blowholes in the seals. Outgassing products can also become permanently sealed into the package. Belt-furnace soldering is generally applicable to metal packages and lids and optimally gold-plated metal. Metal lids can be sealed to ceramic packages if they have metallized seal rings. Belt-furnace sealing has many variables that must be taken into account to get high yields. The eutectic ratio of gold to tin results in the lowest seal temperature with the highest amount of gold. As the solder melts, it takes some of the gold plating into solution which then raises the melting temperature (Figure 7.44). If the plating is too thick excessive gold will dissolve, increasing the melt temperature above the furnace temperature and affecting the reliability of the seal. It is therefore important to control the plating thickness. Other critical parameters include thickness of the preform, flatness of the lid and package, and furnace temperature profile. The preform thickness and flatness of the package and lid are interrelated. If the mating surfaces of the package and lid are flat, the solder does not need to be very thick to thoroughly wet. However, if gaps exist in the seal area, more solder will be needed to make a hermetic seal. Sealing yields will vary widely if flatness is not controlled. The optimum thickness is arrived at through experimentation. In some cases, lack of flatness can be compensated by applying pressure during sealing by clamping the lid to the package.

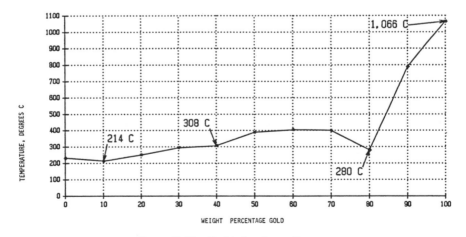

Figure 7.44: Gold-tin phase diagram.

The belt-furnace temperature profile must also be accurately controlled to produce reliable seals. Since each package/lid combination has a different mass, the seal interface temperature will differ from the furnace temperature. Therefore, the furnace temperature must be adjusted to attain the correct interface temperature. A typical furnace profile consists of a minimum of three stages: pre-heat, dwell, and cool-down. The temperature profile for solder sealing a package with a gold-tin preform is shown in Figure 7.45. The parts are slowly heated (to prevent thermal shock) to approximately 25 to 50°C above the eutectic temperature of the solder, kept at that temperature for a short while to allow solder flow and wetting, then slowly cooled to ambient temperature. The peak temperature is a function of the amount of gold available to leach into the interface. It should not be so high that the solder expands and wicks onto the surface of the lid. The furnace should be continuously flushed with dry nitrogen. Any air leaking into the system will cause oxidation and inhibit the wetting of the solder. The effects of various parameters on solderability are given in Table 7.12.[52]

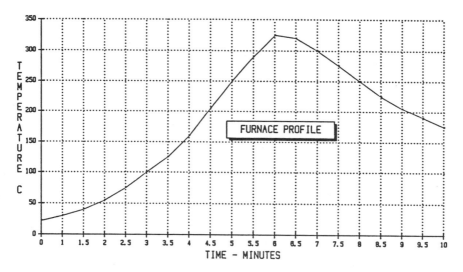

Figure 7.45: Furnace temperature profile for solder sealing with gold-tin preform.

Seam Soldering. Seam soldering utilizes the same equipment as seam welding (see below) except that a hard solder preform such as a gold-tin alloy is used. The solder is reflowed by resistance heating by allowing electrical pulses to travel through the lid or from the lid through the preform between electrodes. Most of the heat is generated at the point of electrode contact because it presents the highest resistance. This process is relatively slow but has advantages over belt-furnace soldering in that heating is localized to the seal area and the parts can be vacuum-baked and sealed in a dry-box which has a controlled nitrogen atmosphere.

Platen Soldering. In platen soldering, the package-preform-lid combination is placed in a fixture which provides for water-cooling. The chamber is evacuated and back-filled with an inert gas. A platen is then brought into contact with the lid and temperature is increased to achieve melting of the

Table 7.12: Effects on Solderability of Metallurgical, Mechanical and Furnace Parameters[52]

Parameter	Too much	Too little	Impact magnitude
Metallurgical			
Lid and package gold plating	Gold leaching raises melting temperature	Difficulty wetting	Moderate
Preform volume	Unnecessary wetting of surrounding areas	Difficulty obtaining complete seal	High
Package/lid cleanliness	Cannot be too clean, but too aggressive cleaning methods can produce corrosion	Solder wetting inhibited	High
Mechanical			
Lid and package flatness	Cannot be too flat	Thicker solder preform and/or use spring clips	High
Package lid and fixture mass	Longer furnace preheat and/or longer furnace dwell	Cannot get it too light	High
Lid compliance	Easy deformation from external forces	Difficult to compensate for lack of flatness	High
Preform attach	Tacked or bonded; means easy handling	Chance of preform misalignment	Moderate
Furnace			
Belt speed	Package/lid mass cannot reach melting temperature	Silicon component damage	High
Peak temperature	Solder crawls up on lid	Proper solder flow not achieved	High
Fillet appearance	Hermeticity probable	Hermeticity questionable	High
Atmosphere: Moisture and oxygen content	Poor wetting	The less the better	High

solder. The chamber pressure during sealing is increased to assist in forcing the solder inward to form a better fillet and to compensate for expansion of gases within the package. Platen soldering is relatively fast requiring only about 2 minutes but subjects the substrate and circuit to high temperatures (about 180-190°C).

Welding. Welding consists in the direct fusion of two metals through the application of heat. Heat may be generated by passing an electric current through high resistance metals (seam sealing), by focusing a laser beam (laser welding), focusing an electron beam (e-beam welding), or by direct heating (projection welding).

Seam Welding. Seam welding, also called seam sealing or resistance welding, is the most popular and widely used process for hermetically sealing hybrid packages. In seam sealing, the lid is positioned on the package and two truncated, cone-shaped copper alloy electrodes (Figure 7.46) are allowed to travel along the top edge surface of the opposite sides of the package lid. These electrodes deliver high current pulses that flow through the lid-package interface, laterally across the lid as well as through the package, and out the other parallel electrode. The travel rate of the electrodes, intensity and duration of the current pulses, and time interval between pulses are controlled to provide the optimum molten region of overlap with the minimum energy input. Seam sealing localizes the heat to the point-contract of the lid and electrodes, focusing sufficient power into the small seal area so that the weld is completed before the rest of the package has time to heat up. The temperature at the point contact (about 1500°C) can be controlled by the current pulse amplitude. By timing the pulses with the advance of the electrodes, overlapping spot welds are formed that produce a continuous weld.

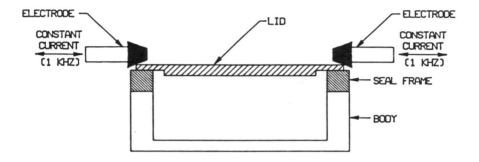

Figure 7.46: Parallel seam sealing (package cross section).

Heat is proportional to I^2R, thus it concentrates at the point of highest resistance—the juncture between the lid and electrode. Step lids in which the thickness at the edges has been reduced to 5 mils and which are 5 mils narrower than the package outside dimensions permit current and heat to concentrate, thus producing a faster, more reliable seal (Figure 7.47). After the electrodes have travelled along the two parallel edges of the package, the package is rotated 90 degrees to seal the other two sides. The key to producing low temperature sealing is to use a minimal total energy input.[53] During sealing, the temperature at the interface must reach at least 200°C if tin-lead eutectic solder preforms are used or 1500°C for direct Kovar welding. The equation that governs the sealing process is:

$$\text{Energy (in watt-seconds)} = [\text{Avg. Power}] \times t$$
$$= [I^2 \times R_c] \times t$$
$$= \frac{I^2_{p-p}}{8} \times R_c \times \frac{PW}{PRT} \times t$$

where: I_{p-p} = peak-to-peak amplitude of 1 KHz constant current (0-1000 amps)

R_c = contact resistance between electrode and edge of lid

PW   = pulse width in milliseconds

PRT   = pulse repetition time in milliseconds—defined as the time between leading edge of successive pulse widths

t   = total sealing time—seal perimeter in inches divided by the sealing rate in inches/sec.

These parameters can be varied to produce an optimum seal. A typical set of parameters for welding a gold-plated Kovar lid to a gold-plated Kovar package is given below.[53]

I_{p-p} = 700 amps
PW   = 20 ms
PRT   = 60 ms
t   = 7.6 sec
Localized seal temperature   = about 1500°C
Package temperature during seal = about 50°C

Figure 7.47: Cross section of lid-package seam weld joint (X100) (Courtesy of Rockwell International).

Figure 7.48 shows a typical seam sealer while Figure 7.49 shows the sealer installed in a dry-box.

Figure 7.48: Parallel seam sealing machine (Courtesy Solid State Equipment Corp.).

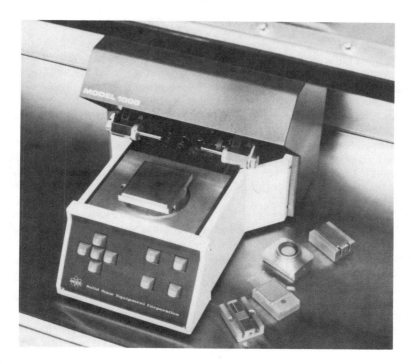

Figure 7.49: Seam sealer installed in dry box (Courtesy Solid State Equipment Corp.).

Commercially available seam sealers can handle square or rectangular packages up to 8 inches on a side and circular packages from 0.125 inches to 8 inches in diameter. Several guidelines in using seam sealers may be listed as follows:

- Thickness of the lid. The thicker the lid the more current must be supplied to form the weld and the hotter the entire package will get. For best results, the thickness of the lid should be reduced to 5 mils at the edges by milling or chemically etching while keeping the center portion at 10-15 mils for strength.

- Electrode wear. Since the current that produces the weld also passes through the electrodes and electrode bearings, these parts have a limited lifetime, should be inspected for wear, and replaced periodically.

- Package material. Seam sealing requires two high resistance materials to form the weld joint. Kovar packages and lids can be easily welded, but not copper or aluminum packages.

- Package shape. Seam sealing machines can only be used to seal parallel-sided or circular packages. Irregular-shaped parts must be brazed, soft solder sealed, laser welded, or, if hermeticity is not a requirement, epoxy sealed.

Electron-Beam Welding. Electron-beam (e-beam welding) involves focusing an electron beam on the parts to be joined, resulting in heating and fusion of the parts. E-beam welding as a method for sealing metal packages is not extensively used because other methods (seam-sealing, belt-furnace soldering) are more economical and adaptable to production. E-beam welding is used in special situations where metals or alloys are difficult to weld (refractory types such as titanium, tungsten, molybdenum, stainless steel) or where the package size is too large or irregular to be sealed by more conventional means. A further advantage of e-beam welding is that the beam can be programmed and focused so that hard-to-get-at surfaces can be joined. E-beams can penetrate very deeply into a structure.

On the negative side, electron-beam welding is expensive, requiring high-vacuum equipment, high initial investment in electron-generating equipment, costly maintenance, and highly skilled operators. Also because of its deep penetrating power, heat can be transferred to and degrade adjacent parts such as glass-to-metal feed-throughs and circuit components. Thus, heat-sinking and shielding may be necessary.[54,55]

Laser Welding. A high power (400 watt) neodymium YAG laser provides a unique source of high-intensity energy that can be applied to many industrial applications, among which are: welding, drilling, brazing, soldering, heat treating, and cutting. In general, the laser is a potentially useful tool in cases where high intensity heat needs to be applied in a precise, controlled manner. Laser welding has many advantages over other welding methods, in particular over e-beam welding. These include:

- Low heat input results in small fusion and heat-affected areas

- A wide variety of similar and dissimilar metals can be welded

- No direct contact with the work piece is required

- No vacuum is required

- There is no thermal damage to heat-sensitive devices or to glass feed-throughs

- Weld characteristics and yields are highly reproducible

As in the case of e-beam welding, the initial investment for laser equipment is high and, in addition, special safety precautions and facilities are needed to operate the laser.

Projection Welding. In projection welding, one of the parts, either the lid or the package case, has a continuous metal projection that makes contact with the opposite part at the weld interface (Figure 7.50). This projection forms a high resistance point contact that melts and forms a weld along the entire seam as the current is passed through. Several guidelines must be considered in the use of projection welding:

- Each package size and style requires a separate set of electrodes and weld parameters. The parameters must be established experimentally.

- The case and lid must have closely controlled tolerances; only small dimensional variations are allowed from one part to another.

- Energy consumption is high and costly, rendering this process economical only for sealing small packages.

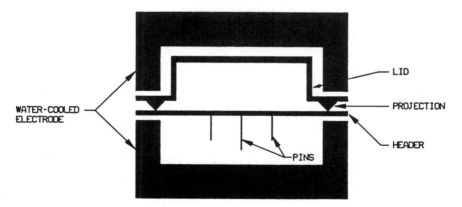

Figure 7.50: Diagram of projection welding.

Glass Sealing. Low-temperature-melting glass with an expansion coefficient closely matching that of the mating surfaces may be used for sealing ceramic or ceramic-metal packages. The glass, usually as a preform or paste, is applied between the surfaces to be bonded, held in a clamping fixture which also provides some pressure, then conveyorized through a furnace. The glass melts and fuses with some of the glass of the ceramic or with oxide of the metal. The coefficient of thermal expansion of glass (about 5 ppm/C is not as close to that of alumina ceramic as it is to Kovar, so some warpage can occur when sealing thin sections of alumina.

Glass is widely used for sealing small ceramic packages that house single devices. It has not found extensive use for sealing large multi-chip hybrid circuit packages because the high temperature required to melt the glass (400-500°C) can degrade devices, wire bonds, and epoxies within the package. Also the low tensile strength of glass and mismatches in expansion coefficients result in microcracks and loss of hermeticity.

Epoxy Sealing. The question is often raised as to whether epoxy adhesives can be used to seal metal or ceramic cavity packages and meet the hermeticity requirements of military specifications. Indeed epoxies both as adhesives and as encapsulants are widely used for sealing devices and hybrid circuits, but these are for commercial and industrial applications where the strict leak rate and moisture requirements of MIL-STD-883 are not necessary. Though not hermetic in the sense of the military specifications, epoxy-sealed circuits have a long history of reliability. Epoxies may be dispensed automatically as liquids, applied by screen-printing, or used as B-staged preforms. In all cases, epoxy sealing is attractive because of its low cost for both material and processing, low curing temperatures (less than 170°C), and ease of repair. Ceramic lids can be attached to ceramic packages with epoxy in a dry-box containing nitrogen, then cured in nitrogen-purged ovens. These packages initially may even have low leak rates of 1×10^{-8} atm-cc/sec or less; however, epoxies—even the best of them—as is true for all other organic polymeric materials, have finite permeabilities to moisture and air oxygen and in due time will allow moisture to enter and accumulate in the package. In one study, moisture sensors were enclosed in epoxy-sealed ceramic packages so that the penetration of moisture could be monitored dynamically while the packages were exposed to high humidity/temperature conditions (98% RH and 60°C). Though the moisture content remained relatively low and constant for the first two to three days, there was an abrupt increase on the third day which then continued to increase linearly.[56] The latent period of 2-3 days may be explained as the time that was necessary for the water to saturate the epoxy, traverse the width and thickness of the bond line, and be released at the other end (Figure 7.51).

Hermeticity and Leak Testing. Generally, metallurgically-sealed, glass-sealed metal, and ceramic packages are considered hermetic if they meet the minimum leak rates established in MIL-STD-883. Epoxy-sealed or plastic-encapsulated packages are non-hermetic because of the inherent permeability of polymers to moisture, air, and other gases and because of moisture penetration along the interface between the plastic and package leads. Non-hermetic packages are used primarily in commercial applications. Because of their lower cost, plastic-encapsulated devices such as epoxy transfer-molded integrated circuits far outnumber the hermetically-sealed packages.

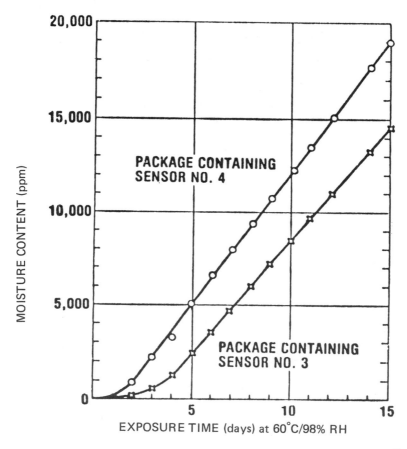

Figure 7.51: Moisture permeation for epoxy-sealed ceramic packages.[56]

Hermeticity testing involves measuring the leak rate of a sealed package that has been pressurized ("bombed") with helium. After removal from the pressure container, a mass spectrometer is used to detect and measure the helium that has penetrated the package. This is referred to as fine leak testing. Fine leak testing is normally followed by gross leak testing in which the sealed package is immersed in a heated inert fluid such as a fluorocarbon fluid. As the package is heated, the internal gases expand, escape from the package into the fluid, and are visible as a stream of bubbles. To increase the sensitivity of this test, the packages may first be immersed and pressurized in a low-boiling fluorocarbon for long periods of time (up to 16 hrs). Once in the package the low-boiling liquid vaporizes when immersed in the heated higher boiling fluorocarbon and creates a high internal gas pressure. Leak rate sensitivity is thus increased and leaks as low as 1×10^{-6} atm-cc/sec can be measured. Leakages can occur at the lid-to-package interface or at the glass-to-metal seals where the external leads emanate from the package.

An alternate leak test utilizes the radioactive isotope of krypton (Kr85). The sealed packages are pressurized for 10-15 minutes with nitrogen gas containing about 0.01% of Kr85, allowing the mixture to enter the package. The parts are then removed and scanned for radioactivity.[57] The radiotracer method has some distinct advantages over the helium mass spectrometer method in the shorter times and lower pressures needed for the gas to penetrate the package but may be outweighed by the licensing and precautions involved in handling radioactive gases. Leak rate test procedures and requirements for both mass spectrometric and radioisotope methods are described in MIL-STD-883, Method 1014.

REFERENCES

1. Weigand B.L. and Caruso, S.V., "Development Of A Qualification Standard For Adhesives Used In Hybrid Microcircuits", *Proc. ISHM*, 1983.
2. Licari, J.J., Perkins, K.L., and Caruso, S.V., "Evaluation of Electrically Insulative Adhesives For Use In Hybrid Microcircuit Fabrication," *IEEE Trans. on Parts, Hybrids, and Packaging*, Vol. PHP-9, 1973.
3. Mitchell, C. and Berg, H., "Use of Conductive Epoxies For Die Attach", *Proc. ISHM*, 1976.
4. Bolger, J.C. and Mooney, C.T., "Volatile Organic and Extractable Ionic Contaminants in Epoxy Die Attach Adhesives", *Proc. NEPCON*, Anaheim, CA, 1983.
5. Planting, P.J., "An Approach For Evaluating Epoxy Adhesives For Use in Microelectronic Assembly", *IEEE Trans. on Parts, Hybrids, and Packaging*, Vol. PHP-11, 1975.
6. Shenfield, D.M. and Zyetz, C.M., "An Investigation of the Effect of 150°C Storage on the Electrical Resistance of Conductive Adhesive Attachment of 2N222A Transistors", *Proc. ISHM*, 1983.
7. Licari, J.J., Weigand, B.L., and Soykin, C.A., "Development of a Qualification Standard For Adhesives Used in Hybrid Microcircuits", *NASA CR-161978*, 1981.
8. IBM Report 74W-00090, "Investigation of Discrete Component Chip Mounting Technology For Digital and RF Hybrid Microcircuits", Final Report, NASA Contract NAS 8-14000/SA2171, 1974.
9. White, M.L., "The Removal of Die Bond Epoxy Bleed Material By Oxygen Plasma", IEEE, *32nd Electronic Components Conference Proc.*, 1982.
10. Sandia Publication, *SAND-80-0834C*, "High Temperature Electronics and Instrumentation Sensing Processes", 1979.
11. Anderson, S.P. and Kraus, H.S., "Heat Aging Characteristics of Polyimide Chip Adhesives", *Proc. ISHM*, 1981.
12. Bolger, J.C., "Long Term vs Short Term Thermal Stability of Polyimide Die Attach Adhesives", *Intl. J. Hybrid Microelectronics*, Vol. 5, 1982.
13. Bolger, J.C., "Adhesive Related Failure Mechanisms in Military Hybrid Packages," *Intl. J. Hybrid Microelectronics*, Vol. 7, No. 4, Dec. 1984.
14. Li, T.P.L. and Chadderdon, G.D., "Cure Schedule and Extended Shelf Life Prediction of Epoxies", *Proc. ISHM*, 1983.
15. Fava, R.A., "Differential Scanning Calorimetry of Epoxy Resins", *Polymers*, Vol. 9, 1968.
16. Dettmer, E.S., et al, "Epoxy Characterization and Testing Using Mechanical, Electrical, and Surface Analysis Techniques", *Proc. ISHM*, 1983.
17. ASTM D 3482, "Determining Electrolytic Corrosion of Copper By Adhesives", 1981.

18. Lee, H. and Neville, K., *Handbook of Epoxy Resins*, McGraw-Hill, 1967.
19. Shukla, R.K. and Mencinger, N.P., "A Critical Review of VLSI Die Attachment In High Reliability Applications", *Solid State Technology*, July, 1985.
20. Dietz, R.L. and Winder, L., "New Die Attach Material For Hermetic Packaging", *EMTAS Conf. Proc.*, Phoenix, AZ, 1983, Soc. Mfg. Eng., EE83-145.
21. Johnson-Matthey Inc., *Electronic Materials Div. Product Bulletin*, "JMI 4613 AuSub Die-Attach Paste."
22. Moghadam, F.K., "Development of Adhesive Die-Attach Technology in Cerdip Packages; Material Issues", *Proc. ISHM*, 1983.
23. Stanley, W.W., "Hybrid Microelectronic Interconnection", *Electronic Packaging and Production*, Oct. 1982.
24. Jones, R.D., *Hybrid Circuit Design and Manufacture*, Marcell Dekker, Inc., 1982.
25. Ginsberg, G.L., "Chip and Wire Technology: The Ultimate in Surface Mounting", *Electronic Packaging and Production*, Aug. 1985.
26. Bonham, H.B. and Plunkett, P.V., "Impact of Plasma Cleaning on Hybrid Bonding", *NEPCON West*, 1978.
27. Lockheed Missiles and Space Co., *Thick Film Microcircuits Notebook*, Sept. 1979.
28. Rodwell, R. and Worrall, D.A., "Quality Control In Ultrasonic Wire Bonding", *Intl. Journal For Hybrid Microelectronics*, June 1985.
29. Carlson, J., "Advances in the Reliability of Ultrasonic Wire Bonding", *Hybrid Circuit Technology*, Dec. 1985.
30. Gehman, B.L., Ritala, K.E., and Erickson, L.C., "Aluminum Wire For Thermosonic Ball Bonding in Semiconductor Devices", *Solid State Technology*, Oct. 1983.
31. Slemmons, J.W. and Woolston, F.J., "Lab Book: Solutions To Unusual Hybrid Circuit Interconnection Problems", *Hybrid Circuit Technology*, Nov. 1984.
32. Oscilowski, A., "Tape Automated Bonding For VHSIC Parts", *Electronic Engineering Times*, July 14, 1986.
33. Warner, R.M. and Fordemwalt, J.N., *Integrated Circuits-Design Principles and Fabrication*, McGraw-Hill, 1965.
34. Harper, C.A., Ed., *Handbook of Electronic Packaging*, Chapter 4, "Welding and Metal Bonding Techniques" by H.F. Sawyer, McGraw-Hill, 1969.
35. Philofsky, E., *Sold State Electronics*, Vol. 13, pp. 1391-99, 1970.
36. Browning, G.V., Colteryahn, L.E., and Cummings, D.G., "Failure Mechanism Associated with Thermocompression Bonds in Integrated Circuits", *Physics of Failure in Electronics*, Vol. 4, RADC Reliability Series, Goldberg and Vaccaro, Eds., 1965.
37. Licari, J.J., *Plastic Coatings For Electronics*, McGraw-hill, 1970: Krieger Publishers, 1980.
38. Swanson, D.W. and Licari, J.J., "Identification and Removal of Contaminants from Hybrid Circuit Packages", *Hybrid Circuit Technology*, June 1985.
39. Trombka, J.A., "Solvent Solutions-Basic Facts for Understanding Defluxers", *Circuits Manufacturing*, Nov. 1985.
40. O'Donoghue, M., "Cleaning Implications in Hybrid Circuit Manufacture", *Hybrid Circuit Technology*, July 1986.
41. Flick, E.W., Ed., *Industrial Solvents Handbook*, Third Edition, Noyes Publications, 1985.
42. Sittig, M., *Handbook of Toxic and Hazardous Substances*, Noyes Publications, 1981.
43. Sax, N.I., *Dangerous Properties of Industrial Materials*, Sixth Edition, VanNostrand-Reinhold, 1984.
44. Gorham, W.F., "A New General Synthetic Method For the Preparation of Linear Poly-p-xylylene", *J. Polymer Science* 4:3027-3039, 1966.

45. Lee, S.M., Licari, J.J., and Litant, I., "Reliability of Parylene Films", *Proc. Met. Soc. Tech. Conf. Defects Electronic Matls. & Devices*, Boston, MA, 1970.
46. Weigand, B.L. and Licari, J.J., "Verification of Selected Solvent-soluble Coatings Using Production Hybrid Microcircuits", *AFWAL TR-80-4048*, 1980.
47. Weigand, B.L., Licari, J.J., and Perkins, K.L., "Manufacturing Technology For Conformal Coatings", *AFWAL TR-80-4139*, 1980.
48. Licari, J.J., Perkins, K.L., and Barnett, B.F., "Solder Mask/Conformal Coating Systems For Improved Maintainability of PWBs", *IPC-TP-304*, Institute of Printed Circuits, San Francisco, 1979.
49. David, R.F. and Bakhit, G., "Block Co-polymer Coating of Hybrid Microcircuits", *Electronic Components Conf. Proc.*, Orlando, FL, 1983.
50. King, J., "Qualification of Dow Corning Q3-6527 As A Particle Getter", *Singer Internal Report YS57A834*, 1985, and "Use of Silicone Gel Particle Getters For Microcircuits Packaging", *Proc. ISHM*, 1985.
51. Duschatko, W.L., "Microcomputer-Controlled-Atmosphere Enclosures", *Hybrid Circuit Technology*, Dec. 1985.
52. Bourdelaise, R.A. and Hill, F.E., "Soldering and Sealing Package Lids", *Electronic Packaging and Production*, Aug. 1986.
53. Solid State Equipment Corp., *Bulletin: Parallel Seam Sealing*.
54. Schwartz, S. (Ed.), *Integrated Circuit Technology*, Chapter 5, "Electron Beam Instrumentation" by J.E. Cline, McGraw-Hill, 1967.
55. Garibotti, D.J., Miller, E.H., and Anderson, P., *First Intl. Conf. on Electron and Ion Beam Science and Technology*, John Wiley and Sons, 1965.
56. Perkins, K.L. and Licari, J.J., "Development of Low Cost, High Reliability Sealing Techniques For Hybrid Microcircuit Packages", *Final Report*, NASA Contract NAS8-31992, Aug. 1977.
57. Neff, G. and Neff, J., "Leak Testing Electronic Devices In Production Quantities", *Microelectronic Mfg. and Testing*, Sept. 1986.

8

Testing

The ultimate requirement for a hybrid microcircuit is its reliable electrical functioning. Thus all tests that are performed, whether during processing or after assembly and sealing, are for the sole purpose of assuring that the intended life-expectancy and electrical performance of the circuit are met. The number, types, and frequency of tests to be performed should be established at the beginning of a program. Sometimes, these are dictated by the customer, while at other times the manufacturer must decide how much testing commensurate with cost and reliability is necessary. Certainly a commercial, low-cost hybrid does not require as much testing as a military, space, or medical product. A hybrid used in a toy, game, or personal computer may only require a short low-temperature burn-in/electrical test compared to a hybrid circuit used in a missile application where over 13 screen-tests in addition to extended high-temperature burn-in and numerous in-process control tests are performed. Thus a large portion of the cost of these hybrids (often over 50%) may be due to testing and its associated traceability and documentation requirements.

Testing may be categorized as: *Electrical* (die level and hybrid level), *Visual Inspection* (internal and external visual), and *Thermal/Mechanical Screen-testing*. Furthermore, testing may be non-destructive, where it may be performed on a 100% basis, or destructive, where it is performed on a small sampling basis for qualification or quality conformance.

ELECTRICAL TESTING

Electrical Testing of Die

Pretesting of semiconductor die (ICs, transistors, and diodes) is essential in producing hybrid circuits with initial high yields. Defective die should be identified and replaced as early in hybrid assembly as possible. The further downstream that a die failure occurs (in a hybrid, a sub-system, a system, or, worse, after delivery to a customer), the more difficult and

249

expensive it will be to isolate and repair. However, if the die can be electrically tested prior to assembly and marginal or failed ones removed, the first-time yield of the hybrid increases and there is less probability of failures later on. The yield of a hybrid, as a function of initial device yield and the number of devices in the hybrid, is shown in Figure 8.1. For example, a hybrid containing 20 devices, each of which has a 90-percent yield, theoretically results in a 12.2 percent overall first-time yield. However, if the devices are pretested, their yield increases to 97.5 percent, resulting in a first-time hybrid yield of 60.3 percent. Calculations for initial yields and first-rework yield are given below.

Initial Hybrid Yield (F_o). Assume 20 devices (M) each having 10% probability of failure (P):

$$F_o = (1 - P)^M$$
$$F_o = (1 - 0.1)^{20} = 0.9^{20} = 12.2\%$$

First Rework Yield (F_1).

$$F_1 = (1 - F_o)(1 - P)^{M_1}$$

where M_1 is the average number of failures per rework

$$M_1 = (M \times P)/(1 - F_o)$$

$$1 - F_o = 1 - 0.122 = 0.878$$

$$M_1 = 20(0.1)/0.878 = 2.28 \text{ average rework failures}$$

Therefore: $F_1 = 0.878(1 - 0.1)^{2.28} = 69\%$

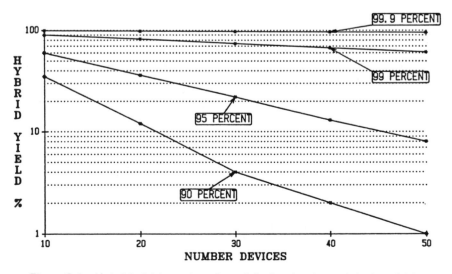

Figure 8.1: Hybrid yield as a function of device density and device yield.

Full functional testing and screening of chip devices are not often performed and may not always be possible. Devices mounted in chip carriers or on tape (TAB) are becoming more popular because they can be electrically tested and even burned-in prior to assembly. If the hybrid is very dense, chip carriers are not the answer due to the increased area that they require. However, if the devices are very complex, such as VLSI or VHSIC, it may not be economically feasible to use untested die in the hybrid. In such cases the die must be assembled in a carrier so that they can be pretested and screened.

Testing of integrated circuits varies with the specific device type; however, all electrical tests fall into three categories: DC parametrics, pulse testing, and function testing.

DC parametric tests are used to measure the worst-case DC parameters by forcing a voltage and measuring the current or forcing a current and measuring a voltage. These tests provide data on device electrical characteristics such as DC current, voltage breakdown, drive capability, leakage current, and power consumption.

Pulse testing provides data on the switching characteristics of a device. A pulse with a controllable rise and fall time is applied to the device and the rise, fall, and propagation times are measured.

Functional tests verify that the many internal transistors or gates within an integrated circuit are operating correctly. This is done by supplying an input signal and observing the output. These tests can also be run at worst-case voltage levels. In analog devices the input and output characteristics are checked. Memory devices are checked for pattern sensitivity to assure that data can be stored and retrieved from the memory cells. For logic or digital devices, functional tests ensure that all the internal switches are operating correctly.

DC parametric tests are the most common tests performed at the die level. The pulse tests are the most costly because of the difficulty in inputting a controlled pulse through a set of probes to the uncased die. Functional testing is likewise difficult and expensive.

In summary, some testing of uncased die prior to assembly in a hybrid is cost-effective. Electrical testing of the die improves first-time-through yield, reduces rework and manufacturing costs, and improves the reliability of the hybrid. If standard untested/unscreened devices are used, one can expect a two percent failure rate within one year.[1] With device testing and screening, this rate can be reduced to 0.01 percent.

Electrical Testing of Hybrids

The trend in hybrids has been toward denser, more complex, and unique custom circuits. Each custom hybrid or hybrid type has a unique set of electrical parameters associated with it. Though many custom hybrids are fabricated and assembled on the same basic equipment, using the same materials and processes, each is unique in the electrical parameters to be measured.

There are four main types of equipment, ranging from manual to automatic, that are used to electrically test hybrids.

1. A small custom test-box that interfaces to a rack of standard test equipment.

2. A custom-built test-box with built-in signal sources and power supplies that provide "go/no-go" test data.

3. A test-card that plugs into a matrix switch and interfaces with a microprocessor that addresses a test.

4. Automated computer-controlled test equipment.

For hybrids produced in small to moderate quantities, the test-box approach is the most cost effective. The average cost of building a custom test-box is about $1,000. The box contains the loads and input-shaping components and is connected to a rack of equipment consisting of power supplies, signal generators, digital voltmeters, and oscilloscopes. The operator inputs the required signal to the unit under test and observes the output. The box contains numerous switches allowing the operator to select the required loads. Hybrid circuits may be tested at high, low, and ambient temperatures. The main drawback of this test equipment is that it is slow. Since it is manual, an average circuit may require one hour to test at each temperature.

The second approach is slightly more expensive because of the added components that are installed in the box; however, it is faster since it is designed to run a test sequence on its own. A key limitation is that it cannot be used to troubleshoot a failed hybrid. It only provides a "pass or fail" indication and does not indicate what specific electrical parameter failed.

The third approach employs a universal test-box that interfaces to a rack of test equipment. It utilizes a test-card upon which the hybrid circuit is mounted and a matrix-switch that interfaces the pins of the hybrid with corresponding pins of the universal test adapter.

The last approach, automated testing, is the most cost effective when a large production rate is involved. A hybrid that takes an hour to test manually, may be tested in a matter of seconds on an automated tester. However, there is a higher initial investment. An automated tester may cost a million dollars or more, in addition to the cost of programming which may exceed $20,000. A trade-off must therefore be made between the time and cost savings of testing and the initial investment in equipment and programming.

Manufacturers of automatic test equipment have not addressed the testing of hybrid circuits directly, but their automated semiconductor testers are extensively used to test both single chip devices and hybrid circuits.

Popular automated testers are the Tektronix 3260 and 3270 testers. Figure 8.2 is a photograph of the Tektronix 3260 system, widely used to test military hybrids. Most automated testers are equipped with a temperature handler so that testing can be performed at various temperatures.

The main features of the Tektronix 3260 tester are:

● 64 pin I/O capability

● 20 MHz burst rate for functional testing

- High speed pattern generator
- DC parametric measuring system
- Delta time measuring system
- Multi-terminals, disk drives, hard copy units, and printers
- Remote test head to interface with handler
- 8 programmable power supplies and 7 programmable clocks

Digital voltmeters and signal generators may be added to perform analog testing.

Figure 8.2: Automated electrical test equipment-Tektronix 3260 (Courtesy Rockwell International).

The hybrid circuit is interfaced to the tester by a so-called "pizza" board (Figure 8.3). This board is a circular-shaped printed circuit card that interfaces the hybrid input/output leads to the electronics of the tester. It is also used to apply loads to the hybrid.

Electrical test equipment may be divided into three groups based on the type of hybrid to be tested: memory, analog, or digital.

Figure 8.3: Interface board for Tektronix tester.

Memory circuit testers—Because of the high density of devices in a memory hybrid, automatic testing is necessary. The tester must be capable of supplying many clocks, data lines, address lines, and power supply lines. If a hybrid is a 64K word by 32 bit memory, the tester requires 13 address lines, 10 clock lines, 32 bi-directional data lines and one power supply line.[2] Because of the mixture of 8, 16, and 32 bit microprocessors on the market, the memory requirements are diverse. The test equipment must be very versatile to adapt to the diversity of the memory hybrids. An ideal memory test system should have a reconfigurable hybrid interface, address generator, and clock generator, and as many flexible features as possible.

Analog and digital testers—Most analog hybrids are tested with manual test equipment. Since hybrids are often partitioned for the convenience of the system designer, they may consist of combinations of digital, analog, power, and microwave functions. This makes the design and manufacture of an automated system even more complicated.

Functional testing of a hybrid microcircuit may be performed at various stages after assembly such as at preseal, preburn-in, post burn-in, and final acceptance. With complex, high density hybrids, open-face burn-in and electrical testing before sealing have been found to be cost-effective in screening out marginal devices.

Preseal functional testing is normally performed at ambient temperature, but in some cases where the hybrid must meet a stringent high-temperature requirement, the hybrid may be tested at that temperature. Hybrids are seldom tested a low temperatures in an unsealed condition because of the risk of moisture condensation with its potential for corrosion. To prevent frost and moisture condensation on unsealed hybrids, the dew point of the air immediately surrounding the part must be lower than the temperature at which testing is performed. To accomplish this, the part is tested in a dry-box using a temperature-controlled chuck—a "hot plate" on which the hybrid is mounted. Low or high temperature testing can also be accomplished by placing a shroud over the hybrid and forcing cold or hot air or nitrogen gas into the shroud. The temperature of the gas is regulated by monitoring the temperature of the hybrid with a thermocouple. Figures 8.4 and 8.5 show examples of both types of temperature systems.

Preburn-in testing is required only if Percent Defective Allowable (PDA) or Parameter Drift Screening (PDS) are specified. PDA is a requirement that a burn-in lot shall be acceptable if greater than 90 percent pass functional testing after burn-in (for Class B) and 98 percent pass (for Class S). PDA is calculated as the number of failures divided by the total number of hybrids that are in a burn-in lot. If the PDA is exceeded, the entire lot should be rejected.

Paramater Drift Screening (PDS) is another requirement often imposed on Class S hybrids. For some hybrids, specific electrical paramaters are critical to their operation. These parameters must be measured before and after burn-in to calculate changes. If the parameter drifts exceed specification requirements, the hybrid is rejected.

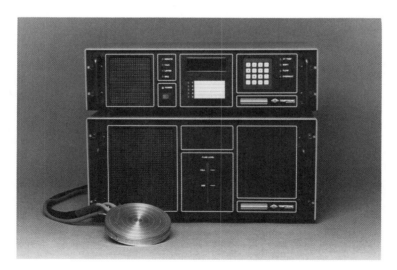

Figure 8.4: ThermoChuck Temperature Controller (Courtesy Temptronic Corp.).

Figure 8.5: Air-Jet temperature system (Courtesy FTS Systems).

VISUAL INSPECTION

Prior to sealing a hybrid microcircuit, visual inspection should be performed. This inspection is done with the aid of optical equipment that has the capability of 30 to 100X magnification. The purpose of this screen is to check the internal materials, construction, and workmanship of the hybrid microcircuit for compliance with requirements. Inspection is performed according to MIL-STD-883 Method 2017 and involves the following:

1. Active Chip Devices—Integrated Circuits, Transistors, Diodes
2. Passive Chip Components—Capacitors, Resistors, Inductors
3. Substrate Defects—Substrate, Metallization, Alignment, Resistors, Resistor Trimming
4. Element Assembly to the Substrate—Solder or Epoxy Attachment, Element Orientation
5. Substrate Attachment to the Case—Solder or Epoxy Attachment, Substrate Orientation
6. Wire Bond—Ball Bonds, Wedge Bonds, Crescent Bonds, Beam Lead Bonds, Mesh Bonds, Ribbon Bonds
7. Internal Wire Bonds (General)—Wires and Beams
8. Through-hole Mounting and Screw Tabs
9. Connector and Feedthroughs
10. Package Conditions—Foreign Material, Stains

After successful completion of the inspection, the hybrid is sealed.

NON-DESTRUCTIVE SCREEN TESTS

Non-destructive screen tests consist of a series of mechanical and thermal tests that are imposed on hybrids on a 100-percent basis. The screens listed in Table 8.1 are considered non-destructive.

Table 8.1: Screening Versus Defects Identified

Failure mechanism

Screen	Wire bonds	Die bonds	Loose particles	Package defects	Package seal	Thermal mismatch	Process induced	Surface effects	Dielectric failure	Metallization	External leads
Precap visual	x	x	x	x			x			x	
Stabilization bake	x	x					x	x			
Thermal cycle	x	x		x	x	x	x				
Thermal shock	x	x		x	x	x					
Centrifuge	x	x	x	x							
Mechanical shock	x	x	x	x							
Vibration	x	x	x	x							
Leak test	x				x						
Power burn-in	x							x	x	x	x
Reverse bias Burn-in (High voltage)	x										
X-ray	x	x	x	x							
External visual	x				x						

Thermal/Mechanical Tests

In addition to electrical testing, hybrids are subjected to varying degrees of mechanical and thermal screen tests. All military hybrids must be subjected to the testing specified in MIL-STD-883. Chapter 11 details the requirements in the military specification while this chapter details the reasons that the tests are required. These screening tests, developed and modified over many years, are an effective means of testing the integrity of

the die and substrate adhesion, and of "weeding out" marginal die that may
fail during the life of the product. Table 8.1 is a matrix of the screen tests
versus the defect that can be identified by each test. It should be noted that
it is not necessary to perform both mechanical shock and constant acceleration but that one or the other must be performed. Table 8.2 provides a
more detailed description of the screen tests.

Burn-In

Burn-in consists of applying power to a circuit while maintaining an
elevated temperature for an extended period of time (160 hours for Class B
and 320 hours for Class S). Raising the temperature and applying power
stresses critical connections and components and accelerates the life
cycle of the hybrid. If there are any weak connections or ionic contaminants,
the device will fail. If it does not fail, the hybrid is considered acceptable
and will remain operational until attrition or normal failure modes occur.

Burn-in is a critical step in the screening of hybrids since it establishes
electrical and thermal conditions that approximate actual operation in
a compressed time frame. As noted in Chapter 12, components, especially
integrated circuits, have a high rate of infant mortality. Thus, if a device is
prone to failure it will fail within the first several months of operation. Burn-in
accelerates this time. A burn-in of 160 hours at 125°C is equivalent to a full
year of operation at ambient temperature. Semiconductor devices are prone
to many types of failures, one of which is ion migration, which generally
occurs in or on the passivation layer, or between metal conductors. Chloride
or sodium ions are the two prevalent forms of ionic contamination. Positively
charged sodium ions under temperature and bias conditions readily migrate
to N-doped regions causing high leakage current and even shorts. Chloride
ions tend to migrate to the P-doped material and may cause emitter-to-
collector shorts in NPN transistors. These defects may not be discernible
for many months, but the combination of high temperature and power that
is provided by burn-in accelerates ionic migration without affecting the
normal failure rate or wear-out rate. Wear-out is associated with metal
migration, long-term threshold drift, and corrosion.[1]

The Arrhenius equation governs the reaction failure rate of an electronic device:

$$F = Ae^{(-E_a/kT)}$$

where:

F   = failure rate
E_a = activation energy (varies from 0.3 to 2.3 ev; if it is not known, MIL-
       STD-883 allows the use of $E_a = 1.0$ ev)
k   = Boltzmann's constant (8.63×10^{-5} ev/K)
T   = junction temperature in degrees Kelvin (Degrees C + 273 =
       degrees K)

To compare the failure rate in normal operation (F_1) to the rate after
burn-in (F_2), the equation can be modified as follows.

Table 8.2: Screen Tests and Detectable Failures

Test	Description	Failure Mechanism Exposed
1 INTERNAL VISUAL (Precap Visual)	A visual inspection before the circuit is encapsulated employing high-power and low-power microscopes. Examines the package, package leads, die bonding, wire bonding, and the topology of the chip.	Defects in the diffusion, oxide, and metalization. Some of the most common faults found are scratched metalization that may open at a later date, contamination (this may induce instabilities) bonding problems that may result in opens and shorts, cracks or chips at the edge of the die, pinholes or diffusion defects caused by dirty or scratched photomasks.
SEM INSPECTION	Scanning electron microscope examination is performed on a sampling basis only.	
2 STABILIZATION BAKE	24 hours at 150°C. Determines the effects of elevated temperature when no electrical stress is applied.	A preconditioning treatment prior to conducting other tests. Improves some characteristics and degrades others as it redistributes ionic contamination.
3 THERMAL SHOCK	(15 Cycles: 100°C, 5 min., 0°C, 5 min. with 10 seconds of transfer time) Determines the ability of the device to withstand sudden exposures to temperature extremes.	Wire bonds that are either mechanically poor or may become intermittent in plastic encapsulated circuits because of the large thermal temperature coefficient of the plastic encapsulation. Cracked silicon die, improperly made die bonds, improper lid seals, defective packages, microcracks in the metalization.
4 TEMPERATURE CYCLING	(65°C, 10 min., +159°C, 10 min. with 5 mins. transfer time, 10 cycles) Determines the ability of the device to survive exposures at extremes of high and low temperatures and to the effect of alternate exposures at these extremes.	Cracking and crazing of the the glassivation, opening of thermal seals and case seams, changes in electrical characteristics due to stresses in the substrate, or rupture of the conductors or insulating materials.
5 MECHANICAL SHOCK	(One of 30,000g's shock of 0.12 millisecond duration in the Y2 plane, or five 1,500g's shock of 0.5 milliseconds in the Y2 plane) Determines the capability of the parts to withstand rough handling encountered in shipment or in field operation.	Detects mechanical weaknesses in wire bonds, die bonds, and package seals. Jars loose mobile contaminant particles in hermetic packages which can then be detected by subsequent electrical testing, X-rays, and/or hermetic seal tests.
6 CONSTANT ACCELERATION	(30,000g's for 1 min. in each of X and Y planes) This test is made to fine mechanical weaknesses that are not already uncovered by the shock and temperature cycling tests.	Weak wire bonds, improperly dressed leads, weak die bonds, cracked die and improper lid seals in hermetic packages. (Ineffective on plastic encapsulated circuits and is the last of the mechanical stresses applied to the circuit.)
7 LEAK TESTS	Detects leaks ranging from 5 x 10⁻⁸ atm cc/sec to gross leaks. Both a fine leak and gross leak test is required. Determines the effectiveness of the seal of cavity devices.	Weeds out defective seals which may become latent failures when exposed to moisture or gaseous contaminants.
8 INTERIM (PRE BURN-IN) ELECTRICAL PARAMETERS	Device dependent.	Uncovers electrical failures which have been caused by mechanical and thermal stresses that have been applied in the previous tests.
9 BURN-IN TEST	(168 hours at 125°C) The burn-in screen eliminates marginal devices that would probably result in infant mortality or early lifetime failures under use conditions.	Detects metalization defects such as intermittent shorts or opens caused by pin-holes in the passivation layer beneath the metalization and corrosion or contamination. Near opens caused by scratches or voids will tend to become permanent opens. Locates circuits which have crystal dislocation, diffusion anomalies, contamination in or on the oxide, improper doping levels, and cracked die.
10 INTERIM (POST BURN-IN) ELECTRICAL PARAMETERS	These electrical tests are made after the burn-in test to determine the number of devices that failed. The result will give an indication of the quality of the entire lot and may serve as a basis for rejection of the lot, if more than the prescribed number of devices fail.	
11 REVERSE BIAS BURN-IN (75 hour power life test at 150°C)	Similar to Test 9, except the temperature is higher and the time is shorter. The electrical stress may also be slightly different.	If further failures can be found, indicates that further burn-in should be performed or the lot should be rejected.
12 FINAL ELECTRICAL TEST	Static and dynamic tests on 100% of the circuits.	Ensures that all the circuits perform the function intended over the voltage and temperature range specified.
13 RADIOGRAPHIC	The circuits are X-rayed.	Detects internal defects such as loose extraneous wire, gold slag, weld residues, improperly dressed wire bonds and voids in the die attach material or in the glass, when glass seals are used. Very effective in most hermetic packages to detect extraneous particles described above, as well as other defects missed by visual examination.

$$\frac{F_1}{F_2} = \frac{Ae^{-E/KT_1}}{Ae^{-E/KT_2}}$$

$$\frac{F_1}{F_2} = e^{-(\frac{E}{K})(\frac{1}{T_1} - \frac{1}{T_2})}$$

For example, a hybrid that is burned-in for 168 hours at a junction temperature of 125°C corresponds to 1.1 years (9,636 hours) of operation at 50°C. This calculation is based on $E_a = 0.6ev$. A small change in the junction temperature produces a drastic change in the failure rate. For instance, if in the above example the part is burned-in at a junction temperature of 135°C for 168 hours, the equivalent operating time would be 1.7 years (14,892 hours). Figure 8.6 is a curve of a junction temperature versus equivalent hours at 50°C. Since junction temperature is critical to the failure rate, it must be precisely controlled during burn-in.

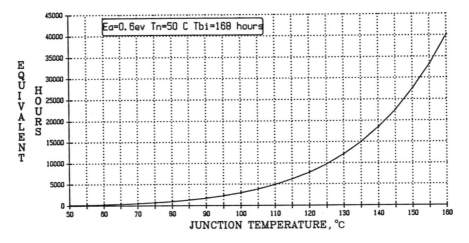

Figure 8.6: Hours at 50°C equivalent to 168 hours burn-in at various junction temperatures.

There are various ways that burn-in can be accomplished:

1. In an oven with heated ambient air or nitrogen circulating around the hybrids.

2. In a chamber filled with a liquid fluid that is heated and circulated over the hybrids.

3. On individual boards with heaters that burn-in one hybrid at a time.

Air Ambient Burn-In. In order to control the junction temperature in an air ambient burn-in oven, a temperature profile of each hybrid type is performed, then the oven temperature adjusted or heat sinks added to the hybrids. If a hybrid generates minimal heat under power, this system is effective. However, if the hybrid generates an excessive amount of heat, burn-in in a liquid medium should be considered. The reasons for this are discussed in the next section. Figure 8.7 provides an example of an air burn-in system.

Figure 8.7: Air ambient burn-in system (Courtesy Aehr).

Liquid Burn-In. The liquid burn-in system consists of a chamber filled with an inert dielectric fluid that may be heated or cooled to maintain the required temperature (Figure 8.8). The fluid circulates over the hybrids, thus heating or removing heat as required. The fluids used must be chemically and electrically stable at the temperatures used and are generally fluorocarbons. A liquid burn-in system is ideal when power devices are being burned-in because the temperature stabilizes and equilibrates better in liquid than in air.

Air vs Liquid Burn-In. Using air-circulating ovens to burn-in power devices creates several problems:

- The need for heat sinks
- Potential hot spots
- Thermal runaway

a

b

Figure 8.8: Liquid burn-in system. (a) console (b) internal chambers (Courtesy FTS Systems Inc.).

Air systems have a poor heat transfer from the hybrids and poor temperature stability throughout the oven. Most ovens can only maintain a 3 degree C variation across the oven. Heat transfer is impeded by the air's low specific heat (0.24 Btu/lb) and low density (0.66 lb/cubic foot). Thus, efficient heat transfer can only be effected by flowing a large mass of air over the hybrids. The equation for the device case-to-ambient thermal resistance (θ_{CA}) in air is:[3]

$$\theta_{CA} = \frac{330(q/A)^{0.048}}{AV^{0.471}}$$

where: q = power generated (watts)
       A = surface area of the hybrid (square inches)
       V = velocity of air (ft/min)

If thermal resistance is high and the hybrid generates a large amount of heat, the temperature of the hybrid increases and may go into thermal runaway. The threshold at which air fails to remove heat effectively lies between 1 and 2 watts/sq inch.[3]

The device-to-ambient thermal resistance for a liquid system is:

$$\theta_{CA} = \frac{10.4}{A[(q/A)^{0.173}](V^{0.724})}$$

The thermal resistances of a hybrid cooled in air and in liquid were calculated using these equations. For a hybrid that dissipates 10 watts per sq inch (at an ambient temperature of 125°C), the case-to-ambient thermal resistance is 27°C/W (v=250 ft/min) for air-cooling and 0.3°C/W (v=70 ft/min) for liquid-cooling.

Generally, liquid burn-in should be considered if the junction temperature of the device is expected to increase 50°C or more above the ambient temperature.

Particle-Impact-Noise Detection (PIND) Testing

Conductive particles trapped in electronic packages have long concerned manufacturers of high-reliability components. There is even greater concern for hybrids because they are more complex; involve more materials, processes, and devices in construction than single devices; and have a higher incidence of loose particles. Hybrid devices are also more expensive than single electronic components, which leads to more concern for yields.[4] There are four ways that the problem of loose particles can be addressed:

1. Reducing or eliminating particles during processing.

2. Protecting the hybrid from loose particles by coating with Parylene or other organic coatings (see Chapter 7).

3. Using particle getters.

4. PIND screen testing to detect loose particles.

No matter how many times a circuit is cleaned or in how clean an area it is assembled, there is no guarantee that it will be particle-free. Thus the first method has never been considered absolutely effective, although with repeated and effective cleaning and handling procedures, it can approach 100-percent yield. Coating effectively immobilizes or traps particles so that they cannot move and cause shorts. However, the extra steps involved in masking, coating, and reworking a hybrid that has been coated are difficult and expensive. Particle getters, consisting of soft silicone gels, are sometimes used and reported to be effective. As with the organic coatings, there is the expense of the extra steps involved and the concern of added outgassing from the silicone.

PIND testing is the test most widely used. It is effective for detecting loose particles in hybrids and can be used as a 100% screen test. PIND testing involves mounting the hybrid on a transducer with an acoustic coupling material, and shocking and vibrating the hybrid while the transducer "listens" for loose particles within the unit under test.

A PIND tester (Figure 8.9) consists of a shaker, an oscilloscope for visual monitoring, a programmable control unit, and a speaker for audio monitoring.

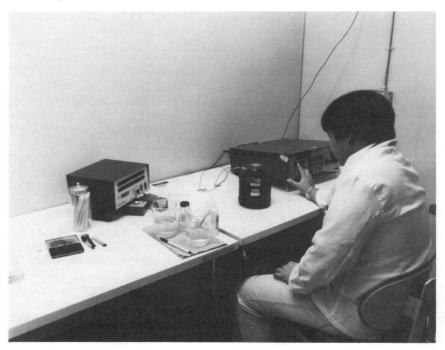

Figure 8.9: PIND Test Station (Courtesy Rockwell International).

Method 2020 of MIL-STD-883 defines the conditions for PIND testing as follows:

1. Preshock at 500-1500 g's.
2. Vibrate for 3-5 seconds at 20 g's, 40-200 Hz.
3. 200-600 g's during vibration.
4. Repeat steps 2 and 3 twice more.
5. Vibrate at 20 g's, 40-200 Hz for 5 seconds.

This sequence is repeated in five independent passes and failures in each pass rejected. To be acceptable, hybrids must survive all the tests.

Infrared (IR) Imaging

Over the past several years, infrared imaging has gained prominence as a non-destructive method for evaluating semiconductor die and hybrid circuits. Thermal imagers detect and measure infrared energy, then convert it to usable temperature information. A hybrid microcircuit can be scanned with infrared equipment to evaluate the integrity of die and substrate attachment and wire bonds, and to produce a thermal profile of the circuit while powered. This thermal profile helps design engineers in locating hot spots and repartitioning the circuit to improve thermal dissipation.

The applications of IR imaging in hybrid processing fall into two categories: IR microscopy and thermography. Figure 8.10 depicts the application of IR imaging for these two categories.[5]

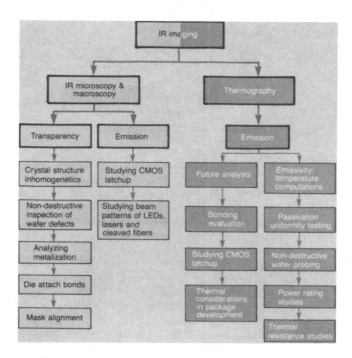

Figure 8.10: The application of IR imaging in semiconductor manufacturing is divided into two distinct categories.

IR Microscopy. IR microscopy is similar to optical (visible light) microscopy, since the instrument that is used is a microscope; however, the image is seen via the detection of IR energy. This energy, operating at wavelengths of 0.7-2 micrometers, is transmitted through or reflected back from the material. Although the IR microscope looks like an optical microscope, it differs in that it includes an IR-sensitive detector. These detectors can be either in the form of an eyepiece (Figure 8.11) or a TV camera with an IR-sensitive photoconductive layer of Si or PbS (Figure 8.12).

An advantage of IR microscopy is that it does not require sample preparation as does X-ray radiography. Another advantage is that, at the wavelength that IR microscopy operates, many semiconductor materials are transparent, permitting inspection of the die-attach medium beneath the die.

This method of scanning is mainly used for the in-process inspection of wafers. Non-uniformity of doping in semiconductors and the crystalline structure of the silicon wafer can be viewed.

Figure 8.11: The Infrared Viewer Model 7310, from Electrophysics, is an infrared to visible image converter that replaces a microscope eyepiece and provides a visual view extending out to a wavelength of 1.2 μm.

IR Thermography. IR thermography is more useful than IR microscopy in the design and testing of hybrid microcircuits. IR thermography provides a color temperature profile of the circuit while the circuit is under power. The scan can be compared with the scan of a failed hybrid to isolate a faulty device or design. In the initial design stage thermal profiles may be used to position and arrange devices so that heat can be more evenly distributed and dissipated. Revised layouts can be evaluated and design changes incorporated before hybrids are committed to production.

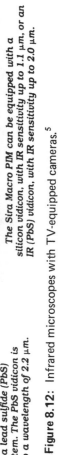

The Sira Macro PIM can be equipped with a silicon vidicon, with IR sensitivity up to 1.1 μm, or an IR (PbS) vidicon, with IR sensitivity up to 2.0 μm.

The Reichert-Jung Polyvar-Infrapol Infrared Research Microscope uses a lead sulfide (PbS) vidicon in its CCTV subsystem. The PbS vidicon is sensitive to IR energy up to a wavelength of 2.2 μm.

Figure 8.12: Infrared microscopes with TV-equipped cameras.[5]

Thermography involves the generation of thermal images of emitted black-body radiation from hybrids. Very few objects are black bodies, and a correction factor (emissivity) must be applied to the black-body equation to get a true temperature. With instrumentation presently available, the detected IR energy is displayed as a digital temperature-profile map of the circuit. This image is presented on the screen of a color monitor, with the various colors (up to 256) indicating temperature differences of 0.1°C. The true temperature of a circuit must take into account the emissivity of the sample. It is possible, with the equipment today, to correct for the differences of emissivity of the various elements in a hybrid. This is accomplished by scanning a non-operating hybrid and storing the IR energy. This is stored in the equipment, and each pixel in the raster can be corrected when a scan of an operating hybrid is input, giving a true temperature picture of the hybrid. Figure 8.13 is a photo of a thermography system manufactured by UTI Instruments.

Figure 8.13: Thermography System (Courtesy UTI Instruments).

Acoustic Microscopy

Another non-destructive testing technique is acoustic microscopy in which a beam of sound is focused by a lens on the device being tested. The sound is reflected from the device and converted by a transducer to an electrical signal which is then processed to form a picture of the device on a monitor in a raster fashion. Acoustic microscopy differs from thermal scanning in that the signal is not dependent on the dielectric properties of the devices but on their physical properties such as density and elasticity.[6] A detailed discussion on the operation of the acoustic microscope may be found in the literature.[7,8,9]

Some specific applications of acoustic microscopy are:

1. Die attach inspection for voids and completeness of coverage of adhesive.

2. Cover seal inspection.

3. Location of cracks in die or substrates.

4. Inspection of multilayer substrates for voids in vias or metallization.

5. Detection of thickness variations in thin-film conductors.

DESTRUCTIVE SCREEN TESTS

Chapter 11 lists the screens that are to be performed on a hybrid microcircuit. In this section a listing of the tests that are considered to be destructive, a brief description of destructive physical analysis (DPA) and a detailed description of moisture analysis (internal water vapor content) will be given. The following is a list of the tests that are considered to be destructive.

Die Shear Strength

Wire Bond Strength

Lead Integrity

Solderability

Salt Atmosphere

Moisture Resistance

Destructive Physical Analysis

Internal Water Vapor Content (Moisture Analysis)

Destructive Physical Analysis (DPA)

Destructive physical analysis is the process of disassembling, testing, and inspecting a hybrid for the purpose of determining conformance with applicable design and process requirements. The purpose of DPA is to verify external and internal physical configuration. While DPA procedures can be performed by the hybrid manufacturer, they are usually performed by the user.

The procedures to be followed are detailed in MIL-STD-883 Method 5009, which requires the following:

1. External Visual

2. X-Ray

3. Particle Impact Noise Detection Test (PIND)

4. Hermeticity

5. Internal Water Vapor Analysis

6. Internal Visual

7. Baseline Configuration

8. Bond Strength

9. Scanning Electron Microscopy (SEM)

10. Die Shear

Moisture and Gas Analysis of Package Ambients

The presence of water, alone or with other gaseous or ionic contaminants, in hermetically sealed hybrid microcircuits or integrated circuit packages is a continuing concern to manufacturers of high reliability systems. The interaction of water with ionic or gaseous residues has been reported to corrode device metallization and electrical interconnections, produce metal migration, induce electrical shorts, create high leakage currents, and produce inversion currents in semiconductor devices. Thus Method 1018.2 of MIL-STD-883C was established to specify test methods and controls for moisture. In spite of the establishment of these requirements and test methods, variations in moisture contents continue to be found among packages of the same lot assembled and processed in the same manner by the same manufacturer. Epoxy adhesives used for substrate and die attachment are key contributors of moisture and other outgassing products, especially if inadequately cured or vacuum-baked. Some epoxy formulations are inherently heavy outgassers and not as thermally stable as others. During the past ten years, considerable progress has been made in understanding and characterizing epoxy adhesives for electronic applications (see Chapter 7). Currently, by following specific guidelines in the selection and qualification of adhesives and by optimizing the cure and vacuum-bake schedules, hybrid circuits that meet the Class B moisture requirement of less than 5,000 ppm can be produced.

In spite of selecting a low-outgassing epoxy and employing optimum processing conditions, some hybrid circuits still show large amounts of water, together with oxygen and argon. Even circuits that have been assembled without organic adhesives sometimes show water, oxygen, and argon. This can only be attributed to the penetration of air into the packages after they have been sealed. The presence of moisture in hermetically sealed hybrid circuits has been correlated with leak rates and with MIL-STD-883 screen tests and burn-in. It has been reported that a relationship exists between the leak rate and moisture content of a package and that, to consistently meet the MIL-STD-883 moisture requirement of 3,000 ppm moisture (for Class S circuits), a leak rate of $<3.0 \times 10^{-9}$ atm-cc/sec after screen tests and burn-in is required.[10] In another study it was found that some packages with very low leak rates before screentests had high leak rates after screens, indicating that stresses imposed on the glass-to-metal seals or lid seal during the screens may have opened the seals.[11]

Industry has standardized on mass spectrometry as the best and most accurate method for the analysis of moisture and other gaseous constituents contained in sealed hybrid circuit packages. Mass spectrometry is a destructive method and thus used only for initial qualification and subsequently, on a sampling basis, to monitor production lots. Non-destructive methods, such as the incorporation of moisture sensors in packages, would be ideal but, to date, none of these methods fully satisfies the requirements of Method 1018.2 of MIL-STD-883. The equipment most widely used for moisture analysis is a computer-controlled quadrupole mass spectrometer. In one version, the equipment contains a carrousel holder to vacuum-bake a number of packages at the same time and a tool to pierce each package sequentially. The packages are first baked at 100°C at a chamber pressure in the low 10^{-8} Torr range for an extended time, usually 12 hours or more. This assures that all volatiles from the outside of the package have been removed and that all adsorbed/absorbed moisture within the package has been outgassed so that it can be measured. In a second version (the rapid-cycle method), a package is externally mated to the instrument through a gasket, then pierced so that the vacuum of the mass spectrometer chamber is not broken. Provision is made for heating the package at 100°C a minimum of 10 minutes prior to piercing. Mass-spectrometer assemblies designed with special features for moisture analysis of electronic packages are commercially available (Figure 8.14).

Figure 8.14: Mass spectrometer for moisture and gas analysis (Courtesy Rockwell International).

Because of the wide variations in moisture values that have been reported by different laboratories and even by the same laboratory for identical packages processed in the same manner, considerable effort has been devoted by both industry and government agencies to standardize the equipment and analysis procedures. Rome Air Development Center assumed the lead for this effort. As a result, equipment-calibration procedures and test procedures (Method 1018.2) have been written and several laboratories have been certified. These actions have resulted in closer correlation of results among laboratories.

Water-vapor-detection sensitivity varies with package volume and water-vapor concentration. The mass spectrometer must be calibrated with samples having known amounts of moisture—generally nitrogen containing 2,000 ppm and 5,000 ppm (by volume) of water. A three-volume calibration valve (TVCV), designed as a package simulator, should be incorporated in the equipment to calibrate the instrument for three volumes, 0.8, 0.1, and 0.01 cc, of nitrogen gas containing known amounts of moisture. The known amounts of moisture may be generated by pressure, divided flow, or cryogenic methods.

After calibrating and preconditioning the equipment, background scans are taken after the valve to the mass analyzer has been open for several seconds. This provides a correction factor for any extraneous outgassing from the chamber walls, sample tooling, and exterior package surfaces (in the case of the carrousel arrangement). Each package is then pierced, generally through the lid, with a sharp piercing tool. The evolved gases are analyzed by the normal mass spectrometric method of measuring mass-to-charge peaks of the spectrum produced.[12]

REFERENCES

1. Sundell, J., Robb, S., Taylor, D., and Crabbe, R., "MIL Screening Boosts Commercial Product Reliability", *Electronic Products*, June 29, 1983.
2. Sergent, J.E., Chiles, H.H., and Power, R., "Test and Inspection of Hybrid Microcircuits", *Electronic Packaging & Production*, February, 1986.
3. Thompson Jr., T.N., "Liquid Burn-In Takes the Heat From High Power Devices", *Electronic Packaging & Production*, September, 1985.
4. David, R.F.S., "Practical Limitations of PIND Testing", 1978 *Proceedings 28th ECC*, 78CH1349-0 CHMT.
5 Burggraaf, P., "IR Imaging: Microscopy and Thermography", *Semiconductor International*, July, 1986.
6. Burton, N.J., "How to Identify Die Bonding Defects", *Microelectronic Manufacturing and Testing*, October, 1986.
7. Smith, I.R., Harvey, R.A., and Fathers, D.J., "An Acoustic Microscope for Industrial Applications", *IEEE Proceedings on Sonics and Ultrasonics* 32(3), 1985.
8. Wickramasinghe, H.K., "Acoustic Microscopy: Past, Present, and Future", *IEE Proceedings*, A 131(4), London, 1984.
9. Bertoni, H.L., "Ray-Optical Evaluation of V(z) in the Reflection Acoustic Microscope", *IEEE Proceedings on Sonics and Ultrasonics* 31(2), 1984, pp 105-116.

10. David, R.F.S., "Effects of Leak Rate and Package Volume on Hybrid Internal Atmosphere", *Intl. Journal Hybrid Microelectronics*, Vol. 6, No. 1, October, 1983.
11. Swanson, D.W. and Licari, J.J., "Effect of Screen Tests and Burn-In on Moisture Content of Hybrid Microcircuits", *Solid State Technology*, September, 1986.
12. Silversteen, R.M., Bassler, G.C., and Morrill, T.C., *Spectrometric Identification of Organic Compounds*, John Wiley, 1981.

9

Handling and Clean Rooms

HANDLING OF HYBRID CIRCUITS AND COMPONENTS

The handling of hybrid circuits and their individual components is a very important aspect of manufacturing. Handling involves keeping parts clean, protecting them from ambient contaminants, protecting them from electrostatic discharge, and avoiding mechanical damage.

Cleanliness of Tools

Tweezers, plastic probes, vacuum probe tips, finger-cots, and plastic gloves can be considered clean when removed from the vendor's package, but contaminated if picked up with bare hands. If so handled, the tools must be cleaned with isopropyl alcohol or Freon TF and dried prior to use.

Table 9.1 details some general handling procedures for hybrids and components.

Storage

As discussed in Chapter 7, cleaning is an important process in the manufacture of hybrid circuits. After cleaning, hybrid components and unsealed assemblies should be stored in an inert nitrogen environment to prevent oxidation, moisture adsorption, and corrosion (Table 9.1). When storing chip capacitors with silver terminations, a sheet of lead acetate paper should be included to prevent sulfide formations on the terminations.

Clean Rooms

Hybrids should be assembled and tested in a clean-room environment in order to keep contamination to a minimum during processing. A single particle of dust can cause an open or a short during the substrate processing. Presently, military specification MIL-M-38510 requires that hybrid manufacture, assembly, and test be performed in a class 100,000 area or better. FED-STD-209B, the current standard that defines the clean room

Table 9.1: Handling and Storage of Circuit Components and Assemblies

Item	Handling Technique	Storage Technique Container	Environment
I Metallized Substrate	1. Tweezers, Broad-Bill 2. Vacuum Probe, Vacuum Cup 3. Teflon or Nylon Probe 4. Film Gloves or Finger Cots	As received from vendor	Class 100,000 per FED-STD-209
II Semiconductor Dice, Chip Resistors, or Chip Capacitors	1. Vacuum Probe, Vacuum Cup 2. Teflon or Nylon Probe 3. Tweezers 4. Brush	As received from vendor, or equivalent	Class 100,000 per FED-STD-209, in a nitrogen purged area
III Metal or Plated Components			
A. Connectors	1. Tweezers 2. Teflon or Nylon Probe 3. Film Gloves or Finger Cots 4. Bare Hands, Avoid Contact Areas	As received from vendor, or equivalent	Class 100,000 per FED-STD-209
B. Package Bases	1. Tweezers 2. Vacuum Probe, Vacuum Cup 3. Film Gloves or Finger Cots 4. Bare Hands, Avoid Seal Areas	As received from vendor, or equivalent	Class 100,000 per FED-STD-209
C. Package Covers and Preforms			
1. Not Cleaned	1. Tweezers 2. Vacuum Probe, Vacuum Cup 3. Film Gloves or Finger Cots 4. Bare Hands, Avoid Seal Areas	As received from vendor, or equivalent	
2. Cleaned	1. Tweezers, Clean 2. Film Gloves or Finger Cots, Clean 3. Vacuum Probe, Vacuum Cup, Clean	Polypropylene, polystyrene, or glass container	Nitrogen Chamber
IV Plastic Coated Components	1. Tweezers, Broad-Bill 2. Teflon or Nylon Probe 3. Film Gloves or Finger Cots, Avoid Devices 4. Bare Hands, Avoid Devices or Exposed Metal Surfaces	As received from vendor, or equivalent	Controlled Area per FED-STD-209
V Unsealed Hybrid Assemblies	1. Tweezers 2. Film Gloves or Finger Cots 3. Bare Hands, Avoid Seal Areas	Polypropylene, polystyrene, or glass container; metal trays, magnets on metal strips, or metal fixture; or equivalent	Class 100,000 per FED-STD-209 NOTE: Store in Nitrogen Chamber
VI Cleaned Unsealed Assemblies – Between Clean and Seal Operations	1. Tweezers, Clean 2. Film Gloves or Finger Cots, Clean 3. Bare Hands, Avoid Seal Area	Polypropylene, polystyrene, or glass container; metal trays, magnets on metal strips, or metal fixture	Nitrogen Chamber
VII Sealed Assemblies	1. Tweezers 2. Teflon or Nylon Probe 3. Film Gloves or Finger Cots 4. Bare Hands	Polypropylene, polystyrene, or glass container; metal trays, magnets on metal strips, or metal fixture; or equivalent	Controlled Area per FED-STD-209

(continued)

Table 9.1: (continued)

Item	Handling Technique	Storage Technique Container	Environment
VIII Static Discharge Sensitive Components and Assemblies			
A. Devices	1. Vacuum Probe, Vacuum Cup 2. Teflon or Nylon Probe 3. Brush	As received from vendor or placed in a Faraday cage bag	Class 100,000 per FED-STD-209. Store in nitrogen chamber.
B. Unsealed Assemblies 1. After wire bonding of integral lead substrates or plug-in lead packages, or 2. After lead trim	1. Tweezers 2. Vacuum Probe, Vacuum Cup 3. Grounded Operator, Bare Hands, Avoid Seal Areas	Faraday cage bag or container	Class 100,000 per FED-STD-209 NOTE: Store in Nitrogen Chamber
C. Clean Unsealed Assemblies, Between Clean and Seal Operations	1. Tweezers, Clean 2. Grounded Operator, Bare Hands, Avoid Seal Area	Faraday cage bag or container	Class 100,000 per FED-STD-209 in Nitrogen Chamber
D. Sealed Assemblies	1. Tweezers 2. Grounded Operator, Bare Hands	Faraday cage bag or container or, if leads are shorted, on an uncoated metal tray, magnets on metal strip, metal box, or metal fixture	Class 100,000 per FED-STD-209
E. Sealed Assemblies, Ready for Delivery to Customer		Faraday cage bag or container.	Class 100,000 per FED-STD-209

classes, defines four classes of air cleanliness: class 100,000; class 10,000; class 1,000; and class 100. These numbers represent the number of particles, 0.5 micron or larger in diameter, in one cubic foot of air. Thus, a class 100,000 clean room contains less than 100,000 particles that are 0.5 micron in diameter per cubic foot of air. It may also contain larger particles up to 100 μm in quantities defined by FED-STD-209B curves (Figure 9.1). Note: a micron is about 1/150th the width of a human hair.

Class 100 is the lowest "officially" defined class (highest cleanliness level), however, the definitions have been used to extrapolate to even lower class levels. For example, a Class 10 area is defined as one in which there are no more than 10 particles of 0.5 micron/cubic foot. While useful, it has been extrapolated unofficially from the Class 100 definition.

Because of the small geometries of the semiconductor devices now being fabricated, clean rooms as low as Class 1 are becoming commonplace. A proposed revision "C" of FED-STD-209 that will extend the definitions down to Class 1 is in draft form. The standard is being drafted by

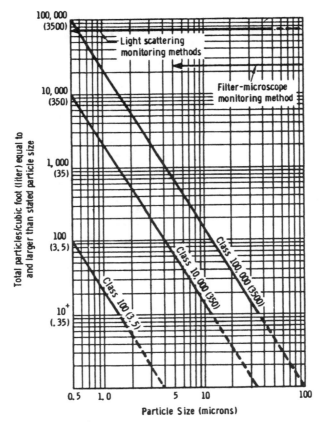

Figure 9.1: Particle Size Distribution Curves (FED-STD-209B).

the RP-50 Committee of the Institute of Environmental Sciences, Mt. Prospect, Illinois. The definition of Class 10 according to the new standard will be: no more than 10 particles that are 0.5 micron or larger, 30 particles that are 0.3 to 0.5 micron, and 350 particles that are 0.1 to 0.3 micron/cubic foot.

Figure 9.2 gives the proposed FED-STD-209C classification curves.

How clean is the air that meets the requirements specified above? As an example, the air in a normal (non-clean) room contains 500,000 particles per cubic foot, and the outside air 1,000,000 particles per cubic foot.[1] When there is a stiff breeze, the dust particles will easily reach 1.5 million particles per cubic foot.

The comparison of the various clean room classes is given in Table 9.2.

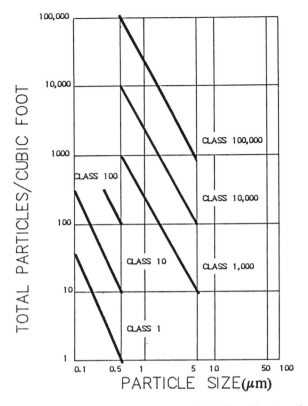

Figure 9.2: Proposed Federal Standard 209C Classification Curves.

Table 9.2: Clean Room Class Comparisons

CLASS	CRITICAL-PARTICLE SIZE (MICRON)	CLEANLINESS IMPROVED BY A FACTOR OF:
100	0.5	1
10	0.5	10
10	0.2	100
1	0.2	1000
SUPER 1	0.1	100000

"Super 1" is 100 times cleaner than the existing Class 1. Within three minutes after the air conditioning is turned on, the number of dust particles that are 0.1 micron in diameter in 30 cubic meters is reduced to one or less.[2]

The biggest contributors of contamination in a clean room are people.

Every move that a human makes inundates the air with a shower of dust and lint. Most of the particulates that come from people are dead skin cells. A single worker can generate up to 300,000 particles per minute simply by sitting and up to 5 million particles per minute by walking.

Cosmetics are a major source of contamination. Millions of particles are released during an application. Lipstick, for example, can generate 1.1 $\times$ 10^9 particles and mascara can generate 3.0 $\times$ 10^9 particles that are greater than 0.5 μm in diameter.

Table 9.3 depicts the sizes of particles arising from some common activities.

Figure 9.3 depicts the relative sizes of human-generated particles.

Table 9.4 depicts the increase of particles in a room arising from human activities.

Table 9.3: Particle Sizes Generated
(Courtesy Rockwell International Semiconductor Division)

SOURCE	SIZE OF PARTICLES
Rubbing skin	4 microns
Scratching vinyl	8 microns
Writing with pen on paper	20 microns
Turning screws	30 microns
Rubbing epoxy coatings	40 microns
Folding paper	65 microns
Sliding metal on metal	75 microns
Rubbing ordinary paint	90 microns

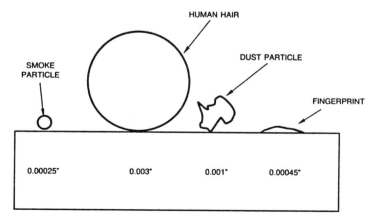

Figure 9.3: Relative sizes of human-generated particles.

Table 9.4: Pollution Comes from People
(Courtesy Rockwell International Semiconductor Division)

ACTIVITY	PERCENTAGE INCREASE OF PARTICULATES
NORMAL BREATH	NO PARTICULATES
SMOKER'S BREATH AFTER SMOKING	500 PERCENT
SNEEZING	2,000 PERCENT
SITTING QUIETLY	20 PERCENT
RUBBING HANDS OR FACE	200 PERCENT
WALKING	200 PERCENT
REMOVING A HANDKERCHIEF FROM POCKET	1,000 PERCENT
STAMPING SHOE ON FLOOR	5,000 PERCENT

ELECTROSTATIC DISCHARGE

Each year millions of dollars are lost by the semiconductor and hybrid industries because of electrostatic damage to devices and circuits. This loss can come from damage to semiconductor devices causing a product to fail or, in a more extreme case, an explosion due to a static spark in a volatile atmosphere. Uncontrolled static charge can cause a variety of problems in industrial operations. They can be grouped into:

Component destruction or damage

Dust attraction

Fire and explosion

Personnel shocks

A static charge develops when two materials come into contact and then separate, causing the transfer of electrons from one surface to the other. The material that has the greater affinity for electrons becomes negatively charged. Thus, when rubber is rubbed against wool, the rubber becomes negatively charged while the wool assumes a positive charge due to the loss of electrons.

Development of Charge

The amount of charge that can build up on a surface depends on the conductivity of the material. Metals, being good conductors, tend to dissipate and lose the charge immediately, whereas nonconductors will retain their charge for a long time. If the conditions are right, plastics can maintain a charge for weeks. The polarity and the magnitude of the charge on a material can be approximated from the "Triboelectric Series". In this chart, the further apart two materials are, the more readily a static charge is generated. The materials at the top of the series acquire a more positive charge relative to materials below it. Thus, if glass is rubbed with cotton, the glass will acquire a positive charge, but if glass is rubbed with asbestos, it becomes negative. Table 9.5 is a listing of the Triboelectric Series.

Table 9.6 gives some typical electrostatic voltages for common everyday activities.[3]

Table 9.5: Triboelectric Series

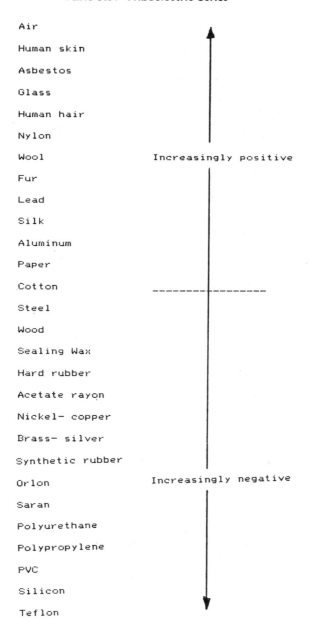

Air

Human skin

Asbestos

Glass

Human hair

Nylon

Wool   Increasingly positive

Fur

Lead

Silk

Aluminum

Paper

Cotton

Steel

Wood

Sealing Wax

Hard rubber

Acetate rayon

Nickel- copper

Brass- silver

Synthetic rubber

Orlon   Increasingly negative

Saran

Polyurethane

Polypropylene

PVC

Silicon

Teflon

Table 9.6: Typical Electrostatic Voltages

Source of Electrostatic Generation	. . .Electrostatic Voltages (VDC). . . .	
	@ 10 to 20% RH	@ 65 to 90% RH
Walking across a carpet	35,000	1,500
Walking over a vinyl floor	12,000	250
Worker at a bench	6,000	100
Vinyl envelopes	7,000	600
Picking up a plastic bag	20,000	1,200
Work chair padded with polyurethane foam	18,000	1,500

Device Susceptibility

Many devices used in hybrid microcircuits are susceptible to damage even if electrostatic voltages are below 500 volts. It must be pointed out that a human does not even feel an electrostatic voltage unless it is at least 4,000 volts, and it takes at least 10,000 volts to cause an audible crackle or spark.

Table 9.7 gives the damage susceptibility of semiconductor devices.[3]

Table 9.7: Damage Susceptibility of Common Semiconductor Devices

DEVICE TYPE	RANGE OF SUSCEPTIBILITY (VOLTS)
MOS FET	100-200
J-FET	140-10,000
CMOS	250-2,000
SCHOTTKY DIODE	300-2,500
SCHOTTKY TTL	1,000-2,500
BIPOLAR TRANSISTORS	380-7,000
ECL HYBRID ON A PCB	500
SCR	680-1,000

There is a plethora of information on the subject of static electricity and the effects of electrostatic discharge on electronic devices and circuits.[4-7]. Suffice it to say that hybrids and hybrid components should be handled in a static-safe manner.

The military documents that the reader is directed to for more information are:

DOD-HDBK-263	Electrostatic Discharge Control Handbook for Protection of Electrical and Electronic Parts, Assemblies and Equipment.
DOD-STD-1686	Electrostatic Discharge Control Program for Protection of Electrical and Electronic Parts and Assemblies and Equipment.
MIL-STD-883	Method 3015, Electrostatic Discharge Sensitivity Classification.
MIL-B-81705	Military Specification, Barrier Materials, Flexible, Electrostatic-Free Heat Sealable.
NAVSEA SE 003-AA-TRN-010	Electrostatic Discharge Training Manual.

Static Damage

Devices may be degraded and even destroyed by ESD (electrostatic discharge) producing excessive voltage or excessive current. Figures 9.4 and 9.5 are examples of static damage on a component.

There are three types of damage to integrated circuits that are commonly caused by ESD: dielectric failure, interconnect metallization failure, and PN junction failure.

Dielectric Failure. This failure mode is an electric field induced action that ruptures oxides on a chip. Figure 9.4 is a representation of this type of damage.

Interconnection Metal Failure. This type of failure is due to the metal reaching its melting temperature. This can be at a lower temperature than expected, since oxides that may be underneath the metal are very poor thermal conductors and therefore all the heat is dissipated in the metal and not conducted away. Figure 9.5 is an example of this type of damage.

PN Junction Failures. This failure mode occurs when the ESD voltage/current causes melting of the silicon or metal "spiking" from one diffusion layer to another.

The above discussions involve the destruction of devices due to ESD; however, devices can also be degraded but not destroyed. Ninety percent of all ESD problems are due to degradation. This can only be detected by observing the I/V curves of the semiconductor junctions. In a part not exposed to ESD, the knee of the curve is very sharp; however, subjecting the part to ESD causes the knee to soften (Figure 9.6).[8]

Coping with ESD

In order to prevent damage or degradation of a hybrid or a hybrid component many safeguards are available.

Personnel Awareness. It is very important that hybrid manufacturers have a training program for their personnel so that they are aware of the dangers of ESD. All personnel who are to be in an area where there are sensitive devices must be trained. After the program, they should be tested and not allowed to handle devices until they pass the test and are certified.

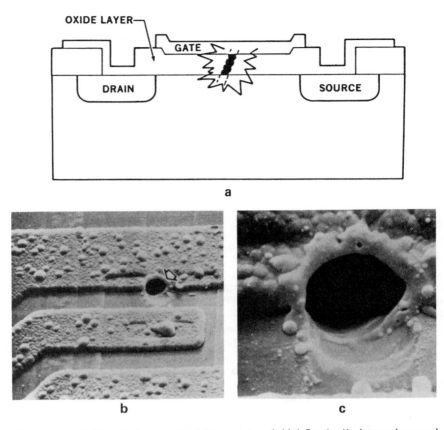

Figure 9.4: (a) Static-damaged MOS transistor. (b)(c) Static-discharge-damaged operational amplifier integrated circuit. (c) is a 5-fold magnification of (b).

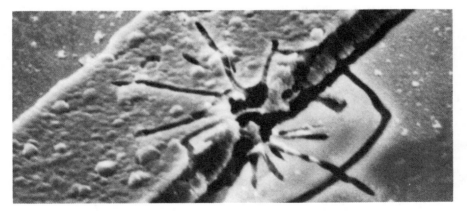

Figure 9.5: ESD discharge damage to aluminum metallization. Note the classic lightning pattern and melting from the thermal energy.

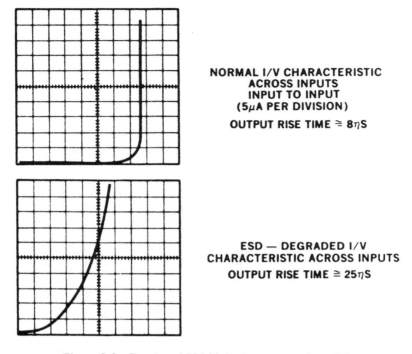

NORMAL I/V CHARACTERISTIC
ACROSS INPUTS
INPUT TO INPUT
(5μA PER DIVISION)

OUTPUT RISE TIME $\cong$ 8ηS

ESD — DEGRADED I/V
CHARACTERISTIC ACROSS INPUTS

OUTPUT RISE TIME $\cong$ 25ηS

Figure 9.6: Results of 500-V discharge to package lid.

Personnel Grounding. When personnel handle ESD sensitive devices, they must be grounded through the use of a wrist strap. The wrist strap that connects the wearer's skin to ground has become one of the most commonly used products in the fight against ESD. There are many designs of wrist straps, but they all rely on a conductive strap that contacts the wearer's wrist and drains static charge to ground. The strap cannot go directly to ground for safety reasons, since a wearer touching test equipment that is not grounded can be shocked. For this reason, a current limiting resistor is added in series with the operator's wrist to ground. This resistor is in the range of 250k ohms to 10M ohms with 1M ohm being the standard. When working with high voltages, this resistor value must be changed to assure that the current the wearer is subjected to is less than 10 to 15 ma. One megohm of resistance has appeal because the decay time is short enough for most circumstances and yet spares the operator from feeling the static charge. When a wrist strap is connected to ground, it must not be connected to a connection point on the work surface, since the work surface adds more resistance to ground thus changing the decay time. Wrist straps are prone to failure due to the flexing of all of the elements in the strap (wrist cuff, resistor, connecting wire). In view of the fact that wrist strap failures are not necessarily observable, the straps must be checked on a regular basis.

Storing or Transporting of Hybrids. Hybrids must be protected from ESD by storing and transporting them in static-safe tote boxes and bags. In order to decide on the type of containers, various definitions must be understood. The following are the definitions from DOD-HDBK-263.

Classification	Definition
Conductive	Material having surface resistivity of 10^5 ohms/square maximum.
Static Dissipative	Material having surface resistivity of $>10^5$ and $<10^9$ ohms/square.
Anti-Static	Material having surface resistivity of $>10^9$ ohms/square and $<10^{14}$ ohms/square.

Components and hybrids must be placed in bags during the times that they are not being worked. In the past "pink-poly" was thought to be protection enough; being an anti-static type, no charge would build up on the bag. However, it would not protect the internal component from discharges from external fields. This means that if a charged person would handle an anti-stat bag, the charge would go through the bag and damage the component inside. The newer bags are a combination of conductive and anti-static materials.

Humidity. The risk of ESD damage is much greater in an area that has low humidity (less than 30 percent) than in an area that is above 50 percent relative humidity.

Labels. Static-sensitive components and hybrid assemblies must be labeled with warning signs. The label shown in Figure 9.7 is described in Electronic Industries EIA Standard RS-471. Doors leading to an area where static-sensitive devices are being assembled should also have a warning sign.

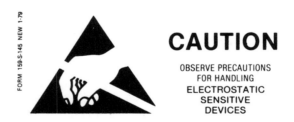

FORM 159-S-145 NEW 1-79

CAUTION

OBSERVE PRECAUTIONS
FOR HANDLING
ELECTROSTATIC
SENSITIVE
DEVICES

Figure 9.7: Static sensitive label.

Static Dissipative Work Bench Tops. All table tops must be of an anti-static or static dissipative material. Bench tops should be grounded through a 250K ohm to 10 megohm resistor with 1 megohm standard. Initially, as in the case of enclosures, a conductive (metal) table top was used; however, this suffered with two disadvantages: it discharged the static too fast and could damage components with a spark, and it was dangerous to personnel if live circuitry was placed on the surface.

Conductive Floor Mats. Floor mats that are grounded should be used at every work station. The conductive floor has conductive elements heat- and pressure-fused into vinyl material. This controls static-charge build-up on personnel and equipment; it also allows a path for static-charge drain. The drawback is that the effectiveness depends upon humidity and the type of footwear that contacts the floor.

REFERENCES

1. Carver, G., National Bureau of Standards
2. "Super Clean Room is 100 Times Cleaner", *Electronic Engineering Times*, August 4, 1986.
3. Frank, D.E., "ESD Considerations for Design, Manufacturing, and Repair", *Worldwide 1982 Electrostatic Discharge Workshop*, February, 1982.
4. Jonassen, N., "The Physics of Electrostatics", Technical Univ. of Denmark, EOE/EOS Symposium Proc., 1984.
5. Quinn, G., "ESD Damage Today: Potentially More Destructive and Costly", *Electronics Test*, Apr. 1984.
6. Turner, T.E. and Morris, S., "Electrostatic Sensitivity of Various Input Protection Networks", EOS/ESD Symposium EOS-2, 1980.
7. Wilson, D.D., Echols, W.E., and Rossi, M.G., "EOS Protection for VLSI Devices," RADC Report MCR-84-506, 1984.
8. Frank, D.E., "ESD Considerations for Electronic Manufacturing", *Westec Conference*, March, 1983.

10

Design Guidelines

Converting design requirements into a producible hybrid microcircuit design involves the application of many diverse engineering disciplines. The disciplines include circuit design and analysis, layout design, thermal analysis, test and test equipment design, process development, and drafting. An effective method of communicating these design requirements to all concerned personnel is through a formal Design Transmittal Document.

HYBRID MICROCIRCUIT DESIGN TRANSMITTAL

The Hybrid Microcircuit Design Transmittal provides information necessary for the applications engineer and the layout designer to proceed with the hybrid design. As a minimum, the transmittal contains the following:

Hybrid parts list

Circuit schematic

Component and total circuit power dissipation

Performance requirements

MIL-SPEC requirements

Documentation requirements

Circuit layout flow, power, and grounding requirements including input/output pin arrangement

The design transmittal may also be used as a change control document to implement design changes that may occur during the hybrid microcircuit development. The transmittal is revised by the circuit design engineer and returned to the applications engineer (Figure 10.1). Figure 10.2 is an example of an alternate transmittal form.

HYBRID MICROCIRCUIT DESIGN TRANSMITTAL

PART TITLE	PART NUMBER
PROGRAM NUMBER ASSIGNED	PROGRAM

CIRCUIT TYPE:	MIL-STD REQUIREMENTS
(1) ANALOG _____	(1) MIL-STD-883 CLASS _____
(2) DIGITAL _____	(2) N/A _____
(3) IF/RF _____	(3) OTHER MIL SPEC REQ'D _____
(4) OTHER _____	

DRAWING REQUIREMENTS:

(1) ESWA _____	(4) PRODUCTION _____
(2) ENGINEERING _____	(5) OTHER _____
(3) DEVELOPMENT _____	

REMARKS:

PREPARED:	APPROVED:
DATE:	DATE: REVISED:

Figure 10.1: Hybrid microcircuit design transmittal form.

CONTENTS:

   ITEM   ADVANCE FIRM   DUE

  (1) PARTS LIST _____ __ __ __   _____

  (2) SCHEMATIC* _____ __ __ __   _____

  (3) SPECIAL REQUIREMENTS_____ __ __ __   _____

  (4) PACKAGE DESCRIPTION _____ __ __ __   _____

  (5) CIRCUIT PERFORMANCE/TEST

      REQUIREMENTS _____ __ __ __   _____

    *RESISTOR POWER DISSIPATION AND TOLERANCES WITH CAPACITOR
    TOLERANCES AND VOLTAGE REQUIREMENTS MUST BE INCLUDED.

  (Item 1) Parts List

Item No.	Qty. Req'd.	Part No. or Identifying No.	Nomenclature/ Description	Reference Designation	Data: Vendor, Size See ◯ General Note

(ITEM 2)
  SCHEMATIC

(ITEM 3)
  SPECIAL REQUIREMENTS

(ITEM 4)
  PACKAGE DESCRIPTION

(ITEM 5)
  CIRCUIT PERFORMANCE/TEST REQUIREMENTS

Figure 10.1: (continued)

PRODUCT CHECKLIST
(Hybrid & High Frequency Products Worksheet)

I.   CUSTOMER INFORMATION: DATE: _____
     Customer: _____
     Address: _____
     City:_____ State:_____ ZIP:_____
     Engineer: _____ Title:_____ Telephone: _____
     Purchasing: _____ Title:_____ Telephone: _____
     Marketing: _____ Title:_____ Telephone: _____

II. APPLICATION AND FUNCTION:_____
     _____
     _____
     _____

     PROGRAM NAME:_____
     DISCLOSURE STATEMENTS IF APPLICABLE:_____
     _____

III. QUALITY REQUIREMENT: ☐ CLASS B ☐ CLASS S ☐ OTHER: _____
     _____

IV. SCHEMATIC (Please describe or attach):_____
     _____
     _____
     _____

     POWER SUPPLY VOLTAGES AND CURRENT REQUIREMENTS (viz. 15V @ 20ma max.):_____
     _____
     CIRCUIT POWER DISSIPATION:_____
     _____
     OPERATING FREQUENCY (and other special considerations such as bandwidth):_____
     _____
     _____

V.   COMPONENTS:
     • ACTIVE (Note Special Requirements):_____
     _____
     • ACTUAL POWER DISSIPATION:_____
     _____
     • SEMICONDUCTOR GATE ARRAY TECHNOLOGY (TTL, CMOS, ECL) PREFERRED:
     _____
     _____
     • RESISTORS: ACTUAL POWER DISSIPATION:_____
     TEMPERATURE COEFFICIENT:_____ PPM/°C FROM: _____ °C to_____°C
     SPECIAL TRACKING OR RATIO MATCH REQUIREMENTS: _____
     _____
     • CAPACITORS: TEMPERATURE COEFFICIENT: _____ PPM/°C FROM:_____ °C to_____°C
     EIA CHARACTERISTICS RATING:_____
     VOLTAGE RATING:_____VOLTS
     • CRYSTALS, CRYSTAL FILTERS (Calibration ±, Temp. Coef., Series or Parallel Mode):_____
     _____
     • SAW DEVICES (Functions and Specification) : _____
     _____

Figure 10.2: Product checklist (hybrid and high frequency products worksheet). (Courtesy of Teledyne Microelectronics)

VI.   PHYSICAL CONSIDERATIONS
       SIZE REQUIREMENTS: LENGTH:_____ WIDTH:_____ THICKNESS:_____
       PACKAGE TYPE: ☐ PLUG-IN ☐ BUTTERFLY ☐ OTHER _____
       LEAD SPACING: _____.050" _____.100" _____OTHER: _____
       _____
       TELEDYNE OPTION:_____
       _____
       _____
       _____

VII.   PACKAGING:
       What areas of modification for packaging design are permissible?
       ☐ CHANGE OR DELETE COMPONENTS:_____
       _____
       ☐ PARTITION SCHEMATIC INTO MULTIPLE PACKAGES:_____
       _____
       ☐ LEAD BREAKOUT:_____
       _____
       ☐ OTHER:_____
       _____

VIII. TESTING REQUIREMENTS (APPLICABLE MIL-STDs, etc., SHOCK, etc.):
       OPERATING TEMPERATURE: MIN:_____°C TYPICAL:_____°C MAX:_____°C
       STORAGE TEMPERATURE: MIN:_____°C TYPICAL:_____°C MAX:_____°C
       OTHER: _____
       _____
       _____
       _____

IX.   ESTIMATED USAGE:
       DEVELOPMENT MODELS:_____ EACH.   DELIVERED BY:_____
       PROTOTYPES:_____ EACH.   DELIVERED BY:_____
       PRODUCTION: _____EACH AT_____ (RATE DELIVERED/MTH), BEGINNING_____
       PRODUCTION TARGET PRICING:_____
       _____

X.   CUSTOMER-FURNISHED MATERIALS/EQUIPMENT:_____
       _____

XI.   DISCUSSION OF DESIGN/PROGRAM GOALS:_____
       _____
       _____
       _____
       _____
       _____
       _____
       _____
       _____
       _____

Figure 10.2: (continued)

SYSTEM REQUIREMENTS AFFECTING HYBRID CIRCUIT DESIGN

After the requirements have been defined in the Design Transmittal
Form, the following sequence must be followed to effectively design a
system using hybrids:

Partitioning the circuit.

Selecting a packaging approach.

Selecting materials and processes.

Defining quality assurance provisions.

Partitioning

Three major factors must be considered when partitioning system electronics into hybrid microcircuits: (1) number and types of input/output pins required, (2) device density, and (3) power dissipation. In addition, consideration must be given to possible circuit commonality (using one circuit for multiple functions) and testability. These factors are interdependent; they may be modified by commonality and testability requirements and usually require several iterations during the partitioning process. Input/output pin and device density requirements are addressed first so that a package and substrate size can be selected. The power dissipation of each candidate hybrid is then calculated by determining the dissipation of each of its components. Partitioning is still largely a trial-and-error application of several rules.

Input/Output Leads

The maximum number of leads that a package can have depends on the package size and lead configuration—the larger the package, the greater the number of leads that can be accommodated. For packages with plug-in leads, the standard pin spacing is 0.100-inch centers. This is to provide sufficient area on the printed wire board (PWB) for plated through-holes for each pin and to allow running conductor traces between the microcircuit pins on the board. Staggered double rows may be used to increase pin density but only at the expense of reducing usable substrate area in the package.

If a "butterfly" or flat-pack-style package is specified, the leads may be on 0.050-inch centers, because leads of this style are usually lap soldered or welded to a multilayer printed circuit board, and conductor traces are not run between the microcircuit leads. Once the package style has been selected, package supplier catalogs should be reviewed to select packages with the desired pin and substrate capabilities.

Testability of a hybrid microcircuit is directly dependent on electrical access to the circuit via the input/output leads. A desirable rule is to dedicate 20 percent of the total available leads as test points, though in high density systems this cannot always be followed.

One common way to increase the number of leads is to have leads on all four sides. However, packages with leads on all four sides are undesirable from the standpoint of heat-sinking capability. Heat-sinking is usually effected through a metallized area on the printed wiring board directly beneath and in contact with the microcircuit. The metallized area conducts heat from the microcircuit to a board side-rail or similar heat sink and requires that at least one side of the microcircuit package be free of pins. A second drawback of having leads on all four sides is the greater care that is required in handling and inserting the package into test adapters without damaging the leads.

Component Density

Several guidelines exist to size substrate area requirements, none of which is exact or without qualifying conditions. One generally practical rule limits the area required for chip components (active and passive) to 20 percent of the substrate area. This rule is valid for substrates of one-half-inch square or larger, but smaller substrates require a preliminary layout for valid sizing. The 20-percent rule is sufficiently accurate for partitioning.

Power Dissipation

In general, some form of heat sinking is required to satisfy the power dissipation requirements of hybrid microcircuits. The heat-sinking technique previously discussed (conduction through a metal pad) can dissipate 1 to 2 watts per square inch of substrate. In the range of 2 to 5 watts per square inch, special heat-sinking techniques must be employed. For dissipation greater than 5 watts per square inch, special materials and processes are required including the use of beryllia substrates, direct bonded copper to alumina or beryllia, and the use of alloy attachment are effective. Where hybrids consist only of CMOS integrated circuits, operating at low speeds (<100 kHz), little or no heat sinking is necessary.

Mechanical Interface/Packaging Requirements

Hybrid microcircuits are normally mounted on printed wiring boards (PWB). If system interconnections require the use of multilayer PWB's with components mounted on both sides, the hybrid package should be of the "butterfly" or flat-pack type (Chapter 6) (leads parallel to the principal plane of the package) so that the leads can be attached by lap-soldering or welding. If the PWB components are to be wave-soldered, or if components are to be mounted on one side of the board only, the hybrid package should be of the plug-in style (leads perpendicular to the principal plane of the package). The leads (pins) project through plated through-holes in the PWB and are soldered in place. In this approach the inspectability of the soldered joint at the pin is limited to one side of the PWB, since the package case obscures the joint on the hybrid circuit side of the PWB.

MATERIAL AND PROCESS SELECTION

Three substrate fabrication processes are available for selection: single-layer thin film, multilayer thick film, and multilayer co-fired thick film. Performance requirements, resistor requirements, component interconnect density, and production quantities are factors to be considered in the process selection.

A guide to process selection is contained in Table 10.1, an overall comparison of co-fired, thick- and thin-films is shown in Table 10.2, and a comparison of the thick-film and thin-film conductor parameters is contained in Table 10.3.

Chip resistors may be used to complement any process; for example, precision thin-film resistor chips may be used with thick-film multilayer circuits, and high-value thick-film chip resistors may be used with thin-film

Table 10.1: Process Selection Guide

Primary System/Microcircuit Requirement	Thick Film	Thin Film	Co-Fired
High-density digital interconnection (require multilayer)	x *	—	x **
Precision resistors (<50 ppm TCR and long-term stability <0.1%/yr)	—	x	—
Low cost-high volume	x	—	—
Low cost-low volume	—	x	—
Integral lead (no package)	—	—	x
Microwave	—	x	—

*Low quantity.   **High quantity.

Table 10.2: Comparison of Thick Films and Thin Films

Thick Film	Thin Film
120,000 Å to 240,000 Å	50 Å to 24,000 Å
Direct Process — Screen and fire (additive)	Indirect (subtractive) process — evaporate or sputter, then photoetch
No chemical etchants used	Problem with disposal and handling of dangerous chemicals, etchants, developers
No problem of recovery of precious metals	Problem with recovery of precious metal from etching solutions
Multilayer process	Multilayering difficult, usually only one layer
Wide range of resistor values by using several pastes with different sheet resistivities from 1 ohm/square to over 1 megohm/square	Limited to low sheet resistivity materials — NiCr and TaN, 100 to 300 ohms/square
Range 50 ohm to 10 megohm	Range 50 ohm to 100 kilohm
Tolerance: 1 Percent Absolute 0.05 Percent Ratio	Tolerance: 0.1 Percent Absolute 0.01 Percent Ratio
No Load Drift: 2500–5000 PPM/C after 1000 Hrs at 125 C	No Load Drift: 500–1000 PPM/C after 1000 Hrs at 125 C
More rugged resistors can withstand harsher environments and higher temperatures	Resistors susceptible to chemical corrosion
TCR's are much higher, 100 to 200 ppm/C	Low TCR resistors 0 ± 50 ppm/C
TCR Tracking 5–30 PPM/C	TCR Tracking 1–5 PPM/C
Line definition 5 to 10 mils 2 mils with etching techniques	Line definition to 1 mil; 0.1 mil possible with sputter etching
Low cost process — continuous conveyorized, automated	Higher cost — batch process
Initial equipment investment low (less than $200,000)	Initial equipment investment high (more than $500,000)
Wire bondability affected by surface roughness of the paste.	Wire bondability better; smoother surface.
Power dissipation 50 watts/sq in	Power dissipation of 30 watts/sq in

Table 10.3: Gold Conductor Current and Resistance

Width (in)	Plated Thin Film 0.01 Ω/□.		Thick Film 0.003 Ω/□	
	Resistance (ohms/in)	Current Rating (amps)	Resistance (ohms/in)	Current Rating (amps)
0.005	2	0.750	0.6	2.25
0.010	1	1.5	0.3	4.5
0.015	0.7	2.25	0.2	6.75
0.020	0.5	3.0	0.15	9.0

interconnect substrates. Chip resistors have been discussed in Chapter 6.

Conductors. The conductivity of substrate conductors is a function of the resistivity of the material and the width and thickness of the conductors. Table 10.3 shows the current-carrying capacity and resistance of thick- and thin-film conductors versus linewidth. Conductors must be designed so that, under the worst-case operating conditions, no gold conductor experiences a current density in excess of 6×10^5 amps/cm^2 (MIL-M-38510).

QUALITY ASSURANCE PROVISIONS

Quality Engineering/Quality Assurance Requirements

Quality Engineering/Quality Assurance (QE/QA) efforts/controls are largely determined by the program requirements. In general, these efforts or controls can vary from good engineering or commercial practice to customer inspection/surveillance of each step in the fabrication, assembly, and test. It is important that the QE/QA requirements be established early in the program and that they be determined for both the hybrid microcircuit and the associated test equipment. The program QE/QA requirements will affect microcircuit design, fabrication, and test; test equipment design, fabrication, and checkout; and documentation and quality assurance. Once the microcircuit QE/QA requirements are agreed upon with the customer, they are incorporated in the specific assembly specification.

Screen Tests

Class S and B levels of quality assurance for military hybrid circuits are established by the screen test procedures of MIL-STD-883, Method 5008. A detailed discussion of screen test methods is presented in Chapter 11.

Preferred Parts List

A preferred parts list is established from which the circuit components are selected. A preferred parts list minimizes the number of part types, with resultant economies in parts procurement, inventory, circuit fabrication, and test. The preferred parts list should be compiled from suppliers that routinely supply chip (uncased) components. The parts must be selected from a supplier's catalog or a source control document, so that he is obligated to deliver devices with a defined and controlled chip size, topology, and electrical performance.

HYBRID DESIGN PROCESS

A hybrid microcircuit design flow diagram is contained in Figure 10.3

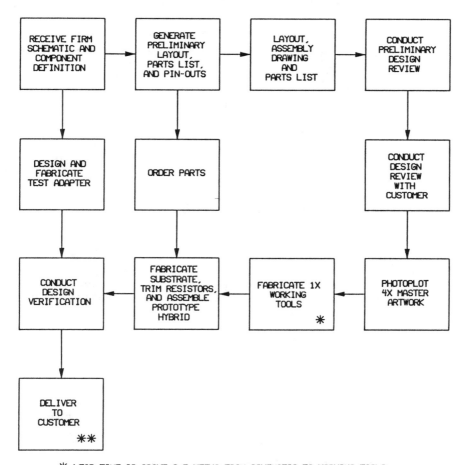

* LEAD TIME IS ABOUT 6-7 WEEKS FROM SCHEMATIC TO WORKING TOOLS.

** IF PARTS ARE AVAILABLE, LEAD TIME IS ABOUT 9-10 WEEKS FROM SCHEMATIC TO DELIVERY OF PROTOTYPE CIRCUITS. HOWEVER, PARTS DELIVERY MAY BE AS LONG AS 20 WEEKS.

Figure 10.3: Hybrid development flow.

Design and Layout

A fully detailed schematic is the data transmitted from the circuit designer to the layout designer. With this schematic, and having selected the hybrid circuit technology, package/substrate size, and following the appropriate design guidelines, the layout designer can proceed with the layout of the circuit. A fully detailed schematic should include:

a.   Resistor values, tolerances, actual maximum continuous power dissipations (no derating or nominal values); special requirements such as temperature coefficient of resistance (TCR), TCR tracking, ratio requirements, and stability.

b.   Capacitor values, tolerances, voltage ratings, and any special dielectric characteristics such as temperature coefficient.

c.   Manufacturer's name and part number for each semiconductor chip.

d.   Conductors with any special requirements identified, such as high current power and ground lines, separate grounds, and conductors sensitive to distributed parameters as in high-frequency circuits.

e.   High dissipation elements identified with expected dissipation specified (not the maximum dissipation capability of the device).

There are several approaches to a hybrid layout; manual methods (single-line and full layout), and computer aided methods.

In the first manual-method layout, a single-line diagram is produced that interconnects the circuit components by single lines denoting conductors. After completing the single-line diagram, the input/output leads are defined. In the second phase of the layout, the single-line diagram is expanded to a full detail drawing from which artwork and 1 X working tools are generated; the drawing is usually 20 times actual size but can be 10 times or 40 times, depending on the actual substrate size and the most convenient working-drawing size. The single-line method is a fast way of obtaining rough pin-outs and component positions. A second manual method involves analyzing the circuit, then laying out full line-widths instead of single-lines.

Computer-Aided Design (CAD)

Automation of electronic designs has become important due to increasing complexity of devices, printed circuit boards, and hybrids. Due to the complexity, the design cycle has lengthened, causing the cost of design to increase. In order for the hybrid industry to become more cost effective the use of CAD became a necessity.

The first stand-alone workstations were introduced in 1981. Prior to that, only a few designs were done on CAD systems such as the large time-shared CALMA. With the introduction of the Personal Computer an affordable CAD became available for hybrid design. This allows the hybrid designer to increase his productivity and the flexibility of changing designs without a cost impact.

CAD equipment varies in price depending on the extent of automation. Lower priced systems will not automatically place the components; the designer still needs to analyze the hybrid, then place the components. The computer will then "rats-nest" connect the components. At this point the designer will move some of the components so that the connections are routed the most efficient way. The more expensive CAD systems have the added capabilities of auto-placing and auto-routing according to the design guidelines entered into the CAD.

Artwork

After the layout is completed, the next step in the design of a hybrid microcircuit is the artwork generation. Care must be taken in preparing and handling of photo-masks due to the further ramifications this artwork has on the substrate processing. Film quality, as far as stability and minimum line widths are concerned, is an important factor. Kodak Precision Line LPD7 (Positive Film) and LP7 (Negative Film) is the standard film for camera reduction of hybrid layouts. Line widths as small as 0.002 are workable with precision developing of this film.

Photo-Plotter. There are several mediums used to generate master artwork from the layout. Presently, the most popular utilizes photo-plotting technology. A photo-plotter is a machine that has a bed for unexposed film which usually is pre-punched for accurate registration. A photo-plotting head, which houses a controlled light source, illuminates the film through apertures which are loaded on a carousel wheel. This wheel is controlled by a motor which drives the light source across the film. A previously encoded set of instructions directs the head, which aperture to select, where to move, when to turn the light on, continue movement to the desired coordinate, stop, and turn the light off. This cycle must be completed and maintain an accuracy of 0.0005 inch over twenty-four inches of film. If the layout has been designed on a CAD system, the data to drive a photo-plotter is readily available. Most photo-plotting is done at four-times size, so when the plot is photoreduced to 1X working tools, the error is 0.0005/4, or 1/8 of a mil.

The photo-plotter described above is the standard type that has been on the market for many years. New plotters are now available that contain a laser as the light source. The laser illuminates a drum which exposes the film in 1/4 mil increments. These laser plotters are able to process a C size piece of film in 10 minutes, which is much faster than the conventional photo-plotters.

Rubylith. Another common method for producing artwork is with "Rubylith" film. Ruby film is a laminate of a hard, clear film on a soft opaque base film. The Ruby is overlayed on the layout and scribed, with an X-Acto knife or cutter on an X-Y guide, around all the perimeters of the desired pattern on the layout. Then, whichever is the least amount of area is peeled away. In this way a positive pattern or negative pattern can be generated. The pattern is as accurate as the skilled person doing the cutting. This cutting is usually done at 20X; therefore, when reduced to 1X, any errors in accuracy are reduced by 20. When generating a thin-film mask, the composite (both conductor and resistor) should be cut and peeled first, shot with a camera, then finished, cutting away the desired resistor pattern for use in the second etch. This method, even though it is done by hand, gives excellent sheet-to-sheet registration.

Other methods for master artwork generation exist, but the two mentioned above are the most popular. Masks have been successfully created using pen-plotters, Printed Circuit Board tape, and even hand drawn masks copied from view-foil transparencies when time constraints so mandate for simpler circuits.

Accuracy of masks is more critical to thin-film circuit generation due to the small precision resistors. As undercutting of the pattern during etching can cause serious yield loss in resistor values, so can undercutting during exposure of the master pattern. Therefore, the emulsion of the artwork should be as close to the substrate as possible for thin film (base side down). In thick-film fabrication, the master artwork is used to fabricate screens. For the maximum accuracy, the artwork is placed against the screen; therefore, the emulusion is placed up (base side down). Standard film thickness, for the master artwork, is in the range of 0.004 to 0.007 inch. If the film is too thin, it is susceptible to stretching or bowing, and if it is too thick, it is hard to keep planar.

Design Review

The completed layout must be reviewed for format, electrical performance, thermal characteristics, producibility, and fabrication and assembly process compatibility. This review is usually held with the designer, applications engineer, production engineering, and the customer. Upon completion of the design review, required design changes are incorporated into the completed layout prior to release for artwork preparation. Checklists, as Figure 10.4 (for thick-film circuits) or Figure 10.5 (for thin-film circuits), are helpful in the design review.

Electrical Performance. The layout must be reviewed for correctness relative to the circuit schematic and component requirements. It should be verified that devices are interconnected per print and that resistor layout and trim capability are compatible with the required values and tolerances.

Conductors must be reviewed for current-carrying requirements. Gold conductors should be designed to carry 6×10^5 amps/cm^2 maximum, in compliance with MIL-M-38510. Conductor widths must be wide enough to eliminate voltage spikes during transient current conditions or voltage offset buildups. While true for signal paths, power and ground lines are particularly important due to generally higher current levels and the potential for crosstalk between devices. The placement of critical circuit components and conductors should be reviewed with respect to parasitic effects.

Thermal Characteristics. Resistor geometries are reviewed for compatibility with power dissipation ground rules. Thin-film resistors, 30 watts per square inch; thick-film resistors, 50 watts per square inch (before trim).

Producibility. The layout design must be reviewed for compatibility with the requirements identified in the layout design guidelines.

Fabrication and Assembly Process Compatibility. The design must be reviewed by Operations personnel for compatibility with on-line fabrication and assembly equipment and tooling. New tooling or special device handling requirements are identified at this step.

1. Substrate
    a. Correct Size _____
    b. Standard Package _____
    c. Registration Marks _____
    d. Orientation marks such as "F" _____
2. Bonding
    a. Beam-Lead Bonding _____
        (1) Bond sites on lowest metallization level _____
        (2) Correct size and spacing _____
    b. Wire Bond
        (1) Size and spacing _____
        (2) Minimum distance when bonding toward package wall _____
        (3) Eliminate die jumpers _____
        (4) Correct size and spacing of opening in insulator (window) _____
        (5) Eliminate two-directional bonding for transistors _____
        (6) Wire bonding around capacitors _____
3. Insulator
    a. Window Around Devices Large as Possible _____
    b. Minimum Spacing from Edge of Substrate _____
    c. Correct Size and Spacing on Vias _____
    d. Vias Parallel to Conductors to be Staggered _____
4. Assembly and Circuit Performance
    a. Minimum Conductor Linewidth _____
    b. Minimum Conductor Spacing _____
    c. Minimum Circuitry Beneath Discrete Parts _____
    d. Die Orientation Square with X-Y Axes _____
    e. Correct Component and Pad Sizes _____
    f. Circuitry Clearance Around Components _____
    g. Noise and Feedback Problems _____
    h. Circuitry Continuity _____
    i. Wire Bonding Around Capacitors _____
    j. Static-Sensitive Devices _____
5. Resistors
    a. Aspect Ratio <10 _____
    b. Minimum size >0.020 Wide and >0.040 Long _____
    c. TCR and Stability are per Requirements _____
    d. Ratio and Trim Requirements _____
    e. Power Dissipation, 50 watts/in.^{2} Before Trim _____
    f. Open Parallel Resistor Paths _____

Figure 10.4: Design review checklist for thick-film circuit.

1. Substrate
    a. Correct Substrate Size _____
    b. Standard Package _____
    c. Orientation "F" _____

2. Bonding
    a. Minimum Pad Provided for Bonding _____
    b. Maximum Lead Lengths _____
    c. Minimum Bond Distance from Edge of Die _____
    d. Minimum Jumpers _____
    e. Minimum Bond to Package Wall Distance _____
    f. Wire Bonding Around Capacitors _____

3. Resistor
    a. Probe Pads Minimum Size and Specified Critical Points _____
    b. Open Parallel Resistor Paths _____
    c. Clearance for Scribing _____
    d. Correct Methods of Trimming _____
    e. Correct Up-Trim for Nominal Value _____
    f. Power Dissipation = 30 w/in.^{2} - Noncritical Resistors _____
    g. Power Dissipation = 15 w/in.^{2} - Critical resistors _____

4. Assembly and Circuit Performance
    a. Minimum Distance from Edge _____
    b. Minimum Spacing Between and Linewidth on Conductors _____
    c. Circuitry Beneath Discrete Parts _____
    d. Die Orientation, Square with X-Y Axes _____
    e. Correct Component Size, Pad Size, Clearance Around Circuitry _____
    f. Noise and Feedback Problems _____
    g. Circuit Continuity _____
    h. Minimum Width and Spacing for Input-Output Pad _____
    i. Static-Sensitive Devices _____

Figure 10.5: Design review checklist for thin-film circuits.

Engineering-Model Design Verification

The first engineering-model circuit is integrated with the test adapter and tested according to the requirements of the Functional Test Specification. Deviations between the test requirements and circuit performance are identified. The test results are analyzed to determine if the deviations are related to marginal circuit design, test adapter design, or unrealistic specification requirements. The system user is requested to evelute the deviations between actual and specified performance with regard to required system performance and, when possible, to integrate the circuit into the next assembly level to verify system performance.

The engineering-model circuit performance should also be verified over the system operating temperature range. Temperature testing is particularly important for those applications requiring production testing at temperature. The test results are used to aid in establishing test limits at temperature.

The design verification is also used to perform any special tests such as design margin, limit testing, and noise testing.

Modification and Redesign

Modification or redesigning the hybrid microcircuit may be necessary because of a system requirement change, adverse results from the design verification testing, or next-level integration testing. The circuit modification may or may not require a redesign of the circuit layout. A simple component change (transistor, chip capacitor, resistor, or, in some cases, an integrated circuit) may suffice to make a correction without a major layout revision. In most cases, layout changes are required when it is necessary to modify circuit interconnect pattern, add components, or change components that are different in physical size or number of pin-outs.

In cases requiring a change in layout, the interconnect substrate may sometimes be altered by scribing conductor lines and using jumper wires to incorporate and verify the change. After being modified, the engineering-model circuit is subjected to design verification testing. When possible, the circuit is integrated into the next level to verify system performance.

When the system user is satisfied with the circuit performance, required changes are permanently incorporated into the layout, assembly drawing, and Functional Test Specification, and the design change can be released.

SUBSTRATE PARASITICS

Capacitance Parasitics

In hybrid microcircuits, the ceramic used as the substrate material is the source of two types of parasitic capacitances:

(1) Interelectrode capacitance, the capacitance between two conductors or resistors on the substrate (X-Y direction).

(2) Capacitance to the case through the substrate (Z direction).

Due to the high dielectric constant of alumina (8 to 10), parasitic capacitances can be large enough to present problems for some circuits.

The capacitance between two parallel plates is given by the equation:

$$C = \varepsilon_o \varepsilon_r A / d$$

where ε_o = absolute permittivity of free space
    = 0.255 picofarads/inch

ε_r = dielectric constant of the material between the plates.

A = Area of the plates (square inches).

d = Distance between the plates (inches)

This equation does not take into account fringing at the edge of the plates and assumes a large area compared to the distance between them. This equation, with some factor, can be used to calculate the capacitance of a hybrid conductor to the case. In order to calculate the capacitance between two conductors on the substrate, the general equation must be modified. To do this, conformal mapping must be performed to transform these lines onto another plane. When these lines are mapped onto a plane where they are in the same configuration as a parallel-plate capacitor, the general equation can be used.

The transformation equations were first introduced by P.S. Castro and P.N. Kaiser in 1962.[1] The derivation is long and will not be presented here. A complete listing can be found in Reference 2. The result of the derivation is shown in the following, which gives the total capacitance between two parallel conductors or resistors.

$$C = \frac{\epsilon_o L}{2} \left[\frac{K(k_1')}{K(k_1)} + \epsilon_r \frac{K(k_2')}{K(k_2)} \right]$$

WHERE:

$$k_2 = \frac{\tanh(\pi W_s/4d_s)}{\tanh\left[\frac{\pi(2w_L + W_s)}{4d_s}\right]}$$

$$k_2' = \sqrt{1 - k_2{}^2}$$

$$k_1 = W_s/(2W_L + W_s)$$

$$k_1' = \sqrt{1 - k_1{}^2}$$

L = Length of line (inches)
ϵ_o = 0.225 pf/inch
W_s = Spacing between conductors (mils)
W_L = Width of conductor (mils)
d_s = Thickness of substrate (mils)

$K(k)$ = elliptical integral of the first kind

Through many laboratory measurements[2] it was noticed that the measured values differed from the calculated values by a multiplication factor. This factor is due to the influence of the case on the lines of flux that are between the conductors. Some of the lines of flux terminate on the case thus lowering the capacitance between the conductors. This factor approaches 1.0 as the line widths become large, which is reasonable because as the line-widths increase the equation approximates the general equation. From the data, a curve-fitting routine was used and a multiplication factor was derived. When the substrates were removed from the case, values calculated more closely approximated the measured values (Figure 10.6).

$$\text{FACTOR (F)} = \frac{1}{1.2455 - 0.013146 \, W_L}$$

$$C = F \left[0.11225 \frac{K(k_1')}{K(k_1)} + \epsilon_r \frac{K(k_2')}{K(k_2)} \right] \qquad \text{pf/inch}$$

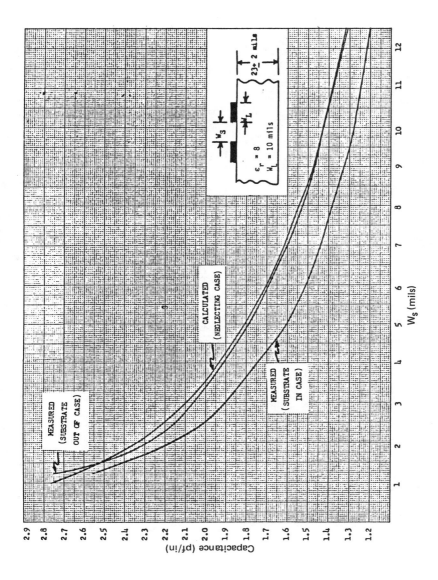

Figure 10.6: Interelectrode capacitance.

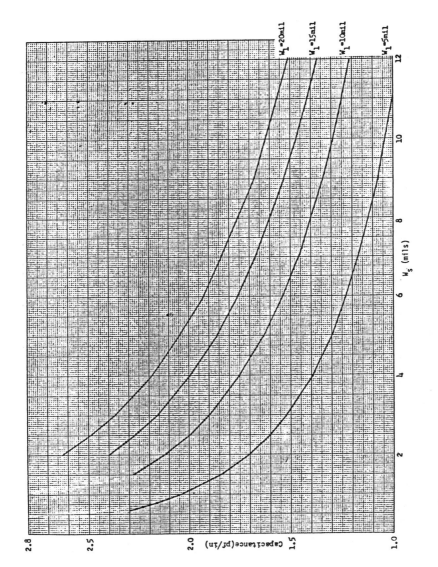

Figure 10.7: Calculated interelectrode capacitance with multiplication factor included.

Figure 10.7 shows the interelectrode capacitance for various line widths.

Conclusions on Interelectrode Capacitance

From the equation and the curves the following conclusions are reached for the interelectrode capacitance between two conductors on a substrate mounted in a metal case.

1. The capacitance is directly proportional to the dielectric constant of the substrate material and to the substrate thickness. Capacitance can be minimized by a thin substrate having a low dielectric constant.

2. Capacitance varies directly with the line width. Minimum capacitance is obtained with narrow lines.

3. Thickness of the conductors does not affect the value of the capacitance as long as it is small compared with the other dimensions.

Capacitance Computer Program

A computer program was written to facilitate the calculation of the interelectrode capacitance between two resistors or conductors. Listed below is an example of the output:

LINE WIDTH = 10 MILS, LINE SPACING = 10 MILS, SUBSTRATE THICKNESS = 25 MILS, DIELECTRIC CONSTANT = 8, INTERELECTRODE CAPACITANCE = 1.30 pf/inch.

Program Description. A computer program was written which will calculate the interelectrode capacitance between conductor or resistor lines on a thin- or thick-film substrate. The program was written in BASIC and run on a Tektronix 4050 series computer.

The data input to the program is: ε_r (dielectric constant of the substrate), W_L (width of conductor or resistor), W_s (spacing between conductors or resistors), and D_s (thickness of substrate). When these parameters are input, the program will calculate the interelectrode capacitance of the resistor or the conductor lines.

It must be remembered that these equations were derived on the assumption that only two lines were present. If more lines are added, the capacitance between different pairs of lines are all connected in series. A listing of the program follows:

```
4 GO TO 100
8 U=51
9 GO TO 460
12 GO TO 660
100 REM ***************************************************************
110 REM ***** THIS PROGRAM CALCULATES INTERELECTRODE CAPACITANCE *****
120 REM *****                        12-23-82                    *****
130 REM ***************************************************************
140 PRINT 'ENTER DIELECTRIC CONSTANT OF THE SUBSTRATEG'
150 INPUT E
160 PRINT 'ENTER CONDUCTOR WIDTH(IN MILS)G'
170 INPUT W1
```

```
180 PRINT 'ENTER CONDUCTOR SPACING(IN MILS)G'
190 INPUT W2
200 PRINT 'ENTER SUBSTRATE THICKNESS(IN MILS)G'
210 INPUT D
220 A=PI*W2/(4*D)
230 B=PI*(2*W1+W2)/(4*D)
240 C=(EXP(A)-EXP(-A))/(EXP(A)+EXP(-A))
250 X=(EXP(B)-EXP(-B))/(EXP(B)+EXP(-B))
260 K2=C/X
270 J2=(1-K2*K2)T0.5
280 K1=W2/(2*W1+W2)
290 J1=(1-K1*K1)T0.5
300 Y=K2
310 GOSUB 520
320 L3=L
330 Y=J2
340 GOSUB 520
350 L4=L
360 Y=K1
370 GOSUB 520
380 L1=L
390 Y=J1
400 GOSUB 520
410 L2=L
420 C1=0.11225*(L2/L1+E*L4/L3)
430 C2=1/(1.24551718264-0.0131467236078*W1)*C1
440 PAGE
450 U=32
460 PRINT @U:'LINE WIDTH= ';W1;'  LINE SPACING= ';W2
470 PRINT @U:'SUBSTRATE THICKNESS= ';D;'  DIELECTRIC CONSTANT= ';E
480 PRINT @U:'JJ'
490 PRINT @U:'INTERELECTRODE CAPACITANCE (PF/IN)= ';
500 PRINT @U: USING '4D.2D':C2
510 END
520 REM *** SUBROUTINE TO CALCULATE COMPLETE ELLIPTIC INTEGRAL ***
530 G1=(1-Y*Y)T0.5
540 A1=1
550 A2=A1
560 T=A2*1.0E-4
570 A1=G1+A1
580 X1=A2-G1-T
590 IF X1<=0 THEN 630
600 G1=(A2*G1)T0.5
610 A1=0.5*A1
620 GO TO 550
630 L=PI/A1
640 RETURN
650 REM ***********************************************************
660 REM ***** ROUTINE TO CHANGE DATA  *****************************
670 REM ***********************************************************
680 PRINT '1....CHANGE DIELECTRIC CONSTANT'
690 PRINT '2....CHANGE CONDUCTOR SPACING'
700 PRINT '3....CHANGE CONDUCTOR WIDTH'
710 PRINT '4....CHANGE SUBSTRATE THICKNESS'
720 PRINT '5....TO END DATA CHANGE'
730 PRINT 'JJ ENTER THE NUMBER OF CHANGE TO BE MADEGG ';
740 INPUT U1
750 IF U1=5 THEN 220
760 GO TO U1 OF 860,830,800,770
770 PRINT 'ENTER THE NEW SUBSTRATE THICKNESSG ';
780 INPUT D
790 GO TO 680
800 PRINT 'ENTER THE NEW CONDUCTOR WIDTHG ';
810 INPUT W1
820 GO TO 680
830 PRINT 'ENTER THE NEW CONDUCTOR SPACINGG ';
840 INPUT W2
850 GO TO 680
860 PRINT 'ENTER THE NEW DIELECTRIC CONSTANTG ';
870 INPUT E
880 GO TO 680
```

Inductive Parasitics[2]

Any conducting path has a self-inductance associated with it. This inductance is a function of the geometry of the conducting path and the permeability of the surrounding material. In the absence of any magnetic material, the inductance becomes dependent only on the dimensions and the permeability of free space ($\mu_0 = 4\,\pi \times 10^{-7}$ henry/meter). The self-inductance of a single, straight conductor line (see Figure 10.8a) is:

$$L = \frac{\mu_0 L_L}{2\pi}\left[\ln\frac{2L_L}{W_L} + \frac{1}{2} + \frac{W_L}{3L_L}\right] \tag{1}$$

$$L = 5.08 L_L\left[\ln\left(\frac{2L_L}{W_L}\right) + \frac{1}{2} + \frac{W_L}{3L_L}\right] \quad nh/in$$

L_L is the length of the line

W_L is the width of the line

$W_L \ll L_L$

The last term in Equation 1 is usually much smaller than the rest of the terms and is often neglected. As the line length is made smaller, the inductance becomes smaller. In fact, the inductance per unit length decreases with decreasing length. As the line is made narrow, the inductance increases.

When another conductor is placed next to the first, there is a mutual inductance which comes into play. The mutual inductance between the two parallel lines (Figure 10.8b) is:

$$L_M = 5.08 L_L\left[\ln\left(\frac{2L_L}{W_L + W_S}\right) - 1 + \frac{W_L + W_S}{L_L} - \frac{1}{4}\left(\frac{W_L + W_S}{L_L}\right)^2 + \frac{1}{12\left(1 + \frac{W_S}{W_L}\right)^2}\right]\frac{nh}{in} \tag{2}$$

W_S is the space between conductors

$S_L \ll L_L$

$W_S \ll L_L$

The last three terms of this equation are usually small and are often neglected. The mutual inductance decreases as the length decreases and the mutual inductance per unit length also decreases as the length decreases. As the width and space width decreases, the mutual inductance increases.

When two conductors are in parallel with current flowing in the same direction, the mutual induction adds to the self-inductance of each line. The total inductance would then be the parallel combination of the two.

$$L_p = \frac{(L_A + L_M)(L_B + L_M)}{L_A + L_B + 2L_M} \qquad L_A = L_B = L \qquad (3)$$

$$L_p = \frac{L + L_M}{2}$$

where L_A, L_B, and L_M are given by Equations 1 and 2.

When two conductors are connected in series (Figure 10.8c) so that current travels in opposite directions, the mutual inductance will subtract from the self-inductance of each line.

$$L_S = L_A + L_B - 2L_M \qquad L_A = L_B = L \qquad (4)$$

$$L_S = 2(L - L_M)$$

If the small terms in Equations 1 and 2 are neglected then added into Equation 4, the total inductance of Figure 10.8c is:

$$L_2 = \frac{\mu_0 L}{\pi} \left[\ln \left(1 + \frac{W_S}{W_L} \right) + \frac{3}{2} \right] \qquad (5)$$

$$L_2 = 10.16 L_L \left[\ln \left(1 + \frac{W_S}{W_L} \right) + \frac{3}{2} \right] nh/in$$

L_L is length of eacn meander

This is the total inductance for a two-meander pattern of Figure 10.8c. Since the logarithm term does not contain the line length, the inductance per unit length of path is independent of length.

This same procedure can be used on the three-meander pattern of Figure 10.8d.

$$L_3 = L_A + L_B + L_C - 2L_{M(AB)} - 2L_{M(BC)} + 2L_{M(AC)} \qquad (6)$$

$$= 3_L - 4L_M + 2L_M^I = 3\left(L - \frac{4}{3}L_M + \frac{2}{3}L_M^I\right)$$

$L_A = L_B = L_C = L \qquad L_M^I$ is given by equation (2) with

$W_S^I = W_L + 2W_S$ substituted in equation (2) for W_S

$$L_3 = \frac{3\mu_0 L}{2\pi} \left\{ \ln \left[\frac{(2L_L)^{1/3} \left(\frac{W_L + W_S}{2} \right)^{2/3}}{W_L} \right] + \frac{7}{6} \right\}$$

$$L_3 = 15.24\, L_L \left\{ \ln \left[\frac{(2L_L)^{1/3} \left(\frac{W_L + W_S}{2} \right)^{2/3}}{W_L} \right] + \frac{7}{6} \right\}$$

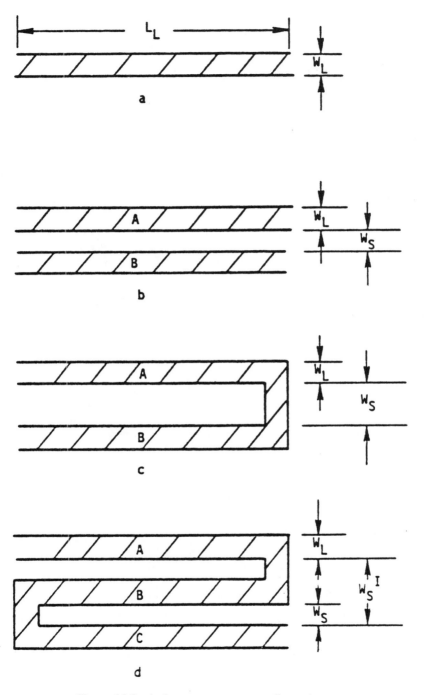

Figure 10.8: Inductance pattern configurations.

This general procedure can be carried out to any number of meanders. The general equation for "n" meanders, in series, is:

$$L_T = nL - 2(n-1)L_M + 2(n-2)L_M{}^I - 2(n-3)L_M{}^{II} + \ldots$$

L_M is mutual inductance between nearest neighbors

$L_M{}^I$ is mutual inductance between next nearest neighbors

where $L_M{}^{II}$ is Equation 2 with $W_M = W_M{}^{II} = 2W_L + 3W_S$

As the number of meanders is increased and more items are added, the smaller terms that were neglected now become important, and error will result if they are not accounted for. For this reason, a computer program was written which takes into account all the terms. Following are the computer results of various configurations along with curves which were drawn from this data. Conclusions concerning this section are listed after the curves.

Figure 10.9. This is a plot of self-inductance of straight conductors on a thin-film substrate (see Equation 1). It gives a family of curves for different line width.

Figure 10.10. This is a plot of mutual inductance between two conductors (see Equation 2). This shows a family of curves for different line width plus line spacing values.

Figure 10.11. This is a plot of total inductance (self and mutual) for a meandering pattern. These curves were drawn for a line width of 5 mils and a total length of the pattern equaling one inch.

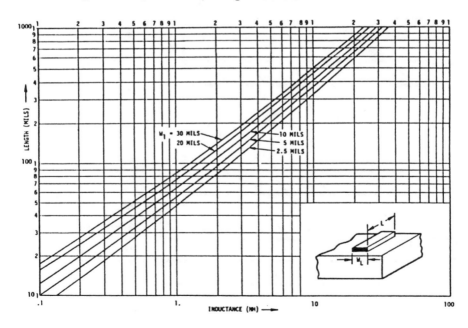

Figure 10.9: Inductance of straight conductors.

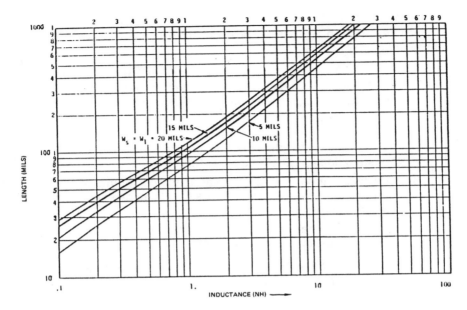

Figure 10.10: Mutual inductance between two parallel conductors.

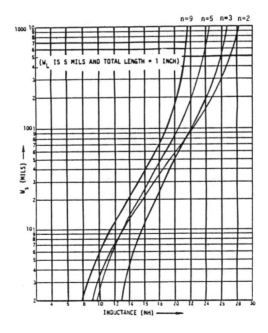

Figure 10.11: Total inductance of meandering resistor pattern.

Conclusions on Parasitic Inductance

The following conclusions can be drawn from the equations presented in this section:

1. The self-inductance of a straight conductor decreases as the line width increases, for a constant line length (Figure 10.9). Therefore, a wide line gives less inductance than a narrow line.

2. The mutual inductance between two conductors decreases as the line width plus the line spacing increases, for a constant line length (Figure 10.10).

3. The inductance is directly related to the length of the pattern. Therefore, the inductance can be made smaller by placing components closer together.

4. The meander pattern is basically a noninductive pattern. It has less total inductance than its straight line counterpart.

In a final conclusion, it can be stated that to minimize the inductance in a hybrid conductor, make the pattern meander or make the conductor short.

THERMAL CONSIDERATIONS

Power dissipation is a major factor in the design of a hybrid circuit. Basically a hybrid circuit must be designed so that the semiconductor devices are maintained at or below their maximum-rated temperature. Since most hybrid circuits are heat sinked in some manner, the temperature of the heat sink plus the temperature rise from the heat sink to the junction of the semiconductors must be less than the maximum-rated device temperature.

Special processes and materials are employed for hybrid circuits having high power dissipation requirements. Alumina, which possesses good thermal properties, is the commonly used substrate material, but for high power-dissipating circuits, beryllia is used because its thermal conductivity is seven times better than alumina. To further enhance the thermal transfer, large-geometry power devices may be attached to beryllia or alumina substrates with eutectic or solder alloys rather than with the conventional epoxy. There are three mechanisms by which heat can be transferred:[3] conduction, convection, and radiation.

Conduction

Conduction, which involves the intermolecular transfer of kinetic energy, is the main means of transferring heat in a solid. For any given temperature difference, more heat can be transferred by conduction than by either convection or radiation. Heat transfer by conduction is governed by Fourier's law, which states that the rate of heat flow through a material is directly proportional to the cross-sectional area of the material, proportional to the temperature difference across the material, and inversely proportional to the thickness of the material. A proportionality constant called the thermal conductivity is used in the equation. Therefore the heat flow, or power dissipated, is equal to the area times the temperature gradient times the thermal conductivity, divided by the thickness.

Convection

Convection involves the transfer of heat through gaseous or liquid fluids. It is the main method of heat transfer between a solid and a fluid in contact with it; for example, a bare die and the surrounding air. Power dissipation by convection is not as easily expressed as that by conduction. The rate of flow by means of convection is mainly a function of the surface area of the solid, the temperature difference between the solid and the fluid, the velocity of the fluid, and some inherent properties of the fluid.

Radiation

Thermal radiation involves the transfer of heat by electromagnetic waves. The rate of flow is directly proportional to the surface area and to the fourth power of the temperature.

Of these three thermal transfer mechanisms, conduction is most applicable to hybrid circuits.

Circuit Design Thermal Criteria

The following criteria should be followed in designing a hybrid that requires a high heat dissipation:

a. Use high-density alumina or beryllia substrates (these have high thermal conductive properties).

b. Power-dissipating devices should be positioned as closely as possible to heat sinks and evenly distributed over the substrate.

c. Use large conductor area contacts to aid in power dissipation for film resistors with aspect ratios <1.

d. Conductors should be as wide as possible.

e. High thermal dissipators should be at least 0.10 inch from the edge of the substrate.

f. Use heat-spreading tabs to attach high-power, small-size transistors. These tabs may be Kovar or molybdenum.

The thermal impedance of the hybrid circuit structure may be calculated to determine the temperature rise from package base to device junction. Figure 10.12 is a representation of the thermal interfaces involved in the attachment of a device and the electrical analog of the heat path. Curves used to calculate thermal resistance are given in Figure 10.13. The thickness of the epoxy joints is difficult to control in production. Consequently, curves representing three thicknesses are given. The curve for 0.001 inch thick adhesive represents a structure containing a device bonded with 0.001 inch of silver-filled epoxy and a substrate bonded with

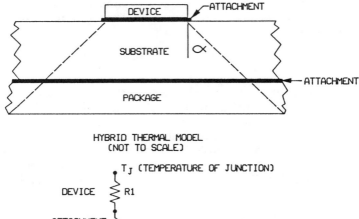

Figure 10.12: Thermal structure of device attachment.

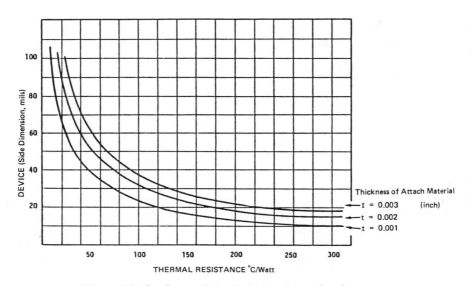

Figure 10.13: Curves for calculating thermal resistance.

0.001 inch of nonconductive epoxy. As a rule of thumb, the 0.002 inch thickness curve should be used. These curves were generated based on a 0.010 inch thick die, 0.025 inch thick substrate, and 0.040 inch thick package.

A matrix as shown in Table 10.4 provides materials data needed to calculate thermal resistance.

Table 10.4: Thermal Properties of Materials Commonly Used in Hybrids[3]

Material	Thermal Conductivity (watt/°C-in)*	Specific Heat(s) (cal/g-°C)	Mass Density (g/cm³)
Silver	10.6	0.056	10.5
Copper	9.6	0.093	8.9
Au-Si eutectic	7.5	–	–
Gold	7.5	0.03	19.3
Beryllia	6.58	0.31	2.8
Aluminum	5.52	0.22	2.7
Molybdenum	3.9	0.066	10.2
Nickel	2.29	0.1125	8.9
Silicon	2.13	0.18	2.4
Silver/glass (die attach)	1.36	–	–
Alumina			
Co-fired	0.37	–	–
94%	0.70	–	–
96%	0.89	0.21	3.7
99.5%	0.93	–	–
Solder (60–40)	0.91	0.04	8.7
Kovar	0.49	0.11	8.2
Thick-film dielectric	0.55	–	–
BeO filled epoxy	0.088	–	–
BeO filled RTV	0.066	–	–
Glass	0.04	0.20	2.2
Epoxy, silver filled	0.04	0.17	3.5
Epoxy, nonconductive	0.01	0.2	2.0

*Watt/°C-in = (0.0255) watts/m-K.

Thermal Performance of a Semiconductor Device. The major characteristics affecting the thermal performance of a semiconductor device are:

   a. Device size
   b. Attachment method (beam lead, adhesive bond)
   c. Attachment material (filled epoxy, eutectic)
   d. Substrate material
   e. Device density
   f. Thermal interface with next assembly
   g. Package thermal design

The active area on simple semiconductor devices (transistors or diodes) range from 10 to 30 percent of the device plan area. Multi-junction devices such as ICs are assumed to have active areas 50 to 70 percent of the plan area.

Thermal Design. Advances in the density and complexity of integrated circuits have resulted in the need to dissipate more power and heat. If the heat is not efficiently removed the life expectancy of a device is shortened. As the junction temperatures of semiconductor devices increase above 125°C there is an increasing risk of thermal runaway.

To assure reliability the thermal impedance of the hybrid structure should be calculated. For this calculation the following parameters are required:

1. Maximum allowable junction temperature of the die (T_J).
2. Maximum power dissipation (P_D).
3. The system ambient temperature (T_A).
4. Thermal conductivities of all materials used in the system (K).

The total thermal resistance of a structure is calculated by summing all the individual thermal impedances, much the same way as an electrical circuit is analyzed. Thermal heat (h) flows down a length (X) of cross-sectional area (A), analogous to a current (I) flowing down a resistance (R) of length (L). In fact, there are many analogies between thermal and electrical mechanisms as can be seen from Table 10.5.

Table 10.5: Thermal Parameters and Their Electrical Analogs

Thermal Symbol	Thermal Parameter	Unit	Electrical Symbol	Electrical Parameter	Unit
ΔT	temperature difference	°C	V	voltage	volt
h	heat flow	calories	I	current	amp
θ	thermal impedance	°C/watt	R	resistance	ohm
γ	heat capacity	watt-sec/°C	C	capacity	farad
K	thermal conductivity	cal/sec-cm-°C	σ	conductivity	mho
Q	quantity of heat	calories	q	charge	coulomb
t	time	second	t	time	second
$\theta\gamma$	thermal time constant	second	RC	time constant	second

The one major difference between the thermal and electrical units is that Q is in units of energy, whereas q is simply a charge. Hence h is in units of power and may be equated to an electrical power dissipation.

Derivation of the Thermal Equations.[4] Fourier's law of Heat Transfer by Conduction states: The rate of heat flow equals the product of the area normal to the heat flow path, the temperature difference along the path, and the thermal conductivity of the material. Therefore:

$$dq/dt = -KA\,(dT/dx)$$

dq/dt is the power (P), expressed in watts, K is thermal conductivity, and T is temperature. To determine the thermal resistance the above equation is integrated.

$$P\int_o^x dx = -KA\int_{T_2}^{T_1} dT$$

$$PX = -KA\,[T]_{T_2}^{T_1} = -KA\,[T_1 - T_2]$$

$$PX = KA\,\Delta T$$

$$PX/KA = \Delta T$$

If the thermal resistance (θ) is expressed as $\theta = X/KA$ an expression is obtained that relates power dissipation to the temperature difference. Therefore:

$$P\theta = \Delta T$$

If the path that the heat travels is short compared to the heat source dimensions, the above equation applies. However, as the path-length increases, heat will spread out laterally (Figure 10.14). The above equation must then be modified to take into account this thermal spreading which lowers the thermal resistance compared to a straight-line path.

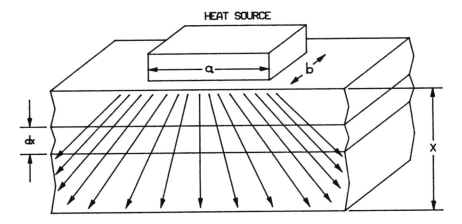

Figure 10.14: Heat source spreading.

The thermal equation then becomes:

$$\theta = 1/K\int_o^x dx/A$$

where A is a function of x and equal to the surface area that the heat traverses at any cross-section. The equations that govern thermal spreading are very complicated. A simplification assumes that the heat spreads at a 45 degree angle corresponding to a truncated pyramid.

Square-Shaped Heat Source.

$$a = b$$

$$\theta = \frac{1}{K} \int_0^X \frac{dx}{A} = \frac{1}{K} \int_0^X \frac{dx}{(a + 2X)^2}$$

$$\theta = \frac{1}{K} \left[\frac{-1}{2(a + 2X)} \right]_0^X$$

$$\theta = \frac{1}{K} \left[\frac{-1}{2(a + 2x)} + \frac{1}{2a} \right]$$

$$\therefore \theta = \frac{X}{Ka(a + 2x)}$$

Rectangular-Shaped Heat Source. (See Figure 10.15.) This is the case that is most often encountered in hybrid thermal analysis.

$$\theta = \frac{1}{K} \int_0^X \frac{dx}{A} = \frac{1}{K} \int_0^X \frac{dx}{(a + 2X)(b + 2X)}$$

$$\theta = \frac{1}{K} \left[\frac{1}{2a - 2b} \right] \ln \frac{(b + 2x)}{(a + 2x)} \Big|_0^X$$

$$\therefore \theta = \frac{1}{2K(a - b)} \left[\ln \left(\frac{a}{b} \right) \left(\frac{b + 2x}{a + 2x} \right) \right]$$

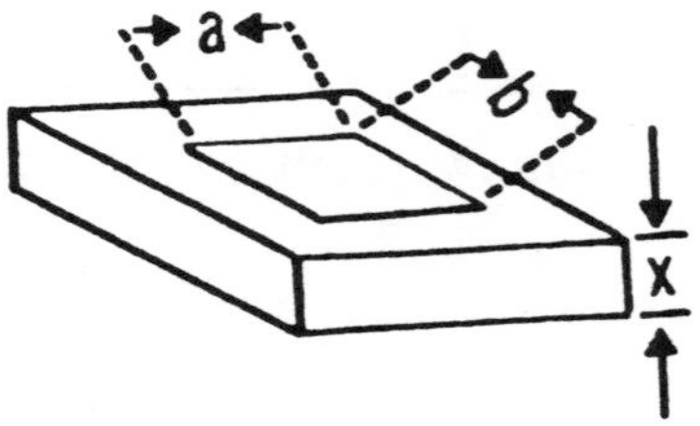

Figure 10.15: Rectangular shaped heat source.

Circular-Shaped Heat Source. If the heat source is in the shape of a circle, the same principles apply to heat spreading. The equation that takes into account the change in geometry is:

$$\theta = X/[K\pi(a^2 + aX)]$$

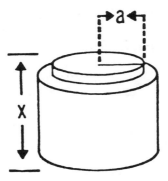

Figure 10.16: Circular shaped heat source.

These equations are first-order approximations and provide a ballpark answer to most hybrid thermal calculations. These equations also only take into account one device. If a more detailed analysis is required one of the computer programs listed at the end of this section should be used.

The "2X" term in the equations is based on the assumption that the heat spreads at 45 degrees. To change that angle one must add $\tan(\alpha)$, the angle from the perpendicular line from the device to the heat-flow lines (Figure 10.12).

Therefore to change the angle of spread from 45 degrees, the "2X" term should be replaced with $(2)(X)(\tan \alpha)$.

Thermal Analysis Computer Program

A thermal analysis program was written in BASIC for the Tektronix 4050 computer using the derived equations.

Data inputs consist of:

Angle of spread

Active area of the chip

Dimensions of the chip

Materials, thickness, and thermal conductivity of each material

Statements 4 through 780 are Tekgraphic commands to draw the initial figure. The program starts at statement 790. It permits entering the information for each layer of the thermal model up to 20 layers. This can be increased by changing the dimension statement in line 890. Any desired angle of spread can be specified. With the change data function, any value can be changed and the program run again without having to re-input all of the data. It should be noted that this program does not take into account heat from sources other than the die.

```
***************************************************************************
    THIS PROGRAM WILL CALCULATE THE THERMAL IMPEDANCE OF A STRUCTURE
                        AS SHOWN BELOW
***************************************************************************
```

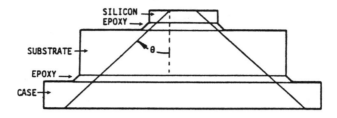

```
TO RUN PROGRAM PRESS  HOME  THEN ONE OF THE FOLLOWING KEYS

KEY 1=INITIALIZE  KEY 2=ENTER DATA  KEY 3=RUN DATA  KEY 4=LIST DATA
KEY 5=CHANGE DATA

COMMON THERMAL CONDUCTIVITY VALUES IN  WATTS/C-IN

ALUMINA    = 0.66 BEO         = 6.50 SILICON = 2.13    KOVAR = 0.40
COND EPOXY = 0.04 NON C EPOXY= 0.01 EUTECTIC= 5.50

4 GO TO 100
8 INIT
9 GO TO 890
12 GO TO 1250
16 GO TO 1450
20 GO TO 1530
24 GO TO 1940
86 GO TO 1250
100 PAGE
110 REM**************************************************************************
120 REM**************************** THERMAL 12-01-82 **************************
130 REM**************************************************************************
140 IMAGE 70('*')
150 PRINT USING 140:
160 PRINT
170 PRINT * THIS PROGRAM WILL CALCULATE THE THERMAL IMPEDANCE *;
180 PRINT 'OF A STRUCTURE'
190 PRINT *              AS SHOWN BELOW*
200 PRINT USING 140:
210 MOVE 70,70
220 DRAW 90,70
230 DRAW 90,66
240 DRAW 70,66
250 DRAW 70,70
260 MOVE 70,66
270 DRAW 68,64
280 MOVE 92,64
290 DRAW 90,66
300 MOVE 50,64
```

```
310 DRAW 110,64
320 DRAW 110,50
330 DRAW 50,50
340 DRAW 50,64
350 MOVE 50,50
360 DRAW 48,48
370 MOVE 110,50
380 DRAW 112,48
390 MOVE 120,48
400 DRAW 120,40
410 DRAW 40,40
420 DRAW 40,48
430 DRAW 120,48
440 MOVE 76,70
450 DRAW 46,40
460 MOVE 116,40
470 DRAW 84,70
480 INIT
490 MOVE 76,70
500 FOR J=1 TO 10
510 RDRAW 0,-1
520 RMOVE 0,-1
530 NEXT J
540 MOVE 56,68
550 PRINT "SILICON"
560 MOVE 56,64.5
570 PRINT "EPOXY"
580 MOVE 34,56
590 PRINT "SUBSTRATE"
600 MOVE 36,49
610 PRINT "EPOXY"
620 MOVE 32,43
630 PRINT "CASE"
640 MOVE 76,57
650 SET DEGREES
660 FOR I=90 TO 45 STEP -5
670 ROTATE I
680 IF I=>60 AND I<=75 THEN 710
690 RDRAW 0,1
700 GO TO 720
710 RMOVE 0,1
720 NEXT I
730 RDRAW 1,-2
740 RMOVE -1,2
750 RDRAW -1,-2
760 MOVE 70.5,56.5
770 PRINT "0"
780 MOVE 0,33
790 PRINT "TO RUN PROGRAM PRESS  HOME   THEN ONE OF THE FOLLOWING KEYSJ"
800 PRINT "KEY 1=INITIALIZE  KEY 2=ENTER DATA  KEY 3=RUN DATA  ";
810 PRINT "KEY 4=LIST DATA"
820 PRINT "KEY 5=CHANGE DATAJJ"
830 PRINT "COMMON THERMAL CONDUCTIVITY VALUES IN  WATTS/C-INJ"
840 PRINT "ALUMINA    = 0.66","BEO       = 6.58","SILICON = 2.13";
850 PRINT "  KOVAR = 0.40"
860 PRINT "COND EPOXY = 0.04","NON C EPOXY= 0.01","EUTECTIC= 5.50"
870 INIT
880 RETURN
890 DIM T(20),C(20),A(20),W(20),C$(10),Z$(200)
900 FOR J=1 TO 200 STEP 10
910 G$="..........."
920 Z$=REP(G$,J,0)
930 NEXT J
940 PRINT "THE ANGLE OF SPREAD IS = G";
950 INPUT D
960 PRINT "THE ACTIVE AREA OF CHIP (ONE SIDE OF SQUARE AREA) IS =G";
970 INPUT S
980 PRINT "THE DIMENSIONS OF THE DIE(ONE SIDE SQUARE DEVICE)= G";
990 INPUT S9
1000 PRINT "ENTER THE TITLE OF THIS ANALYSISG"
```

```
1010 INPUT B$
1020 PRINT "ENTER TITLE OF THE MATERIAL    "
1030 PRINT "ENTER THICKNESS (IN) , THERMAL CONDUCTIVITY (WATT/IN C)"
1040 PRINT "ENTER ZERO TO END THE DATA"
1050 A(1)=S
1060 T1=0
1070 SET DEGREES
1080 N=0
1090 FOR I=1 TO 20
1100 PRINT "G"
1110 INPUT C$
1120 IF C$="0" THEN 1250
1140 INPUT T(I),C(I)
1150 Z$=REP(C$,10*(I-1)+1,10)
1160 IF C$="0" THEN 1250
1170 N=N+1
1180 W(I)=T(I)/(C(I)*A(I)*(A(I)+2*T(I)*TAN(D)))
1190 T1=T1+W(I)
1200 A(I+1)=A(I)+2*T(I)*TAN(D)
1210 IF A(2)<S9 THEN 1230
1220 A(2)=S9
1230 NEXT I
1240 RETURN
1250 PAGE
1260 PRINT
1270 PRINT B$
1280 PRINT "J THE ANGLE OF SPREAD IS= ";D;"J"
1290 IMAGE 70("-")
1300 PRINT USING 1290:
1310 PRINT "TITLE          O C/WATT    K(WATT/IN C)";
1320 PRINT "      THICKNESS(IN)        A(IN2)"
1330 PRINT USING 1290:
1340 FOR I=1 TO N
1350 IMAGE 10A,4D.2D,7X,3D.2D,11X,1D.3D,8X,1D.3D,1X,1A,1X,1D.3D
1360 F$=SEG(Z$,10*(I-1)+1,10)
1370 PRINT USING 1350:F$,W(I),C(I),T(I),A(I),"X",A(I)
1380 NEXT I
1390 IMAGE 70("*")
1400 PRINT USING 1390:
1410 PRINT
1420 PRINT "TOTAL O (C/WATT) = ";
1430 PRINT USING "4D.2D":T1
1440 END
1450 REM************ROUTINE TO LIST DATA****************************
1460 PRINT "J    DATA LISTING    "
1470 PRINT "JTHE ANGLE OF SPREAD IS = ";D
1480 PRINT "THE AREA IS = ";A(1),"J"
1490 FOR I=1 TO N
1500 PRINT "T(";I;")=";T(I),"C(";I;")=";C(I)
1510 NEXT I
1520 RETURN
1530 REM************ROUTINE TO CHANGE DATA****************************
1540 PRINT "JTO CHANGE ANGLE OF SPREAD   ENTER 1"
1550 PRINT "TO CHANGE THICKNESS         ENTER 2"
1560 PRINT "TO CHANGE THERMAL COND      ENTER 3"
1570 PRINT "TO CHANGE AREA              ENTER 4"
1580 PRINT "TO CHANGE THE TITLE         ENTER 5"
1590 PRINT "TO END DATA CHANGE          ENTER 0"
1600 INPUT X
1610 IF X=0 THEN 1840
1620 GO TO X OF 1630,1660,1710,1760,1810
1630 PRINT "ENTER NEW ANGLE OF SPREADG";
1640 INPUT D
1650 GO TO 1540
1660 PRINT "WHICH THICKNESS REQUIRES CHANGING?G";
1670 INPUT M
1680 PRINT "ENTER NEW VALUE FOR T(";M;")G";
1690 INPUT T(M)
1700 GO TO 1540
1710 PRINT "WHICH THERMAL CONDUCTIVITY REQUIRES CHANGING?G";
1720 INPUT M
```

```
1730 PRINT 'ENTER NEW VALUE FOR C('#M#')G'#
1740 INPUT C(M)
1750 GO TO 1540
1760 PRINT 'ENTER NEW VALUE FOR AREAG'#
1770 INPUT A(1)
1780 PRINT 'THE DIMENSIONS OF THE DIE(ONE SIDE SQUARE DEVICE)= G'
1790 INPUT S9
1800 GO TO 1540
1810 PRINT 'ENTER THE NEW TITLEG'
1820 INPUT B$
1830 GO TO 1540
1840 PAGE
1850 T1=0
1860 FOR I=1 TO N
1870 W(I)=T(I)/(C(I)*A(I)*(A(I)+2*T(I)*TAN(D)))
1880 T1=T1+W(I)
1890 A(I+1)=A(I)+2*T(I)*TAN(D)
1900 IF A(2)<S9 THEN 1920
1910 A(2)=S9
1920 NEXT I
1930 GO TO 1250
1940 REM************* PRINT ON PRINTER **************************
1950 PRINT @51:
1960 PRINT @51;B$
1970 PRINT @51:'J THE ANGLE OF SPREAD IS= '#D#' DEGREES'#'J'
1980 IMAGE 70('-')
1990 PRINT @51: USING 1290:
2000 PRINT @51:'TITLE      O'#'H'#'/ C/WATT    K(WATT/IN C)'#
2010 PRINT @51:'      THICKNESS(IN)      A(IN2)'
2020 PRINT @51: USING 1290:
2030 FOR I=1 TO N
2040 IMAGE 10A,4D.2D,7X,3D.2D,11X,1D.3D,8X,1D.3D,1X,1A,1X,1D.3D
2050 F$=SEG(Z$,10*(I-1)+1,10)
2060 PRINT @51: USING 1350:F$,W(I),C(I),T(I),A(I),'X',A(I)
2070 NEXT I
2080 IMAGE 70('*')
2090 PRINT @51: USING 1390:
2100 PRINT @51:
2110 PRINT @51:'TOTAL O'#'H'#'/ (C/WATT) = '#
2120 PRINT @51: USING '4D.2D':T1
2130 END
```

Following is an example of the computer output where θ is the thermal impedance.

TITLE	θ (C/WATT)	K (WATT/IN C)	THICKNESS (IN)	AREA (IN x IN)
Silicon	0.39	2.13	0.010	0.100 x 0.100
Conductive Epoxy	3.36	0.04	0.002	0.120 x 0.120
Alumina Substrate	1.76	0.66	0.025	0.124 x 0.124
Insulative Epoxy	6.46	0.01	0.002	0.174 x 0.174
Kovar PAckage	2.18	0.40	0.040	0.178 x 0.178

****

TOTAL θ (C/WATT) = 14.15

Ideally, for power devices, the total θ should be reduced to as close to zero as possible. In the above example, an improvement can be made by using eutectic or alloy attachment in lieu of epoxy.

The conversion of K to other units is:

$$watt\text{-}in/in^2\text{-}°C \ = \ 10.67 \ x \ cal\text{-}cm/cm^2\text{-}sec\text{-}°C \ = \ 0.00366 \ x \ Btu\text{-}in/hr\text{-}ft^2\text{-}°F$$

Thermal Testing

The thermal performance of a hybrid may be evaluated by means of thermographic testing or computer simulation.

1. An instrument such as the UTI thermal imaging system or the Hughes Probeye thermal video system can be used to evaluate a hybrid. More discussion of these systems is given in Chapter 8.

2. Computer simulation of a hybrid can be performed using existing software such as TXYZ, a program available from the National Bureau of Standards.

LAYOUT GUIDELINES COMMON TO BOTH THICK- AND THIN-FILM HYBRIDS

The need to produce a low-cost system requires that hybrids be economical to produce. Innovative steps must be taken at the circuit design and layout phases to simplify fabrication, assembly, and testing. The following procedure applies to both thick- and thin-film hybrid technologies.

Preliminary Physical Layout

The circuit engineer and layout designer must work closely during this phase to assure that both the functional and physical design goals can be met. An important function of the layout designer is to determine whether a particular function will physically fit a given substrate area.

Estimating Substrate Area

Using a listing like the Hybrid Master Parts List (Figure 10.17), list all discrete devices, transistors, diodes, ICs, chip capacitors, and chip resistors, and record maximum size, value, tolerance, and power dissipation. Calculate and list the area of each circuit element. Add the individual component areas and multiply the resulting sum by ten. This figure accounts for the area of the lead pads and all interconnections. This calculation gives the total substrate area for a given number of devices.

Final Physical Layout

To finalize the layout:

1. Analyze and redraw the circuit schematic to eliminate or minimize the number of cross-overs, to position the external leads at the substrate edge, and to place all components in the relative positions they are to occupy. All power components should be evenly distributed and input and outputs well separated.

2. Determine size and configuration of all deposited resistors.

3. Locate all key elements.

4. Locate external-lead bonding pads; usually a package master outline is used.

5. Locate other circuit elements and draw interconnect patterns.

6. Orient all chip elements, circuit-element characteristics, trimming direction, and materials and processes necessary to fabricate the design.

7. Place conductors parallel to the edges of the substrate wherever possible.

8. Keep conductors as short and wide as possible to minimize added circuit resistance, stray capacitance, and increases in TCR, particularly when terminating low-value resistors and for ground or transistor collector paths.

9. Design wire bond pads for bonding from the substrate to the package pin-outs to be 0.015 × 0.015 inch minimum. Wire bond pads for substrate-to-device and substrate-to-substrate jumpers shall be 0.012 × 0.012 inch minimum (0.015 × 0.015 inch preferred).

10. Avoid running circuitry beneath discrete parts (i.e. capacitors).

11. Test Points: When the need for test points occurs they should be designed to be 0.015 inch square pads and offset to prevent damage to conductors when probing (internal test points are provided only if package is pin limited).

12. After the layout is completed, number the components (R1, R2, Q1, Q2, etc.) on the layout from left to right. Then transfer the designations to the schematic. This helps locate a component on the substrate during rework and troubleshooting.

Assembly Aids

Assembly aids are designed into the layout, where space permits, and consist of bond site locators and chip orientation locators. This aid takes the form of a half-circle for wire bond sites. Whenever size permits, the part number shall be added to the layout.

TITLE _____ PART NO. _____
DESIGNER _____ DATE _____
        PROGRAM _____

QTY REQD	PART NUMBER	PART NAME	VENDOR	REF DESIGNATION	L MAX	W MAX	H MAX	VALUE, TOL, POWER	REMARKS

Figure 10.17: Hybrid master design parts list.

Device Placement

1. Semiconductor devices are attached to metallized pads on the substrates. The normal pad potential necessary for proper device operation is listed below. Any uncertainty about pad potential can be resolved by consulting the specific device manufacturer.

    PMOS - most positive chip supply

    COSMOS - isolated

    Transistors - collector contact

    Diodes - cathode contact

    Op-Amps - most negative chip supply

    TTL - isolated

    ECL - most negative chip supply

    These should be considered as guidelines, but some changes may be required depending on the sensitivity of the circuit design.

2. Components of the same type should be oriented in the same direction where possible. This will lend itself to fewer errors during assembly.

3. All die should be oriented square with the edges of the substrate.

4. Die-pad dimensions for epoxy die attach shall be equal to the maximum die size, plus 0.005 inch on a side, with a 0.010-inch minimum clearance between pad edge and adjacent circuitry when the electrically conductive epoxy is applied by screening. If the epoxy is to be hand applied, the die-pad dimensions for die greater than 0.05 inch on a side shall be the maximum die size plus 0.010 inch on a side. A tab opposite pad 1 of an integrated circuit die can be included for die orientation purposes were possible.

5. For eutectic die attach, the pad dimensions shall be 0.020 inch longer along each edge than the maximum dimension of the die. Where space permits, the die-pad area should be large enough for two die to facilitate rework.

6. The minimum dimensional requirements for installation of chip capacitors are as shown below. Use more liberal pad area wherever available. Lay out using maximum capacitor size as specified in the vendors' catalog.

7. Minimum spacing between integrated circuit die is 0.040 inch.

8. Unused inputs to devices should be connected as specified on the applicable circuit diagram.

9. Transistors should be positioned for unidirectional wire bonding whenever practical. Figure 10.19 illustrates acceptable bonding directions.

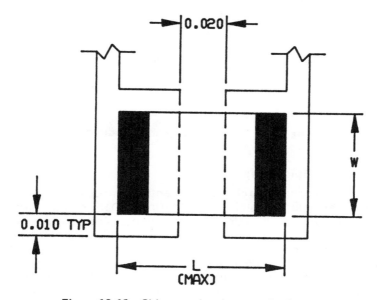

Figure 10.18: Chip capacitor layout criteria.

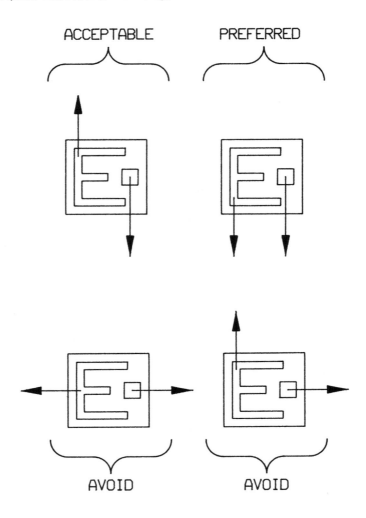

Figure 10.19: Transistor wire bonding directions.

Wire Bonding Guidelines

Circuit connections to chip components are made by wire bonds. The wire is either gold or aluminum. Gold wire 0.0015 inch in diameter is typically used for substrate-to-substrate jumpers and substrate-to-package pin-out connections. Aluminum wire 0.001 inch in diameter is typically used to make connections from chip components to thin-film substrates. Gold wire 0.001 inch in diameter is typically used to make connections from chip components to thick-film substrates. Both types are ultrasonically bonded. Table 10.6 lists the conductive characteristics of the wire bonds. See Chapter 7 for a detailed discussion on wire bonding.

Table 10.6: Wire Bond Characteristics

Wire Type	Size (in)	Rated Current (ma)	Resistance (ohm/in)	Fusing Current	Single Bond Resistance (mΩ) To Thick Film Au	To Thin Film Au	To Device Aluminum
Gold	0.0015	400	0.52	1.0 amp	–	–	–
Gold	0.001	250	1.20	0.5 amp	5	45	45
Aluminum	0.001	200	1.50	500–600 ma	20	30	60

Following is a list of guidelines that should be followed for wirebonding:

1. Wire bond lengths shall be 0.100-inch maximum, measured point-to-point on the layout.

2. The wire-bonding-machine operating clearance determines how close to relatively high components (package wall, chip capacitor, inductor, etc.) a wire bond can be made. When the bonding direction is perpendicular to such a component, the bond site must be as shown in Figure 10.20. The minimum distance can be significantly reduced by bonding the wire at an angle to the wall. All wires should be laid out parallel to package walls if possible.

3. Substrate bond sites for die-to-substrate wire bonds shall be a minimum of 0.020 inch from the edge of the die.

4. Wire bonding from die to die should not be permitted; all device wire bonds must terminate on the substrate. This is to allow the probing of any circuit node for fault isolation.

5. Only one lead may be bonded to a die pad; stitch bonding is not permitted.

6. Crossing over die with flying leads should not be permitted.

7. Wire bond sites should have a minimum dimension of 0.010 inch square.

Conditions to Avoid

Experience has shown that some assembly techniques should be avoided. Among these are:

a. "Piggy-back" mounting of components on substrate resistors.

b. Use of conductive epoxy to make electrical connections, in lieu of wire. Long add-on conductors runs should be made by using Kovar tabs mounted with nonconductive epoxy and/or Teflon-coated gold wires, or using jumper wire (0.0015 gold wire 0.100 inch long or less) ultrasonically bonded.

c. Use of solder-plated-pad thick-film chip resistors.

d. Handling devices and hybrids without proper grounding techniques.

e. Use of tin-lead solder and flux on circuits that have gold-aluminum wire bonds.

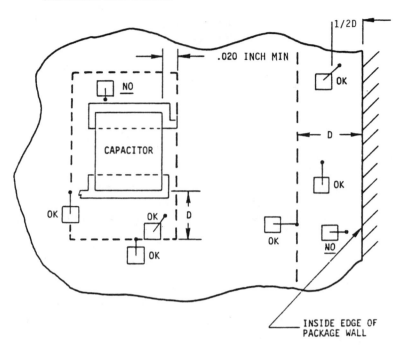

Minimum dimension "D" is established for bonding directly (perpendicular) toward a device or package wall.

Capacitor

"D" varies according to the capacitor height.

IF	height	= 0.040 inch	D = 0.070 inch
		= 0.050 inch	D = 0.085 inch
		≥ 0.060 inch	D = 0.100 inch

Package

"D" varies according to the package wall height.

IF wall height	≤ 0.040 inch	D = 0.100 inch
	≥ 0.126 inch	D = 0.085 inch

Figure 10.20: Wire bonding guidelines.

THICK-FILM MATERIALS AND PROCESSES DESCRIPTION

Thick-film circuits utilize screen-printed and fired conductors, resistors, and insulators on ceramic substrates. They are considered to be relatively rugged, have excellent power dissipation capability, and possess good operational characteristics. Two key advantages of thick-film over thin-film circuits are the capability of multilayering thick film to give high-density interconnects with z-direction conductor vias and the ability to apply a wide range of resistor values well into the megohm range.

Design guidelines generally tend to deal in terms of maxima or minima, but usually the optimum design is not based on either. For example, for a 1-inch-by-1-inch substrate which utilizes 0.020-inch-wide conductor lines and 0.020-inch-wide spaces, yields are close to 100 percent and the labor costs extremely low. However, for a 2-inch-by-2-inch substrate which utilizes multilayer 0.0075-inch-wide conductor lines, 0.0075-inch-wide spaces, and numerous resistors, yields are lower, perhaps by 20 percent, and labor costs increase sometimes by an order of magnitude. On a practical basis, the preferred minimum line width and spacing is 0.010 inch.

The following sections contain design guidelines which have been developed for a variety of situations. Naturally, each packaging layout contains its own constraints and, in these cases, the guidelines must be modified. These guidelines are intended to be compatible with the requirements for automatic resistor trimming and wire bonding.

Thick-Film Substrates

Normally, thick-film circuits are fabricated on substrates made from 96-percent alumina (Al_2O_3). The 96-percent-alumina content represents only the nominal percentage of alumina, with the actual content varying from slightly under 95 percent to slightly over 97 percent, depending on the supplier. All paste manufacturers use 96-percent-alumina substrates to check out their pastes, unless specifically requested to do otherwise. The 96-percent-alumina substrate is compatible with most thick-film pastes, and has a high flexural and compressive strength and good thermal conductivity—approximately one-seventh the thermal conductivity of aluminum. Where greater thermal dissipation is required, 99.5-percent beryllia should be used because of its high thermal conductivity (about the same as aluminum).

Substrates are available in either the "as-fired" condition or with ground surfaces. Substrates with all surfaces ground may cost up to five times as much as "as-fired" substrates. But with dense, fine-line circuitry, high resistor density, or because of system packaging considerations, grinding may be necessary, at least on the surface side to be screened.

A more detailed discussion of properties of thick-film substrates may be found in Chapter 2.

Substrate Holes and Machining. When required, the substrate supplier can install holes in the substrate prior to firing and delivery; the holes are punched into the "green" ceramic before firing. After firing, holes may have changed their relative position from a base line by as much as ±1 percent. For example, two holes located two inches apart before firing of a "green" ceramic could have the relative spacing between them change by as much as 0.040 inch. This change in relative position is caused by material shrinkage during firing, and this shrinkage is difficult to control uniformly across the substrate. For a premium, most suppliers will supply to ±1/2-percent tolerance on hole locations, depending on the configuration of the particular part. Hole diameters may vary ±10 percent except that the maximum tolerance should not exceed ±0.005 inch. Holes with better dimensional tolerance can be produced by drilling on fired ceramic using a diamond drill, a CO_2 laser, or an ultrasonic tool.

Diamond Drills. Diamond drills are relatively expensive and tend to degrade quickly. Small-diameter diamond drills are usually of solid construction, and the diamond tips break off rather easily. Above 0.040 inch diameter, diamond core (hollow) drills that have a longer life are used. Obviously, tool life and thus cost per hole is also a function of the depth of the holes (thickness of substrate) being drilled.

Ultrasonic Drilling. When an ultrasonic tool is used to drill fired ceramic, the costs are only slightly less expensive than when a diamond drill is used. However, when a large number of holes are required, or when precision hole locations (±0.005 inch) are necessary, the use of a multiple drill head on an ultrasonic tool is a good method. For example, 25 to 30 holes can be drilled through 0.020-inch alumina in approximately 30 seconds, while 200 holes can be drilled through 0.050-inch-thick alumina in 3 minutes. Because of expensive tooling costs, this method is economical only for high production runs.

Airbrasive Drilling. Airbrasively drilled holes will have an 8 to 10 degree taper.

Laser Machining and Drilling. Laser machinery and hole drilling have been particularly successful on many military substrates requiring numerous through-hole configurations. Laser-scribed substrates are commonly used so that several patterns can be processed at one time ("multiple" ups).

Occasionally it is necessary to radius the edge of a substrate or the circumference of a hole. This can be accomplished by tumbling (edges only), laser, liquid honing, or sand blasting (air abrading). Except for laser machining, these methods of material removal do not lend themselves to great accuracies, so care should be exercised when specifying their use.

Conversely, when radiused edges on substrates are not desirable, the maximum acceptable radius should be specified on the procurement document.

Substrate Dimensions. The following are guidelines for selecting substrate sizes:

a.   Use of substrates thinner than 0.020 inch is not advisable due to fragility and difficulty in handling. Thicknesses of 0.025 inch to 0.060 inch are generally used; selection of a specific thickness may depend upon the size (area) and shape required to resolve a specific packaging problem.

b.   Preferred size of substrates is 2-1/2 inches × 2-1/2 inches × 0.025 inch. However, 4 × 4 inch substrates are becoming popular. Off-the-shelf substrates as large as 5-1/2 inches × 7.0 inches are available. Larger area and/or thicker substrates can be purchased from suppliers; however, special runs and tooling are required.

c. A practical rule-of-thumb for determining the substrate thickness required for multilayer thick-film circuits is to allow a minimum of 0.025 inch of substrate thickness for each inch of length or width, whichever is longest, e.g., 0.025 inch thick when the longest substrate dimension is 1 inch, 0.050 when the longest substrate dimension is 2-1/2 inches.

Thick-Film Conductor Materials

Although a variety of conductor compositions are commercially available, only silver, silver alloys, gold, platinum-gold, and palladium-platinum-gold are extensively used. Silver thick-film conductor compositions, although relatively inexpensive, are in disfavor for military and high-reliability applications because of the potential for silver migration. Gold conductor compositions are used in those applications requiring eutectic and epoxy die bonding, thermocompression, thermosonic, or ultrasonic bonding. When tin-lead solder attachment is required, palladium-platinum-gold alloys and/or copper compositions must be used. It is important that these compositions have good adhesion to the substrate and exhibit a very low sheet resistivity. Typical tensile adhesion strengths range from 1,000 to 4,000 psi. The sheet resistivity of fired gold and copper compositions is less than 0.005 ohm per square, and the sheet resistivity of solder tinned platinum gold is less than 0.01 per square. For comparison, the resistivities of selected conductors are as follow:

Copper	2 to 5 milliohms/square
Gold	3 milliohms/square
Palladium-Gold	5 milliohms/square
Platinum-Palladium-Gold	25-60 milliohms/square
Solder-Coated Pt-Pd-Au	Less than 10 milliohms/square
Copper-Coated (plated) Pt-Pd-Au	5 milliohms/square

Thick-Film Resistors

Resistor pastes are available from many suppliers in sheet resistivities from 1 ohm/square to above 300 megohms/square. Sheet resistivities are available in increments of one and three decades. Sheet resistivities best suited for use are 100, 300, 1 K, 3 K, 10 K, 30 K, and 100 K. Higher (300 K, 1 M, 3 M, 10 megohm, 100 megohm, 300 megohm) and lower (1, 3, 10, 30) sheet resistivities are also available, but they have less desirable characteristics, so they should only be used where absolutely necessary. The one-decade sheet resistivities are off-the-shelf items, whereas the 3×10^x values must be blended and are subject to added lot charges.

Resistor tolerances should be as wide as possible though resistor pastes with tolerances tighter than ±1 percent can be requested because special stabilization techniques are required which greatly increase costs and lengthen fabrication time.

Resistor pastes can be reliably screen-printed and fired to approximately 20 percent of nominal value. For most applications, the resistors are purposely printed to a nominal value that is 20 to 30 percent below the required final value. The resistance is then increased by removing resistor material so as to decrease the effective width, either by air abrasive methods or by laser-beam vaporization. Typical thick-film resistor characteristics are given in Table 10.7.

Table 10.7: Thick-Film Resistor Characteristics

OHM/SQUARE/MIL	TCR (PPM/DEG C)		QUAN-TECH NOISE
	Type I	Type II	Max (db)
5	>+100		−10
10	±250		−20
30	±150		−20
100	±125		−20
300	±100		−15
1K	±100		−10
3K	±100		− 5
10K	±100	±150	0
30K	±125	±150	+10
100K	±150	±150	+15
300K	±350	±150	20
1M	±600	±150	+25
3M	±750	±250	
10M		±750	

Overcoating Resistor Surfaces. Resistors shall not be covered with fired dielectric and shall not be positioned under any component. Resistors may be coated with overglaze (fired) and/or approved organic coatings.

Overglaze Design Guidelines

Overglaze dielectric is noncrystalline, low-temperature-firing glass with or without a color pigment that remains visible after firing. This is the last material to be fired during the fabrication of the substrate.

Overglaze is used for many functions, some of which are listed below.

Circuit Identification. By screening overglaze with a color pigment on the circuit for identification purposes, the following applications are possible:

a. Print the part number

b. Identify a pad number

c. Identify a bonding direction

d. Identify a bonding pad

e. Show the routing of a specific conductor

f. Show the end points of a multilayered conductor and components with bodies which are not mounted directly to the substrate.

g. Identify static-sensitive devices.

Circuit Protection. Overglaze can provide environmental protection for single and multilayer circuits by acting as a barrier against solder, fluxes, solvents, adhesives, humidity, and other contaminants.

Protection for Gold Conductors. If a substrate is to be immersed in solder, a layer of dielectric or glaze shall be screened over gold conductors which are to be protected from the solder. An alternative approach would be to use platinum-palladium-gold conductors for areas to be soldered or exposed to solder.

Resistor Protection. Overglaze can be screened onto resistors when the designer wants to protect the resistor from the environment and/or from solder and flux. Overglaze should overlap all resistor and conductor protected areas by a minimum of 0.0075 inch. Figure 10.21 shows the preferred overlap.

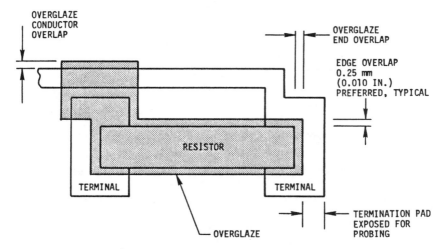

Figure 10.21: Overglaze overlap for resistor and conductors.

Solder Control. Overglaze can be used to control the flow of solder during solder screening and flowing operations, preventing shorting of adjacent conductors.

Sealing Surface. Overglaze used to provide a sealing surface for epoxy mounting a cover shall have dimensions as shown in Figure 10.22. The width of the overglaze surface insures the placement of the cover without assembly difficulty. Tolerances on the dimensions of the cover must also be included when defining the sealing surface width.

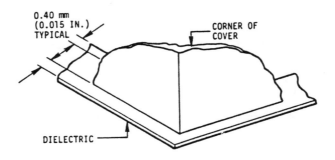

Figure 10.22: Sealing surface.

Short Prevention. Overglaze can be used on chip and wire circuits to prevent bond shorts as shown in Figure 10.23.

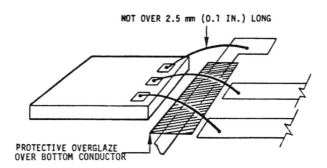

Figure 10.23: Prevention of electrical shorting.

Chip-Carrier Overglaze Design. Figure 10.24 illustrates the use of the overglaze around a chip-carrier component. The conductor patterns are shown for reference. The overglaze is used to prevent the solder from flowing away from the chip-carrier pads.

By controlling the solder to the region of the pad, uniformity in the size of the solder fillets can be maintained.

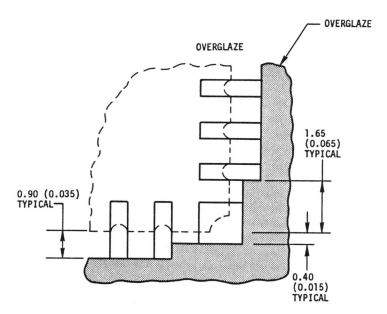

Figure 10.24: Chip-carrier overglaze design.

Figure 10.25 shows a poor overglaze design. Here, the solder is allowed to flow away from the three pins. Thus, the fillet size and reliability will not be the same as on the other pads.

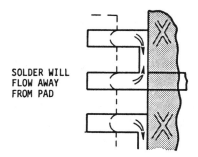

Figure 10.25: Poor overglaze design.

Solder Application

Solder can be added to thick-film substrates by dipping, by hand application, and by printing. The following presents the design guidelines for the printing of solder.

Solder Printing Line Width—Uniform solder wetting of a conductor line depends upon the width of the solder paste patterns, the emulsion thickness, conductor composition, and the flux in the solder paste, as well as the reflow parameters.

Artwork for depositing screenable solder should be the same as for the metallization artwork, perhaps 0.0025 inch less.

Thick-Film Dielectrics

High dielectric constant (K) insulator compositions (as high as K = 1,200) are used to make capacitors, and low K insulator compositions (K = 9 to 15) are used to provide insulation between conductors.

Although the thick-film process provides good general-purpose capacitors, it is usually not practical to screen thick-film capacitors; they utilize too much substrate area and require considerable additional processing. Usually chip capacitors are more suitable. Thick-film capacitors are normally utilized only where the end product would consist primarily of capacitors, such as in delay lines.

Low K dielectric compositions are used to form either insulated crossover or multilayer circuitry. The use of numerous individual insulated crossovers, in lieu of multilayer approach, is not recommended when resistors are to be used because irregular topography results, causing large dispersions in resistance values.

THICK-FILM DESIGN GUIDELINES

When making a thick-film layout, various items must be added to the artwork to make fabrication easier. These items, such as alignment marks and artwork identifiers, are described in the following sections.

Artwork and Drawing Requirements

Alignment. Alignment marks are specified to cover screen pattern alignment and alignment of deposition levels during circuit fabrication. A "+" shall be used in diagonally opposite corners of all artwork circuit areas to align successive levels (see Figure 10.26). The dielectric artwork(s) shall have corresponding "+" shaped windows in two corners. It is preferable that the "+" be designed using 0.010-inch-wide lines and a total length of 0.040 inch, if space permits. Artwork for each sheet resistivity shall have an alignment square that fits into one corner of the conductor "+" alignment mark. The sheet resistivity, to be screened first, shall have a square that conforms to the No. 1 space in the bottom left corner of the crosses. The second sheet resistivity square will fit into the No. 2 corner of each cross.

Center Lines for Emulsion Patterns. Center-line marks approximately 0.003 inch wide by 1.0 inch shall be placed on each piece of artwork as shown in Figure 10.26. These marks are for locating artwork when making exposed emulsion patterns on screens. They should be positioned at least 1/2 inch away from the active substrate area.

"F" Indicator and Part-Number Alignment System. The letter "F" indicator is a nonsymmetrical letter which provides a quick way to insure correct orientation of 1 X artwork and screen relationships on all levels. The

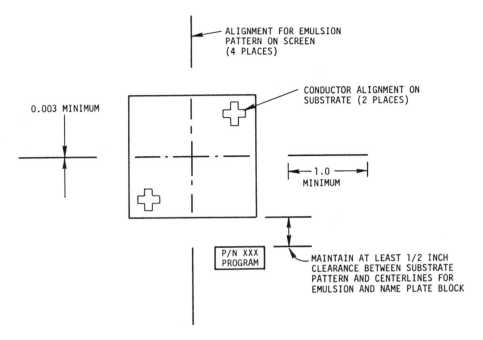

Preferred Conductor Alignment Method and Alignment for Positioning Artwork on Screen Emulsions

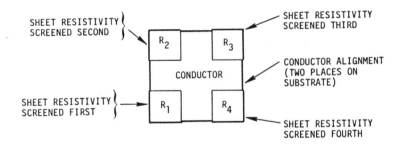

Figure 10.26: Alignment marks.

"F" indicator is designed into each level of artwork at the same location (for that design). The "F" is not used for critical level-to-level alignment during printing. As each level is overprinted on the "F", the edge definition and the subsequent alignment become poorer. No "F's" need be used on the resistor or dielectric artworks.

As an alternative, the part number could be used in place of an "F" on the level 1 conductor on the substrate and used in combination with a "+" to provide alignment for conductors.

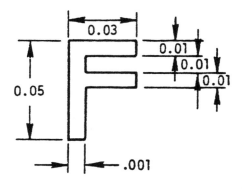

(All dimensions are in inches.)

Figure 10.27: Letter "F" alignment indicator.

Artwork Identification and Requirements. Separate artwork is required for each level of conductors, dielectrics, and resistors (i.e., for each sheet resistivity) and for solder paste, conductive epoxy, nonconducting epoxy, overglaze, and via fills. All artwork shall have the correct part number and configuration, as well as program identification. Each sheet of artwork for a particular circuit shall be numbered consecutively, starting with Sheet 1, Sheet 2, etc., in the order required for processing as shown in the following examples. The sequence for resistor processing normally is most economically performed when the sheet resistivity with the largest number of resistors is processed first and the last sheet resistivity to be processed has the fewest number of resistors. One-to-one artwork shall be "read right emulsion" (RRE) positives and shall be on 0.004-inch thick polyester film. The black areas of the positive artwork are the areas in which material is deposited onto the substrate. Two sets are required for each sheet. The black areas on the artwork should be on top of the polyester sheet.

Examples:

   (a) *Conductor on Substrate*

       P/N XXXX-X Sheet 1
       Name: Level 1 Conductor Au
       Date: _____
       Program: HYBRID

   (b) *Dielectric over Level 1 Conductor*

       P/N XXXX-X Sheet 2
       Name: Level 1 Dielectric
       Date: _____
       Program: HYBRID

(c) *Conductor over First-Level Dielectric*

> *P/N XXXX-X Sheet 3*
> Name: Level 2 Conductor Pt-Au
> Date: _____
> Program: HYBRID

(d) *Resistors*—Separate artwork is required for each sheet resistivity. The sheet number for the first resistor sheet resistivity value would follow that of the top dielectric level over the top conductors. Example for resistors of sheet resistivity of 30 kilohms/sq:

> *P/N XXXX-X Sheet 4*
> Name: Resistor 30K ohms/sq
> Date: _____
> Program: HYBRID

As mentioned earlier, the sequence for resistor sheet numbers would be to start with that sheet resistivity that has the largest number of resistors and to finish with the sheet resistivity that has the fewest number of resistors. Following this sequence will help improve the yield on the parts.

(e) *Conductive Epoxy for Device Attachment*

> *P/N XXXX-X Sheet _____*
> Name: Conductive Epoxy
> Date: _____
> Program: HYBRID

(f) *Epoxy for Solder Masking and/or for Insulating Material*

> *P/N XXXX-X Sheet _____*
> Name: Nonconductive Epoxy
> Date: _____
> Program: HYBRID

(g) *Solder Paste*

> *P/N XXXX-X Sheet _____*
> Name: Solder Paste
> Date:
> Program: HYBRID

Multilayer Yields

Although there is no theoretical limit to the number of levels of conductor metallization that can be fabricated, judgment should be exercised because as the number of conductor levels increase, yields decrease.

Multilayer process yield is dependent upon each of the following items *and* upon the interrelationships among these items:

Number of conductor levels

Via size

Via space

Conductor width

Conductor space

Conductor Patterns—General Considerations

(a) Maximize conductor densities on the lowest possible conductor level so as to minimize complexities on the upper conductor levels.

(b) Avoid adjacent parallel conductor runs on different levels to reduce parasitic capacitance effects.

(c) During design and layout of bond sites, adequate bonding tool clearance between the bond sites and surrounding obstructions must be provided for assembly and rework.

(d) Minimum width and spacings shall be utilized only where necessary.

Conductor Orientation—For best results, all conductors less than 0.010 inch wide should be oriented parallel or perpendicular to each other and to the edges of rectangular substrates. Conductor angles at 45 deg may also be used, but are difficult to consistently print and should be avoided.

Conductor Width—Minimum conductor width of most hybrid houses is 0.0075 inch (0.010 inch preferred). Conductors to be soldered should not be less than 0.010 inch wide in order to minimize leaching and enhance reworkability.

Conductor Spacing—Minimum spacing between conductors should be 0.0075 inch (0.010 inch preferred). Conductor spacing should be equal to, or greater than, any adjacent conductor width that is less than 0.010 inch. Conductors running parallel to the edge of a substrate should be at least 0.015 from the edge of the substrate. Conductors running perpendicular to the edge of the substrate should be at least 0.0075 inch from the edge.

Conductor Layer Limits. Multilevel substrates without resistors should be limited to four conductor levels. Increasing the number of levels will typically lower the substrate fabrication yields. However, in some cases, an extra level may be warranted as a "trade-on" to avoid a situation with an even more negative yield impact (such as use of 0.010-inch vias versus the use of preferred 0.015-inch vias, with the extra conductor level). Generally, multilayers with resistors should be limited to a maximum of two levels of dielectric.

Dimensional Relationships. Table 10.8 presents the dimensions for improved yield and for multilayer design minimums.

Parallel Conductors on Adjacent Levels. Parallel conductors on adjacent levels are not permitted. Avoid stacking conductors in the same direction on adjacent levels. If it is not possible to design the circuit with perpendicular conductors on adjacent levels, then parallel conductors should not run on top of one another in the same direction.

Table 10.8: Multilayer Dimensional Relationships

	Improved Yield Designs	Absolute Design Minimums
Conductor Levels	3 or less	
Via Size	0.020 inch	0.015 inch
Via Configuration	Square	Rectangular
Space Between Via and Any Other Conductor Feature	0.015 inch and greater	0.010
Conductor Width		
Level 1	0.015 inch	0.0075 inch
All other levels	0.015 inch	0.0075 inch
Conductor Space	0.015 inch	0.0075 inch

Ground and Voltage Planes and Inner Level Metallization. Design ground and voltage planes using a grid pattern rather than a solid layer. This saves a significant amount of conductor paste. It is preferable to keep inner metallization levels as flat as possible in order to have reasonably flat surfaces on the top level. An air-fired multilayer, which will be subsequently soldered, should use one of the gold-platinum-palladium alloy pastes for the top conductor (or any areas exposed to solder). The inner protected conductor layers should use gold alloy pastes in order to provide the lowest electrical resistances.

Conductor Spacing from the Edge of a Dielectric. The spacing between a conductor and edge of the dielectric on which it is screened, or along the edge of a window in the dielectric (device installation area and/or resistor deposition area), should be 0.0075 inch minimum. This spacing decreases the possibility of conductor material flowing into the window and safeguards the conductor from damage during device installation, testing, or removal.

Conductor Lines from One Level to the Next Lower Level. When a continuous conductor is required to go from one level to the next lower level, the upper conductor width should be decreased at the edge of the dielectric, preferably to 0.0075 inch if the upper conductor is at least 0.010 inch wide. If the conductor width cannot be decreased, consider insulating the lower conductor's lines from adjacent conductors with dielectric in between them, or maintain 0.010-inch separation (Figure 10.28).

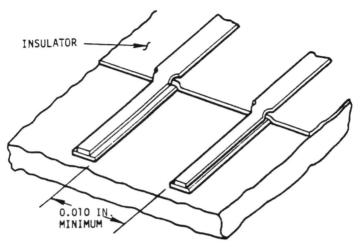

INSULATOR

0.010 IN. MINIMUM

Figure 10.28: Interlevel connections at edge of insulator.

Overlapping Different Conductor Metallizations. It may be occasionally necessary to use a conductor line composed of different metallizations in different areas, e.g., gold (for wire bonding) overlapping a platinum-gold conductor to be used for soldering attachments. In this case, one conductor should overlap the other by 0.0075 to 0.010 inch. The gold conductor should be protected from solder by a dielectric or overglaze insulating layer at least 0.0075 inch from the exposed gold overlap areas (Figure 10.29).

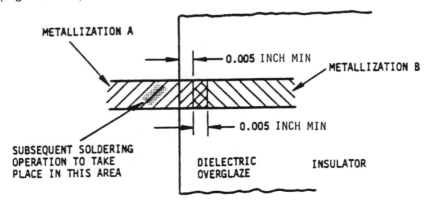

METALLIZATION A

0.005 INCH MIN

METALLIZATION B

0.005 INCH MIN

SUBSEQUENT SOLDERING OPERATION TO TAKE PLACE IN THIS AREA

DIELECTRIC OVERGLAZE

INSULATOR

Figure 10.29: Transition between different metallizations.

Double (Superimposed) Conductor Layers. In order to decrease conductor line resistances, and/or increase resistance to solder exposure, it may be desirable to have one conductor layer superimposed on a lower layer. Except at vias, the upper conductor layer should be 0.003 to 0.004 inch less in width than the lower conductor layer.

Probe Pads. Resistor Probe Pads (Automated Probing)—Probe pads will be provided where necessary for resistor trimming or checking of the circuits (Figure 10.30). Where possible, probe pads will be located toward the outside perimeter of the circuit so that probing of one pad will not be prevented by the space required by another probe. Probe cards are used for automatic high-speed resistor trimming. The probe card leads have minimum dimensional requirements and require positioning so that they do not shadow (interfere with) the laser beam during trimming.

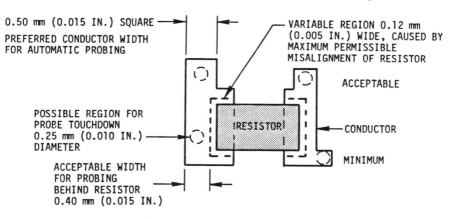

0.50 mm (0.015 IN.) SQUARE

PREFERRED CONDUCTOR WIDTH
FOR AUTOMATIC PROBING

VARIABLE REGION 0.12 mm
(0.005 IN.) WIDE, CAUSED BY
MAXIMUM PERMISSIBLE
MISALIGNMENT OF RESISTOR

ACCEPTABLE

POSSIBLE REGION FOR
PROBE TOUCHDOWN
0.25 mm (0.010 IN.)
DIAMETER

RESISTOR

CONDUCTOR

MINIMUM

ACCEPTABLE WIDTH
FOR PROBING
BEHIND RESISTOR
0.40 mm (0.015 IN.)

Figure 10.30: Four possible probing locations.

Vias—Interlevel Conductor Connections Through Multilayer Dielectric

Vias are used for interlevel metallization connections through dielectric. Vias should not be used for wire bond sites.

Noncongruent Vias. Interlevel connections (vias) shall be made only to adjacent levels of metallization, i.e., two separate conductor levels with just one dielectric level between them. Vias making a connection through more than one dielectric level shall be staggered vertically from level to level (see Figure 10.31). Vertical vias through more than one layer of dielectric can be accomplished if a separate screening (via fill) is performed to fill the vias.

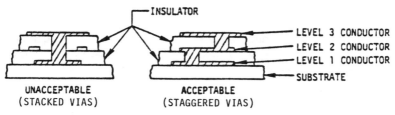

INSULATOR

LEVEL 3 CONDUCTOR
LEVEL 2 CONDUCTOR
LEVEL 1 CONDUCTOR
SUBSTRATE

UNACCEPTABLE
(STACKED VIAS)

ACCEPTABLE
(STAGGERED VIAS)

Figure 10.31: Interlevel via design.

Minimum Size of Vias. The minimum size of vias should be 0.015 inch × 0.015 inch square. Round vias shall not be used. Square vias are preferred over rectangular shapes.

Staggered Vias on the Same Level. Vias on parallel conductors should be staggered rather than arranged in even steps from conductor to conductor on the same level. Vias on separate conductors shall be separated by at least 0.010 inch. This rule applies to vias in the same conductor level, adjacent levels, and levels two away from it. Once separated by more than two levels, the rule in inapplicable. Staggered vias are illustrated in Figure 10.32.

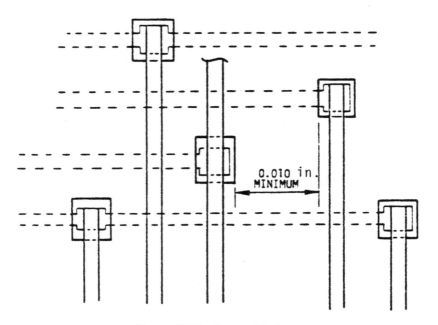

Figure 10.32: Staggered vias.

Minimum Conductor Size at Base of Via. The underlying conductor at the base of a via should be at least 0.005 inch wider than the vias, or 0.0025 inch away from the via on each side.

Substrate Holes. Holes may be machined in substrates, or punched in the "green" tape before the substrate is fired, to electrically interconnect circuitry on both sides, or to mechanically attach the substrate. For electrical connection the holes are filled with conductor metallization. Conductive through-holes shall be at least 0.005 inch greater than component leads (maximum tolerance) to allow for the thickness of metallization (including solder, when used). Conductor pad areas around holes in the substrate shall be at least 0.020 inch greater in diameter than the hole.

Wire and Die Bonding Pads

Positioning of Wire-Bonded Parts and Leads. During layout for wire-bonding (flying lead) parts, care should be exercised that the bond sites are positioned so that the leads will have minimal possibility for electrically shorting out circuitry on the dice. Also, to facilitate bonding, bond sites should be positioned to minimize the number of directions of wire lay.

Orientation of Bonding Pads. Die bonding pads should be designed to have the pad edges parallel to the edges of rectangular substrates. This facilitates assembly and rework time by greatly simplifying alignment; thus, costs are reduced.

Bonding Tool Clearance. During design and layout of bond sites, adequate tool clearance between the bond sites and other obstructions must be provided for assembly and rework.

Minimum Size of Wire Bond Pads. When possible, make wire bond pads 0.015 X 0.015 inch, minimum. The absolute minimum for automatic wire bond pads is 0.012 X 0.012 inch.

Minimum Dielectric Openings for Bonding Pads. Openings in a dielectric to expose a bonding pad shall be 0.020 X 0.020 inch minimum.

Minimum Distance of Bonding Pads from Vias. A bonding pad which is connected directly to a via metallization shall be placed a minimum of 0.005 inch away from the outer edge of the via.

Minimum Spacing Between Edge of Pad and Component. The near edge of a bonding pad should never be closer to any component or package wall than the height of that component.

Bonding Pads and Package Pins. Bonding pads to be interconnected to the output pins of the package should always be placed on the perimeter of the circuit substrates.

IC Wires. Where possible, the wires from ICs should be radial from the perimeter and of approximately the same length.

Design of Conductor Pads. Conductive Pads for Components with Conductive Bases. Discrete components requiring direct attachment and having a bottom electrical contact shall be mounted on a conductive pad designed for conductive epoxy, solder, or eutectic mounting. Conductive pads over dielectric shall have at least two or three separate layers of dielectric under the pad for air- or nitrogen-fired systems, respectively.

Sealing-Ring Metallization for Cover (Lid) Attachment. If the thick-film substrate is to serve as the base of the package, a seal ring can be screened on the substrate so that a metal cover can be attached directly to the substrate. Design dimensions for sealing rings are a function of the cover being installed and the proximity of the adjacent circuitry. Best results are obtained when the sealing-ring sides are oriented parallel to the sides of rectangular substrates. The sealing ring should be wider than the flange of the cover to allow for adequate solder filleting, perhaps as much as 0.010 inch on both the inside and the outside edges of the cover flange.

Sealing rings are generally fabricated from platinum-gold conductor compositions. Where possible, the sealing ring should enclose all circuitry to minimize possible damage to circuitry during lidding and delidding operations. Also, to avoid damage from solder splatter, the sealing ring should be separated from adjacent circuitry by 0.040 inch minimum, or adjacent circuitry may be covered with insulator.

Terminal Pad for Soldering. Pads for attaching external leads to a substrate that serves as the package base shall be as large as real estate permits. Pad widths shall be 3X the lead diameter or 1.9 mm (0.075 inch), whichever is greater. Preferred width is 2.5 mm (0.1 inch). Pad lengths shall be a minimum of 1.9 mm (0.075 inch). Preferred length is 2.5 mm (0.1 inch). The minimum spacing between the edges of adjacent lead pads shall be 0.5 mm (0.020 inch). Preferred spacing is 0.75 mm (0.030 inch). Terminal pads are illustrated in Figure 10.33. Terminal pads over dielectric shall have at least two layers of dielectric underneath for air-fired systems and at least three layers for nitrogen-fired systems.

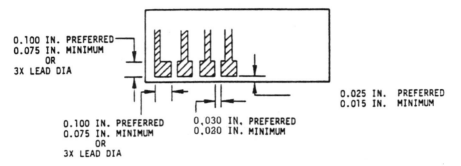

0.100 IN. PREFERRED
0.075 IN. MINIMUM
OR
3X LEAD DIA

0.025 IN. PREFERRED
0.015 IN. MINIMUM

0.100 IN. PREFERRED
0.075 IN. MINIMUM
OR
3X LEAD DIA

0,030 IN. PREFERRED
0,020 IN. MINIMUM

Figure 10.33: Terminal pad for soldering.

Axial Lead Component Pads. Axially leaded component pads shall be designed per Figure 10.34. Note the acceptable alternate pad position for axial leads bent in opposite directions.

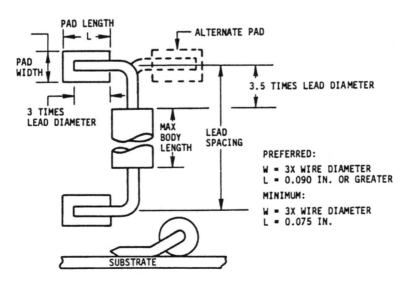

Figure 10.34: Axial lead component pad.

Chip-Carrier Pads. Figure 10.35 shows the plan view of a chip-carrier component. Figure 10.36 represents the design of the conductor pads on the substrate for the component. Even when a pin-out is not required, the conductor pads will be designed into the circuit.

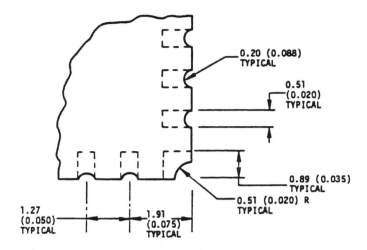

(Numbers in parentheses are in inches; numbers not in parentheses are in mm)

Figure 10.35: Chip-carrier component.

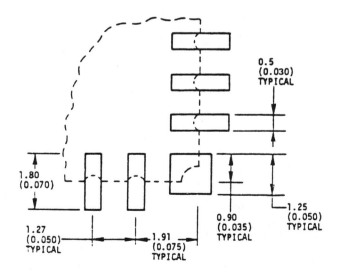

(Numbers in parentheses are in inches; numbers not in parentheses are in mm)

Figure 10.36: Chip-carrier pads on substrate.

Chip-Carrier Conductor. Figure 10.37 illustrates the conductor design for a chip-carrier component. Important items in the design are:

a.   Side pad conductors are at least 1.78 mm (0.070 inch) long.

b.   Corner pad conductors can proceed outward in any direction, but the 1.27-mm (0.050-inch) pad size *must* be maintained.

c.   Pads may be connected. However, dielectric or overglaze *must* be used to limit solder flow, and possible shorting, between pads under the base of the chip carrier.

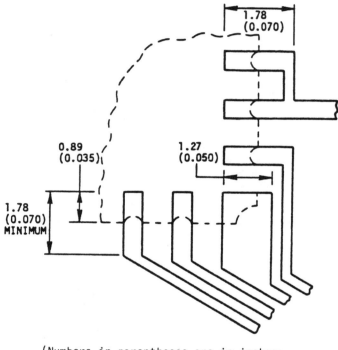

(Numbers in parentheses are in inches;
numbers outside of parentheses are in mm.)

Figure 10.37: Chip-carrier conductor example.

Thick-Film Resistor Design Guidelines

Resistor Configurations. The most useful and easy-to-process resistor has a rectangular configuration. "Top hat" (lobe type) resistors are often used where space limitations prohibit adequate resistor length in the direction of current flow, or where dynamic trimming over a significant range is required. "U" (folded resistor) and "L" shaped resistors are not recommended because they are more difficult to deposit and to adjust.

Aspect Ratio. The resistance of a specific resistor with a given resistivity is dependent on its aspect ratio, which is the ratio of its length to its width, expressed as a number of "squares." The resistor "length" is always the dimension of the resistor which is parallel to the current flow (Figure 10.38). The aspect ratio of most thick-film resistors generally range from 1/6 to 4, which limits the spread of available resistor values to 24:1, for rectangular resistors of a given resistivity.

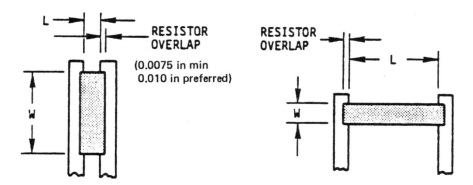

Figure 10.38: Rectangular resistors.

Rectangular Resistors—

$$\text{Aspect ratio} = \frac{\text{Length}}{\text{Width}}$$

Length is always that direction which is parallel to the current flow.

*Top-Hat (Lobe Type) Resistors—*Avoid designing a "top hat" resistor except in the following situations:

a. When the resistor tolerance is greater than ±5 percent, e.g., ±10 percent, ±20 percent, etc.

b. When a power level of 25 watts/inch^2 can be used without sacrificing space.

When used, a "top hat" resistor shall be designed as indicated in Figure 10.39. Top-hat resistors should be designed at 25 watts/inch^2 with tolerances of ±5 percent or greater.

The advantage of a top hat is its trim range. Shown is a 5-square top-hat resistor. The disadvantages are: hot spots and, sometimes, extra stabilization time is required during the trimming operation (especially for tighter tolerances).

By trimming the resistor as shown in Figure 10.39, a 9-square resistor can be made from a 5-square resistor. The resistance has been increased by 80 percent by removing only 15 percent of the total area. The trim areas in Figure 10.39 are exaggerated for illustrative purposes.

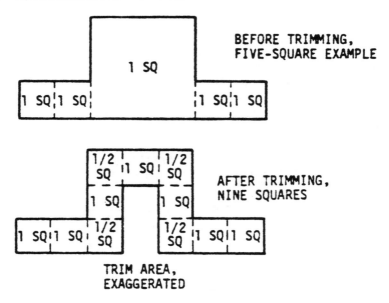

Figure 10.39: Top-hat resistor.

Bent (L) Design—The best resistor is a square or rectangle. However, if a resistor of greater than ±1 percent tolerance cannot be fitted into a specific area, it may be bent. In this situation (to be avoided if possible), allow the corner area to count one-half the resistivity of a straight area. Reference Figure 10.40.

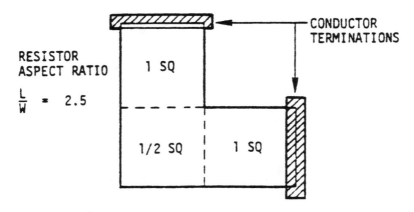

Figure 10.40: Bent resistor design.

Number of Sheet Resistivities. Design the resistors using sheet resistivities such that the number of sheet resistivities required on a substrate is minimized. It is preferable to limit the number of sheet resistivities on one substrate to four. Although geometric similarity among resistors decreases the variance in resistor values of the as-fired (unadjusted) resistors, it is more important to keep the number of sheet resistivities to the lowest number possible. This reduces the number of screening operations which decreases fabrication costs and increases yield.

Minimum Length. The minimum length of a resistor shall be 0.040 inch. Resistors should not utilize an aspect ratio greater than 10 because the aspect ratios are difficult to adjust. On extremely short resistors, it is difficult to maintain the uniform sheet resistivity expected from aspect ratio calculations. This is due to the greater relative effect of the contact resistance at the resistor terminations. This is attributed to material diffusion processes between the resistor material and the resistor termination material.

Minimum Width. Minimum resistor width is 0.020 inch; the preferred minimum width is 0.040 inch. On extremely narrow resistors, it is difficult to maintain the uniform thickness required for use of aspect ratio calculations. The rounded edges of the resistors form a greater percentage of the cross-sectional area as the width decreases.

Incremental Steps in Dimensions. A resistor should have its dimensions as multiples of 0.127 mm (0.005 inch). This is to simplify the layout and artwork generation. The larger dimension of the resistor will be determined by the aspect ratio which is required to achieve the resistance needed for the paste being used.

Resistor Clearance. The area on the base substrate within 0.015 inch of a resistor shall be free of all insulator and conductor materials except for the resistor terminations and the resistor test probe locations. The clearance between a resistor and the edge of the substrate shall be 0.75 mm (0.030 inch) minimum.

Resistor Overlap. The minimum resistor/conductor overlap shall be 0.0075 inch. Preferred overlap is at least 0.010 inch. Do not completely overlap the conductors at the end of the resistor deposition because the resistor composition will flow down and spread out, possibly causing shorting to adjacent metallization. Resistor patterns shall be designed so that different sheet resistivities do no touch each other.

Orientation. Orient all resistors, without exception, in either the X or Y direction parallel to the substrate edges (Figure 10.41). Attempt to orient all resistors in the direction of squeegee travel (from termination to termination). Any resistor whose long axis is perpendicular to the direction of squeegee travel will have a higher fired value.

Conductor Terminations. Conductor terminations of resistors shall always be on the same level as the resistors (all level 1 conductors, i.e., on the substrate).

Resistor Windows in Multilayer and Cross-Over Circuits. A resistor window in the dielectric level(s) shall be at least 0.010 inch wider than each side of the resistor. The spacing helps resistor trimming and deposition.

Height of Surrounding Topography. The height of resistors can be no less than the surrounding topography formed by the rest of the circuit materials (plus the offset of the screen). It is very important that the height of all materials around the resistors be kept at the same value. For example,

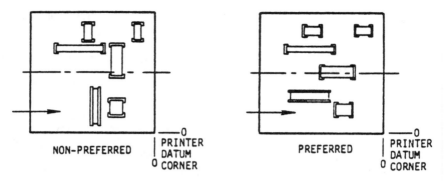

Figure 10.41: Resistor orientation.

on multiple substrates, previously laser scored, there are surrounding substrate areas that will be thrown away once the score lines are broken. A resistor next to a score line, adjacent to the throw-away ceramic, would be without a buildup on one side. In this case, uniform resistor thickness can be provided by depositing a dielectric layer on the disposable ceramic surface, which would be outside the normal artwork areas for the circuit.

Maximum Multilayer Levels with Resistors. When resistors are screen-printed last through apertures in multilayer dielectric, the thickness of the resistor increases with each added level of dielectric. A two-level multilayer is generally considered to be the limit with respect to keeping resistors at a thickness which is easy to trim and does not require added stabilization. Dried thicknesses of resistors (before firing) are approximately 40 microns (1.5 mils) for a two-layer circuit.

Closed Loops. Closed resistor loops should be avoided so that resistors can be measured and trimmed individually. If a resistor loop is necessary, the loop must be left open until all resistors in the loop have been trimmed. It may then be closed with a wire bond, a small-gauge wire or ribbon, or a soldered jumper. The metallized pattern on each side of the gap must be designed to provide an acceptable pad for whichever bridging technique is selected. See Figure 10.42. Resistor patterns shall be designed so that different sheet resistivities do not touch each other.

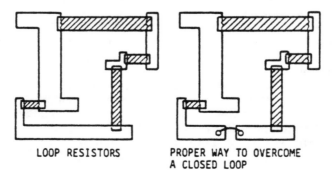

LOOP RESISTORS   PROPER WAY TO OVERCOME
                    A CLOSED LOOP

Figure 10.42: Trimming closed loops.

Resistor Calculations. The basic equation for calculating resistance is:

$$R = \frac{sL}{W}$$

where

  R = resistance, ohms
  L = length of resistor
  W = width of resistor
  s = sheet resistivity of paste, ohms/square/mil (resistor paste sheet resistivity is usually based on 1-mil thickness). Note: The term "square" (or "sq") is dimensionless.

This equation can also be expressed as:

$$R = \frac{\rho L}{tW}$$

where

  t = thickness of resistor, inches
  ρ = bulk resistivity, ohms-inches

Figure 10.43 depicts the relations of the above dimensions. This resistor has four squares of resistive material. If the resistivity of the film is 1,000 ohms per square, then the resistance is 4,000 ohms.

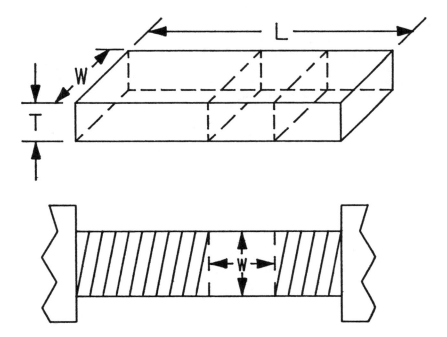

Figure 10.43: Four-square resistor configuration.

Calculation Example—Assume an ink of sheet resistivity 1,000 ohms/ square (s = 1,000 ohms/sq/mil), a length of 0.100 inch, and a width of 0.050 inch (L = 0.100 inch, W = 0.050 inch). Calculate the resistance

$$R = \frac{sL}{W} = \frac{1,000 \text{ ohm/sq} \times 0.100 \text{ in}}{0.050 \text{ in}} = 2,000 \text{ ohm}$$

Resistance and Thickness Considerations. The previous calculation assumes that all resistors have a dried thickness (before firing) of 0.001 inch. This is reasonably true only for a nonmultilayer circuit. The present state of the art requires that resistors be deposited only on the ceramic substrate and that they be fired only once. As a consequence, resistors shall be limited to one side of the substrate and, for multilayers, they must be deposited onto the ceramic through openings (windows) in the surrounding dielectric, which will always result in a dried thickness greater than 0.001 inch. Resistor paste compositions are blended to compensate for added thicknesses in order to provide the design resistance values. These blends are controlled and tested, as are the original "as received" pastes.

TCR Tracking. Worst-case TCR tracking is the sum of the TCR tracking of the resistor members. Tracking between different resistivities must be experimentally determined and is subject to lot-to-lot variation in the paste within specification limits. To obtain best tracking characteristics, resistors should be deposited from one sheet resistivity in tandem (in series, in a straight and continuous pattern laid end to end) with a common center conductor. Tracking between resistors of the same sheet resistivity on the same substrate will be within ±5 ppm/deg C. Special designs will track within ±1 ppm/deg C.

Resistors which must track each other must be designed with as low a power dissipation factor as practical and having the same geometry. The following equations show the relationship between the resistors so that they have the same geometry/square area.

Assume two resistors R_1 and R_2

$$R_1 = \Omega/\square \times L_1/W_1$$

$$R_2 = \Omega/\square \times L_2/W_2$$

Divide the first equation by the second.

$$\frac{R_1}{R_2} = \frac{L_1/W_1}{L_2/W_2} = \frac{L_1 W_2}{L_2 W_1} \tag{1}$$

Since the areas must be equal

$$L_1 W_1 = L_2 W_2 \tag{2}$$

Let n = the ratio of the value of R_1 and R_2.

$$n = \frac{L_1W_2}{L_2W_1} \qquad (3)$$

$$nL_2W_1 = L_1W_2$$

from Equation 2

$$W_1 = L_2W_2/L_1$$

Substituting into Equation 3

$$\frac{nL_2L_2W_2}{L_1} = L_1W_2$$

$$nL_2^2 = L_1^2$$

or
$$L_2 = L_1/\sqrt{n}$$

and
$$W_2 = W_1\sqrt{n}$$

Example

R_1 = 10K	L_1 = 0.5 inch	W_1 = 0.010 inch
R_2 = 20K	L_2 = unknown	W_2 = unknown

$$n = \frac{10}{20} = 1/2$$

Therefore, $L_2 = \dfrac{L_1}{\sqrt{n}} = \dfrac{0.5}{\sqrt{0.5}} = 0.707$ inch

and $\quad W_2 = 0.010\sqrt{0.5} = 0.00707$ inch

This example represents two resistors, one twice the value of the other but designed to cover the same area, thus giving the best tracking stability between the two.

Power Rating. Screened resistors (5 ohms to 10K ohms) can be rated for a dissipation of 50 watts/square inch, but they shall be rated at 25 watts/square inch for design applications to permit a 50-percent reduction in the trimmed path during resistance adjustment. See appropriate resistor specifications for power ratings of resistors above 10K ohms.

$$P \text{ (watts)} = L \text{ (in.)} \times W \text{ (in.)} \times 25 \text{ watts/sq inch}$$

where

L = the useful resistive length measured along the direction of current flow

W = the resistor width

A more useful version of the above formula might be

$$P_d = \frac{P}{L \times W}$$

where

P_d = the rated power dissipation in watts/square inch (25 watts/ square inch)

P = the actual resistor power dissipation in watts

Densities. The maximum resistor density per substrate depends directly on the allowable power dissipation per substrate. It is generally considered that 25 resistors per square inch, or 50 resistors per substrate are reasonable maxima.

Three-Year Operating End Points. Three-year operating end-points in a temperature range of −55 to +125°C are normally expected to remain within +1 percent/−0.5 percent of the initially adjusted resistance value.

Five-Year Non-Operating End Points. Five-year non-operating end-points in a temperature range of −55 to +125°C are normally expected to remain within ±0.2 percent of the initially adjusted resistance value.

Preproduction Design Verification. It may be difficult to design resistors to attain the desired resistivities because of contact resistance, the effect of widely varying widths, and non-uniform substrate topography created by conductors, insulators, and other resistors. Usually, a few circuits of any layout can be made in the laboratory using special techniques, but when production quantities are anticipated, it is recommended that time be allowed for redesign of resistor geometries. This redesign period is best accomplished prior to the official release of drawings.

Deposition Surface. Resistors may be deposited either on the substrate or top dielectric. Where power dissipation is an important concern, it is desirable to screen print the resistors directly onto the substrate. Generally, resistor characteristics (TCRs, value, etc.) may change when deposited on dielectric and should be verified during the engineering phase. Multilayers with resistors should be limited to a maximum of two levels of dielectric.

THIN-FILM GUIDELINES

Standard Practices

Layout Preparation. Layouts are designed on a rectangular grid, usually at 20 times actual size. The lines on the 20X grid fall on 0.1-inch centers, which produce an actual-size minimum dimensional resolution of 0.005 inch. The layout is a manually prepared or computer assisted drawing on which chip components are positioned and interconnected and substrate resistors and conductors are designed.

Artwork and Tool Generation. The artwork tools for substrate fabrication are made by photographically reducing a digitized circuit layout to actual size. A digitized layout is a computer-aided design tool. It is prepared from the layout design by encoding the layout in a digital format. An interim artwork tool, usually at 4X, is generated directly on Mylar film from the digitally formatted layout. The 4X master is photographically reduced to 1X to provide the working tool. The advantage of the digitized layout process is flexibility when changes are required. The software layout design can be iterated until the design is satisfactory, then committed to artwork.

Substrates. Thin-film circuits are fabricated on substrates made from 99.5 percent alumina (Al_2O_3). Standard substrate thickness is 0.025 $\pm$ 0.002 inch. Standard substrate size is 2.000 X 2.000 inches with a tolerance of 0.003 inch. Surface roughness should not exceed 4 microinch center line average (CLA) on the "A" or front surface, and camber should be no greater than 0.002 inch/inch.

Design Limitations

Thin-Film Conductors. The preferred conductor width and spacing is 0.005 inch. The minimum distance between a conductor and the edge of the substrate shall be 0.010 inch; however, the preferred distance 0.020 inch. All thin-film circuits should be gold plated to provide a maximum sheet resistivity of 0.01 ohm/square. Thin-film conductors are not normally multilayered.

Thin-Film Resistors. *Layout Design*—The resistance value for a thin-film resistor is determined by its aspect ratio and the sheet resistivity of the nichrome film in the same way as discussed in the thick-film section of this chapter. Resistors designed in a top-hat configuration are more widely used in thin-film than in thick-film circuits. Due to current density phenomena, the estimation of the effective number of squares in a top-hat configuration is not straightforward. Figure 10.44 may be used to predict the effective square count of a top-hat resistor configuration.

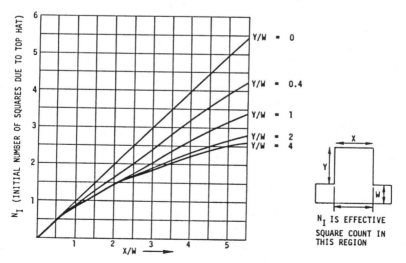

Figure 10.44: Initial square count of top-hat resistor.[5]

To compensate for errors inherent in the fabrication process, resistors with aspect ratios less than one shall be designed 10 percent low. Resistors less than 0.020 inch in width shall be designed in accordance with the nomograph shown in Figure 10.45.

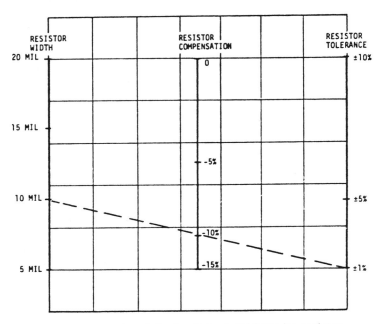

NOTE: The broken line indicates that a 0.010-inch-wide resistor trimmable to ±1 percent should be designed 10 percent low.

Figure 10.45: Nomograph for determination of resistor design values.

Resistor Density. Maximum resistor density per substrate depends on resistor size, area required by other circuit components, and circuit power dissipation limitations. Figure 10.46 shows how an example resistor value varies in size as a function of power dissipation and trim requirements.

Resistor Trimming. Figure 10.47 shows various methods for trimming resistors. The trim capability must be designed so that trimming will not reduce the resistor width to less than one-half the narrowest section.

Uptrimming (increasing the ohmic value) is the normal trim method. However, some downtrim capability is desirable on close-tolerance resistors. In the case of body-trimmed resistors, the design should include a conductor segment in series to provide this downtrim capability.

Power Dissipation. Thin-film resistors on alumina substrates are designed to dissipate 30 watts/square inch maximum. Although actual power dissipation capability is much higher, they are derated for design applications to permit a 50-percent area reduction during resistance adjustment in the manufacturing area.

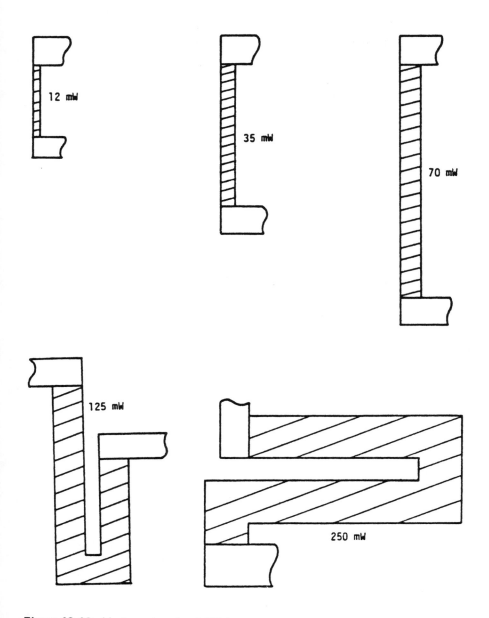

Figure 10.46: Various sizes for 1.8Kohm resistors, depending on power dissipation.

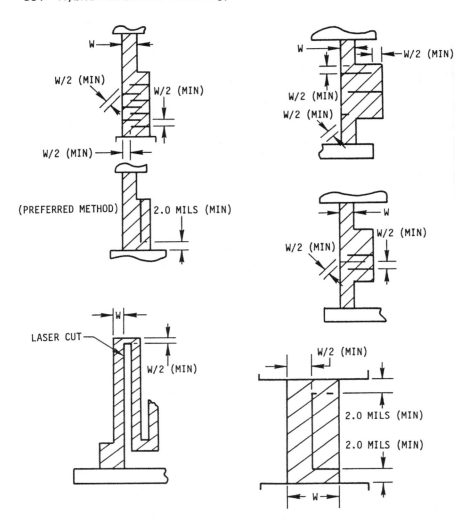

Figure 10.47: Typical laser cuts.

Power dissipation is determined using the same equation discussed in the thick-film section.

$$P_d = \frac{P}{L \times W}$$

where P_d = the rated power dissipation in watts/square inch
  P = the actual resistor power dissipation in watts
  L = the useful resistive length measured along the direction of current flow in inches
  W = the resistor width in inches

The nomograph shown in Figure 10.48 is a handy aid that may be used when determining the aspect ratios for resistors. The procedure for its use is as follows:

GIVEN RESISTOR—10K ohms ± 5 percent, 200 mw, maximum power dissipation to be 15 watts/square inch.

STEP 1   Determine the number of squares required for a 10K resistor. At 200 ohms/square, the number of squares equals 10000/200, or 50.

STEP 2   Trace line A from 0.200 watt on the resistor-dissipation scale through 15 on the watts/square inch scale to determine area.

STEP 3   Trace line B from the area scale through 50 on the number-of-squares scale to determine resistor width.

The resistor width calculated for this example is slightly more than 16 mils. Following standard design practice of working to a 5-mil grid, the resistor should be designed with a 20-mil width (next highest grid increment) so as not to exceed maximum power dissipation.

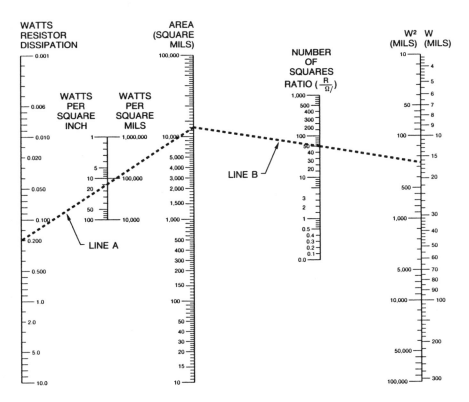

Figure 10.48: Nomograph for determining resistor width vs power dissipation.

REFERENCES

1. Castro, P.S. and Kaiser, P.N., "Capacitance Between Thin-Film Conductors Deposited on a High Dielectric Constant Substrate", *Proceedings*, IEEE, October 1962.
2. Enlow, L.R., "Parasitic Reactances Associated With Conductors and Resistors in Hybrid Thin-Film Microcircuits", Masters Thesis California State University Long Beach, 1971.
3. Dryden, W., "Thick-Film Thermal Design", *Circuits Manufacturing*, January 1979.
4. National Beryllia Corp., *Designing with Beryllia*, 1979.
5. Jones, R.D., *Hybrid Circuit Design and Manufacture*, Marcel Dekker, Inc., New York, 1982.

11

Documentation and Specifications

DOCUMENTATION

The extent and depth of documentation required to produce and screen hybrid microcircuits depends on specific program requirements. The quantity of circuits to be produced and reliability expected are controlling factors. For example, if a small quantity is to be produced, it is cost effective to use engineering documentation. If high-production quantities are involved, production-type documentation should be generated and released. For circuits that require high reliability (manned spacecraft, strategic defense, satellite, and medical implants), more detailed specifications covering certification, qualification, quality assurance, and testing are necessary.

Hybrid microcircuit documentation may be divided into two basic categories: circuit and circuit test equipment. The test equipment group includes both functional test and screen test equipment (test adapters and burn-in boards). The following list details documentation required for most hybrid circuits:

Hybrid Circuit Layout—A layout converts schematic and/or logic diagrams into circuit patterns from which artwork is produced (usually drawn at 20X).

Master Artwork—A photo-plot generated from the layout (usually 4X). It consists of one or more sheets, depending on the number of layers.

Assembly Drawing—Book-form drawing consists of title page, case outline, general-notes page, parts list, schematic, and assembly view (usually 10X). The assembly view is usually a half-size of the layout. This package can also consist of single sheets of various size drawings.

1X Working Tools—The master artwork that has been photo-reduced to actual size.

Assembly and Screen Specification—A specific process specification for a given program detailing how a circuit is to be fabricated, tested, inspected, and screened to meet specific customer requirements.

Functional Test Specification—Details specific electrical test requirements for a hybrid.

Test Adapter Drawing—Defines the adapter required to interface the hybrid with the manual test station.

Performance Board Drawing—Defines the adapter required to interface the hybrid with automated test equipment.

Burn-in Board Drawing—Defines the adapter required to interface the hybrid with the burn-in oven.

Operating Instruction Specification (OIS)—Converts engineering documentation and customer requirements into "how to" information for testing and screening a circuit.

Work Instruction (WIs)—Converts engineering documentation into how-to-build information for Operations. Usually consists of, but is not limited to:

(1) Assembly
(2) Bonding
(3) Passive test and trim (Only required for deposited or screened resistors)
(4) Marking
(5) Burn-in
(6) Lead forming

Resident Order (RO)—Details specific steps to be followed in fabricating, processing, testing, and screening for a given program.

Manufacturing Order (MO)—Each operator must sign and date this form, which follows the steps detailed in the RO, as each step is completed.

Process Work Instruction (PWI)—General information applicable to a specific process setup or operation of a specific piece of equipment.

Materials and Processes Specifications—General process specifications controlling typical hybrid processing steps.

Source Control Document (SCD)—Component Specification used for component procurement to control specific characteristics of a device. These are required by MIL-STD-883 for Class S hybrids.

Program Listing Document (PLD)—Listing of magnetic/paper tape used on auto tester.

Qual Plan—Detailed instructions on tests and requirements to qualify a hybrid or set of hybrids for a particular program.

Additional Specifications—Some programs may require other specifications which are unique, such as for inspection, X-rays, and rework.

MILITARY AND GOVERNMENT SPECIFICATIONS

The tendency to prepare more documentation than is needed should be avoided. Many military, federal, and other government-generated documents already exist and should be used before in-house documentation is generated. It is not cost effective to generate in-house documents simply to "parrot" existing documentation.

The following is a list of publications that are applicable to hybrids:

1. MIL-STD-100 — Engineering Drawing Standards
2. MIL-STD-129 — Marking for Shipment and Storage
3. MIL-STD-202 — Test Methods for Electronic and Electrical Component Parts
4. MIL-STD-750 — Test Methods for Semiconductor Devices
5. MIL-STD-883 — Test Methods and Procedures for Microelectronics
6. MIL-STD-1331 — Parameters to be Controlled for the Specification of Microelectronics
7. MIL-STD-1772 — Product Assurance Provisions for Custom Hybrid Microcircuits Line Certification of Fabrication Processes
8. MIL-Q-9858 — Quality Program Requirements
9. MIL-S-19500 — Semiconductor Devices, General Specification for
10. MIL-C-26074 — Military Specifications, Coatings, Electroless Nickel, Requirements for
11. MIL-M-38510 — Microcircuits, General Requirements for
12. MIL-G-45204 — Military Specification, Gold Plating, Electrodeposited
13. MIL-I-45208 — Inspection System Requirements
14. MIL-STD-45662 — Calibration System Requirements
15. MSFC-STD-587 — Design and Quality Standard for Custom Hybrid Microcircuits

16. MSFC-SPEC-592		Specification for the Selection and Use of Organic Adhesives in Hybrid Microcircuits
17. DOD-STD-1686		Electrostatic Discharge Control Program for Protection of Electrical and Electronic Parts, Assemblies and Equipment
18. MIL-HDBK-217		Military Handbook, Reliability Prediction of Electronic Equipment
19. MIL-H-38534 (PROPOSED)		Military Specification, Hybrid Microcircuits, General Specification for
20. FED-STD-209		Federal Requirement, Clean Room and Work Station Requirements, Controlled Environment
21. QQ-N-290		Nickel Plating, (Electrodeposited)

Of this list, MIL-M-38510, MIL-STD-883, and MIL-STD-1772 are most important for hybrid manufacture.

MIL-M-38510 is the current military specification for the procurement of hybrid microcircuits. It deals with documentation, rework, qualification, and screening. This specification requires hybrids to be tested according to MIL-STD-883, which details all the test procedures. MIL-STD-1772 is a recently issued specification which gives the requirements for the audit procedure of a hybrid facility. It is the document that assures that all the controls imposed by the above two specifications are being followed.

MIL-M-38510 General Requirements for Microcircuits

Originally this specification was written for semiconductor devices but, for lack of a specification on hybrid circuits, it was also applied to hybrids. This caused some confusion as to which portions of the specification were applicable to hybrids. In August 1977 Appendix G was added to MIL-M-38510 to define hybrid procurement procedures. However, it only addressed hybrid microcircuit procurement procedures. Finally in December of 1981 revision "E" was issued which completely revised Appendix G and entitled it "General Requirements for Custom Hybrid Microcircuits". With this revision hybrids became officially recognized as a reliable technology through a defined specification. At this writing MIL-M-38510 has gone through an "F" revision. The scope of MIL-M-38510 states, in part:

This specification establishes the general requirements for monolithic, multichip, and hybrid microcircuits and the quality and reliability assurance requirements which must be met in the acquisition of microcircuits......

The scope of Appendix G states, in part:

This appendix details the procedure to be used in documenting the requirements for custom hybrid microcircuits and specifies the quality and reliability assurance requirements which shall be met in the acquisitions of such microcircuits. The types of devices covered by this appendix include hybrid and multichip microcircuits, microwave/integrated circuits and surface acoustic wave (SAW) devices.....

The main requirements of the specification for a supplier of military or space-grade hybrids are paraphrased below. The reader is referred to the specification for complete wording.

Country of Manufacture. All hybrid microcircuits shall be manufactured, assembled, and tested within the US and its territories.

Product Assurance Requirements. Two levels of quality assurance are provided for in the specification (Class S and B). A Class S (space rated) hybrid is subjected to more stringent screening procedures than Class B hybrids. These procedures will be discussed in the section of MIL-STD-883. Any devices delivered to this specification shall have been subjected to and passed all requirements and Method 5008 of MIL-STD-883.

The custom hybrid manufacturer shall establish and implement a product assurance program as defined in Appendix A of this specification and MIL-STD-1772. After a manufacturer has been certified to MIL-STD-1772, they shall not implement any change in the certified material, process, or control without concurrent change to the process control or quality control documents listed in the approved product assurance program plan. When other than major changes are made to process control or quality control documents, the manufacturer shall notify the Government certifying/qualifying activity at least annually.

Screening. All custom hybrid microcircuits delivered in accordance with this specification shall have been subjected to and passed all screen requirements of MIL-STD-883 Method 5008. Delta measurements are a requirement of Class S hybrids.

Quality Conformance Inspection. Each inspection lot must pass qual conformance testing before any hybrids of that lot can be delivered.

Traceability.

(1) All adhesives and coatings shall be traceable to a lot.

(2) All process steps shall be traceable to an operator and a date that the process was performed.

(3) Records must be retained for a minimum of five years for Class B and 7 years for Class S.

Design And Construction.

(1) All hybrids shall be hermetically sealed in a metal or ceramic package.

(2) Class S hybrids shall include a tracer gas during seal.

(3) No adhesive or polymeric material shall be used for lid seal or repair.

(4) Class S internal moisture content shall not exceed 3000 ppm at 100°C.

(5) Class B internal moisture content shall not exceed 5000 ppm at 100°C.

Polymeric Materials.

(1) All polymeric materials shall be approved by the acquiring activity.

(2) All polymeric materials shall not exceed their cure temperature for a period longer than 10 minutes.

(3) All adhesives shall meet Method 5011 of MIL-STD-883

Documentation. All hybrids manufactured to this specification shall have complete documentation that details the processes, topologies, schematic, and test requirements. This documentation shall be submitted to the acquiring activity for approval.

Device Elements. All active and passive elements and packages used in the manufacture of hybrids shall conform to the requirements of MIL-STD-883 Method 5008.

Thermal Design. A thermal analysis shall be performed to prove that no elements in the hybrid are operating above their rated design when the package is at its maximum rated temperature. In addition, Class S elements shall be derated to the requirements of MIL-STD-975.

Electrical Circuit Design. A worst-case circuit analysis shall be performed which includes stress over temperature.

Serialization. Prior to the first recorded electrical measurement, each hybrid shall be marked with a unique serial number.

Workmanship. Each hybrid shall be manufactured, processed, and tested in a careful and workmanlike manner. These hybrids shall be built using work instructions, inspection instructions, training aids, and any other paperwork required to produce a highly reliable hybrid.

Environmental Control. All fabrication, assembly, and testing of hybrids prior to precap visual shall be in a Class 100,000 environment as defined in FED-STD-209. Photolithographic operations for Class S devices must be performed in a Class 100 environment.

Rework and Repair. All temperature excursions during rework shall not exceed the epoxy cure temperature (except in the local area of the rework). For Class S hybrids the local rework/repair temperature shall not exceed 165°C, and, if gold-aluminum bonds are present, that area must be flooded with nitrogen whenever the rework temperature exceeds 125°C.

(1) Touch-up of the plating on the package sealing surface of delidded packages is not permitted.

(2) The minimum distance from the package sealing area to the top of the glass-to-metal seals must be 0.050 inch. This applies to seam welding only.

(3) Any hybrid that is reworked/repaired after precap visual shall be subjected to complete rescreening, including the full burn-in of Method 5008 MIL-STD-883.

(4) If flux is required during rework/repair, its use and cleaning procedures must be approved.

(5) If a protective coating has been applied, no rework shall be permitted without approval of the acquiring activity.

(6) Replacement elements shall not be bonded onto the element they are to replace.

Element Wire Rebonding.

(1) No conductor patterns shall be repaired on an element by the use of a bond wire.

(2) All rebonds shall be placed on at least 50 percent undisturbed metal.

(3) No more than one rebond attempt, per wire, shall be permitted at any pad.

(4) No rebonds shall touch an area of exposed oxide caused by lifted metal.

(5) For Class S the total number of rebond attempts shall be limited to 10 percent of the total number of wire bonds in the hybrid. New wire bonds resulting from the addition of another element do not count toward the 10 percent.

Substrate Wire Bonding.

(1) Scratched, open, or discontinuous metallization paths, not caused by poor adhesion, may be repaired by the addition of bonded conductors that have at least 3.5 times the operating load current of the original conductor. The quantity of these repairs shall not exceed one for each 0.5 square inch of substrate area.

(2) No rebonds shall be made over an area in which the top layer metallization has lifted, peeled, or been damaged in such a way that underlying metallization or substrate is exposed.

Element Replacement.

(1) An epoxy-attached element may be replaced twice at a given location for a Class B hybrid.

(2) An epoxy-attached element may be replaced once at a given location for a Class S hybrid.

(3) A metallurgically attached element may be replaced once at given location.

(4) A substrate may be removed and put into a new package once.

Seal Rework. It is permissible to reseal hybrids that fail fine leak test, one time only and only if the hybrid was originally sealed with a tracer gas.

Delidding of Hybrids. Delidding of hybrids for rework and reseal is not permitted for Class S and permitted only once for Class B.

Product Assurance Provisions.

(1) A product assurance plan shall be established as defined in Appendix A of MIL-M-38510.

(2) Quality Conformance testing shall be performed.

(3) Screening shall be performed per Method 5004 or Method 5008 of MIL-STD-883. The method used depends on the size of the hybrid.

MIL-STD-883 Test Methods and Procedures for Microelectronics

This military standard is a collection of test methods and procedures to be followed by the hybrid manufacturer to verify the integrity and reliability of the hybrid.

Purpose. MIL-STD-883 establishes uniform methods, controls, and procedures for designing, testing, identifying and certifying microelectronic devices suitable for use within military and aerospace electronic systems. This includes basic environmental tests to determine resistance to deleterious effects of natural elements and conditions surrounding military and space operations. This standard also includes physical and electrical tests; design, package and material constraints; general marking requirements; workmanship and training procedures; and such other controls and constraints that are necessary to ensure a uniform level of quality and reliability suitable to the intended applications of those devices. For the purpose of this standard, the term "devices" includes items such as monolithic, multichip, film and hybrid microcircuits, microcircuit arrays, and the elements from which the circuits and arrays are formed. This standard is intended to apply only to microelectronic devices. The test methods, controls, and procedures described herein have been prepared to serve several purposes:

a. To specify suitable conditions obtainable in the laboratory and at the device level which give test results equivalent to the actual service conditions existing in the field, and to obtain reproducibility of the results of tests. The tests described herein are not to be interpreted as an exact and conclusive representation of actual service operation in any one geographic or outer-space location, since it is known that the only true test for operation in a specific application and location is an actual service test under the same conditions.

b. To describe in one standard all of the test methods of a

similar character which now appear in the various joint-services and NASA microelectronic device specifications, so that these methods may be kept uniform and thus result in conservation of equipment, manhours, and testing facilities. In achieving this objective, it is necessary to make each of the general tests adaptable to a broad range of devices.

c. The test methods described herein for environmental, physical, and electrical testing of devices shall also apply, when applicable, to parts not covered by an approved Military/NASA specification, Military/NASA sheet-form standard, or drawing.

d. To provide for a level of uniformity of physical, electrical and environmental testing; manufacturing controls and workmanship; and materials to ensure consistent quality and reliability among all devices screened in accordance with this standard.

The above stated purpose is from MIL-STD-883C Notice 4 dated November 29, 1985. The test methods in the standard are divided into four classes:

1. Methods 1001 to 1999 Environmental Tests
2. Methods 2001 to 2999 Mechanical Tests
3. Methods 3001 to 4999 Electrical Tests
4. Methods 5001 to 5999 Test Procedures

Figure 11.1 forms the basis for axis definitions to be used in the application of forces.

Now that the purpose of the specification and the definitions of the applied forces have been established, Method 5008 Test Procedures for Hybrid and Multichip Microcircuits can be discussed.

Method 5008. This method established screening and quality conformance procedures for the testing of hybrid, surface acoustic wave (SAW), and multichip microcircuits and microwave hybrid/integrated circuits to assist in achieving two levels (Class S and Class B) of quality and reliability commensurate with the intended application. It shall be used in conjunction with other documentation such as MIL-M-38510 and an applicable detail specification to establish the design, material, performance, control, and documentation requirements which are needed to achieve prescribed levels of device quality and reliability (from MIL-STD-883C Notice 4). This method subjects the hybrid to four phases as defined in Table 11.1.

The first requirement is that every element (Chip, IC, Package) used in the completed hybrid must be subjected to certain tests either at the element supplier or at the hybrid manufacturer. The element requirements are divided into active die, passive die, and packages. Each one of these will be discussed separately.

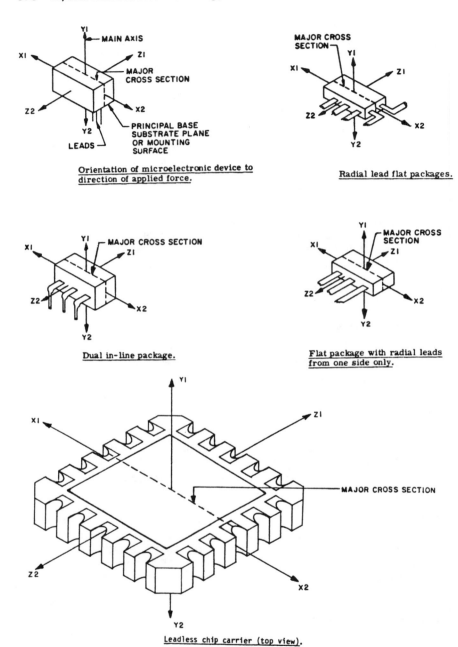

Orientation of microelectronic device to direction of applied force.

Radial lead flat packages.

Dual in-line package.

Flat package with radial leads from one side only.

Leadless chip carrier (top view).

NOTE: The Y_1 force application is such that it will tend to lift the die off the substrate or the wires off the die. The reference to applied force actually refers to the force which operates on the device itself and may be the resultant of the primary forces applied in a different manner or direction to achieve the desired stress at the device. (e.g., constant acceleration).

Figure 11.1: Axis definition.

Table 11.1: Device Evaluation Summary

Paragraph Numbers and Tables Refer to MIL-STD-883

Requirement	Paragraph	Table
Element evaluation	3.2	II
Process control	3.3	VI
Device screening	3.4	VII
Quality conformance evaluation	3.5	IX

PHASE I: DIE EVALUATION

Before hybrid assembly, the element characteristics shall be evaluated per Table 11.2

Table 11.2: Element Evaluation Summary

Element	Paragraph	Table
Microcircuit and semiconductor dice	3.2.2	III
Passive elements	3.2.3	IV
Packages	3.2.4, 3.5.5	V

Active Die Evaluation.

(1) All die shall be electrically tested, which may be done at the wafer level, and must include as a minimum Group A testing Subgroup 1. Group A tests consist of 11 electrical subgroups as shown in Table 11.9.

(2) All die shall be visually inspected to assure conformance with the applicable related requirements of Method 2010 (Internal Visual Monolithics).

(3) From each wafer lot, a sample shall be packaged in a suitable package that simulates the assembly methods and functional conditions of the intended application and then evaluated per Table 11.3.

Subgroups 1 and 2—Subgroup 1 checks for die diffusion, oxide, and metalization faults. Subgroup 2 checks electrical parameters after being subjected to environmental and mechanical screens. Class S: 3 die from each wafer for a total of 10 samples shall be used. Class B: 10 die from each wafer lot shall be used. Minimum electrical testing of Subgroup 2 shall be per Subgroups 1, 2, and 3 of Table 11.9.

Subgroup 3—This subgroup is to evaluate the device metalization. From each wafer lot, a sample of at least 5 die with 10 bond wires shall be selected. Each wire bond will be nondestructively tested then a minimum of 10 bonds shall be destructively tested.

Table 11.3: Active Die Evaluation Requirements

Sub-group	Class S	Class B	Test	MIL-STD-883 Method	MIL-STD-883 Condition	Quantity (accept no.)
1	x	x	Internal visual	2010 2072* 2073*		10 (0)
2	x		Stabilization bake	1008	C	10 (1)
	x		Temperature cycling	1010	C	
	x		Mechanical shock	2002	C, Y1 axis	
			or			
			Constant acceleration	2001	B, Y1 axis	
	x		Interim electrical			
	x		Burn-in	1015	**	
	x		Postburn-in electrical			
	x		Steady state life	1005		
	x	x	Final electrical			
3	x	x	Wire bond evaluation	2011		10 (0) wires or
				2023		20 (1) wires
4	x		Scanning electron microscope (SEM)	2018		See method 2018
5	x		Radiation			
	x		Dose rate and latchup	1020		10 (0)
	x		Total dose	1019		5 (0)
	x		Neutron irradiation	1017		5 (0)

*MIL-STD-750 methods.
**240 hr at 125°C.

Subgroups 4 and 5 are applicable to Class S devices only. Subgroup 4 provides a method of testing the quality of the element metalization. Subgroup 5 is required only when applicable to the microcircuit device.

Passive Element Evaluation.

(1) Each element shall be electrically tested at 25°C as specified in the detail/procurement specification.

(2) All die shall be visually inspected to assure conformance with the applicable related requirements of Method 2010 (Internal Visual Monolithics) and Method 2017 (Internal Visual Hybrid).

(3) From each inspection lot, a sample shall be packaged in a suitable package that simulates the assembly methods and functional conditions of the intended application and then evaluated per Table 11.4. The sample must contain at least 20 wire bonds.

Table 11.4: Passive Element Evaluation Requirements

Sub group	Class S	B	Test	MIL-STD-883 Method	Condition	Quantity (accept no.)
1	x		Visual inspection	2010		100%
		x		2017		22 (0)
2	x		Stabilization bake	1008	C	10 (1)
	x		Temperature cycling	1010	C	
	x		Mechanical shock	2002	C	
			or			
	x		Constant acceleration	2001	B	
	x		Voltage conditioning			
			or			
	x		Aging (capacitors)			
	x		Visual inspection	2017		
	x	x	Electrical			
3	x		Wire bond evaluation	2011		10 (0) wires or 20 (1) wires

Package Evaluation. From each package inspection lot, a randomly selected sample shall be evaluated per Table 11.5. Subgroups 1, 2, and 3 shall be accomplished for each lot. The rest of the table shall be accomplished periodically at intervals not to exceed 6 months. Subgroups 2, 3 and 4 apply to cases only. This completes the requirements of the elements that go into make up a hybrid.

Table 11.5: Package Evaluation Requirements

Sub group	Class S	B	Test	MIL-STD-883 Method	Condition	Quantity (accept no.)
1	x	x	Physical dimensions	2016		15 (0)
2	x	x	Solderability	2003	Soldering temperature $245 \pm 5°C$	3 (0)
3	x	x	Thermal shock	1011	C	3 (0)
	x	x	High temperature bake	1008	2 hr @ $150°C$	
	x	x	Lead integrity	2004	B2 (lead fatigue) D (leadless chip carriers)	
	x	x	Seal	1014	A_4 unlidded cases	
4	x	x	Metal package isolation	1013	600 Vdc, 100 nA max.	3 (0)
5	x	x	Moisture resistance	1004		5 (0)
6	x	x	Salt atmosphere	1009		5 (0)

In summary, when hybrid manufacturers purchase parts for a hybrid design, they must first prepare a Source Control Document for each part detailing the previously stated requirements. This document is then sent to the parts supplier. Depending on the vendor, the tests could be performed at the vendor's location or at the hybrid manufacturer's location.

PHASE II: PROCESS CONTROL

The next phase of device control deals with the controls that are placed on the hybrid manufacturer's processes. The processes indicated in Table 11.6 must be controlled.

Table 11.6: Process Control Summary

Operation	. . Class . .		.MIL-STD-883 Method	
	S	B	Method	Condition
Wire bonding	x	x	2011	
	x	x	2023	
Seal	x		1014	A, 1×10^{-8} atm/cc He

A wire bonding machine/operator evaluation shall be performed after any one of five events occur:

1. When a bonding machine is placed into operation.

2. Periodically while in operation, not to exceed four hours.

3. When an operator is changed.

4. When a bonding machine part or an adjustment has been made.

5. When a spool of wire is changed or a new lot of packages is started.

The wire bond evaluation consists of nondestructive and destructive bond strength tests.

Destructive. A minimum of 10 wires total from three devices (test samples may be used in lieu of a hybrid) shall be tested for Class B. Class S requires 15 wires. These 15 wires must contain one wire from each transistor, diode, capacitor, and resistor chip, and five wires from the substrate to header. If more than one size of wire is used at least 4 samples of each type and size shall be tested.

These samples shall be bond pull tested and recorded. Evaluation results are acceptable if no failures occur; however, if a failure occurs, the bonder shall be deactivated and not returned to service until corrective action has been taken. The machine will be subjected to another sample and will be returned to production when it passes the tests.

Nondestructive. From each wire bonding lot a sample of at least two devices shall be removed and subjected to nondestructive bond pull testing. A wire bonding lot consists of devices bonded by the same machine/operator using the same wire, during the same time period not to exceed 4 hours. These two devices shall contain at least 15 wires. These 15 wires must contain one wire from each transistor, diode, capacitor, and resistor chip, three wires from each type of integrated circuit, and five wires from the substrate to header. The wire lot will be acceptable if no failures occur. If one wire fails, another sample of two devices can be tested, and if no failures occur the lot is acceptable. If more than one failure occurs, or the second sample contains a failure, the bonder shall be removed from operation. The failures shall be investigated and corrective action taken. After the correction action, another sample shall be tested.

The other process that must be controlled, per Table 11.6, is seal testing. This control is for Class S hybrids only. All Class S hybrids shall be fine leak tested, without pressure bomb, immediately after seal and before any other testing.

PHASE III: DEVICE (HYBRID) SCREENING

Preseal Burn-In—This is optional and shall be approved by the procuring activity. The oven must be purged with dry nitrogen and the hybrids must be covered to avoid contamination.

Internal Visual—Prior to and after inspection, the hybrids must be stored in a dry, controlled environment until seal.

Stabilization Bake—Stab bake may be divided into two areas: (1) Preseal bake, occurring immediately prior to seal, and (2) Postseal bake which is 16 hours minimum.

Mechanical Shock/Constant Acceleration—Mechanical shock is not permitted on hybrids with an inner seal perimeter of less than 2 inches. If the inner seal perimeter is less than 2 inches and weighs less than 5 grams, the constant acceleration level shall be Condition E (30,000 g's).

Particle Impact Noise Detection Test (PIND)—With procuring activity approval, this test is not necessary if the hybrid has an internal protective coating applied. The test shall consist of five independent passes with the failures in each pass rejected. The survivors of the last pass are acceptable.

Seal Tests—These tests shall be performed after all shearing and forming operations have been completed.

Preburn-in Tests—These tests are only required if a Percent Defective Allowable (PDA) or delta parameters have been specified.

Burn-In—The burn-in period is 320 hours for Class S and 160 hours for Class B hybrids. For Class S the burn-in period may be divided into two successive 160-hour burn-in periods. Electrical testing shall be performed after the first period and failures shall be removed from the lot.

PDA—If a PDA is required, but a number is not specified, Class B shall use a PDA of 10 percent and Class S shall be 2 percent. PDA is calculated as the failures divided by the total number of hybrids that enter burn-in. For Class S the PDA is calculated across the second 160-hour burn-in period. A failure is defined as any departure from specified parameter limits or a delta measurement. The period of failure accountability shall include burn-in through final functional test. If the PDA is exceeded, the entire lot shall be rejected.

PDA Resubmittal—A failed burn-in lot can be resubmitted (one time only) to burn-in if: (1) the PDA was not more than twice the stated amount, (2) the cause of the failure has been evaluated, (3) the failure was due to random causes, (3) appropriate preventative action has been initiated, and (4) the resubmitted lot shall be equal to the next lower number in the LTPD series table.

Final Electrical Tests—Final electrical shall be as specified in the procurement document. The testing shall include, as a minimum, Group A, Subgroups 1, 2, 3, and 4 or 7.

Table 11.7: Device Screening

Test Inspection	. . .MIL-STD-883. . . Method	Condition	. . Requirements. . . Class S	Class B
Preseal burn-in	1030		Optional	Optional
Nondestructive bond pull	2023		100%	N/A
Internal visual	2017		100%	100%
Stabilization bake	1008	C	100%	100%
Temperature cycling or	1010	C	100%	100% or
Thermal shock	1011	A, min	N/A	100%
Mechanical shock or	2002	B	100% or	100% or
Constant Acceleration	2001	A	100%	100%
		(Y_1 only)		
Particle impact noise detection (PIND)	2020	A or B	100%	N/A
Electrical			100%	Optional
Burn-in	1015	125°C	100%	100%
Final electrical test	Per applicable device specification		100%	100%
Seal a. Fine b. Gross	1014		100%	100%
Radiographic	2012		100%	N/A
External visual	2009		100%	100%

PHASE IV: QUALITY CONFORMANCE EVALUATION

Each inspection lot of hybrids shall be evaluated in accordance with Table 11.8. All hybrids in each test sample of Table 11.8 shall have completed the screening requirements of Table 11.7 and Group A testing, except where nonfunctional samples are allowed. The nonfunctional samples must still have been subjected to all screening including burn-in.

Group A. The hybrids shall be tested per the requirements of Table 11.9. If the sample size exceeds the lot size, 100 percent testing is allowed.

Group B. Sample shall be tested to requirements of Table 11.10.

PIND Test—Lots failing this test shall be subjected to 100 percent testing. The failures shall be analyzed and corrective action taken.

Bond Strength—The samples shall be preconditioned at 300°C for one hour. The sample of wires, from the two hybrids, shall be selected to an LTPD of 5 for Class S and 10 for Class B. In each sample hybrid, at least 22 wires for Class S and 11 wires for Class B shall be tested. These sample wires are made up of one wire from each type of transistor, diode, capacitor, and resistor chip, three wires from each type of integrated circuit, and five wires from the interconnecting package pin.

Die Shear Strength—This test shall be performed to an LTPD of 10 of the elements in the hybrid. The sample shall be uniformly divided among all element types in the hybrid and performed on a minimum of two hybrids.

Solderability—At least 15 leads shall be selected and tested.

Electrostatic Discharge—This is performed on the initial qualification and any redesign.

Group C. A sample shall be tested in accordance with Table 11.11.

Internal Water Vapor Content—The sample shall be selected from Subgroup 1 and tested. The maximum allowable water content at 100°C shall be 5,000 ppm for Class B and 3,000 ppm for Class S.

Group D. Samples shall be tested to the requirements of Table 11.12.

Lead integrity—Fifteen leads minimum shall be tested.

Table 11.8: Quality Conformance Evaluation Summary

Test Inspection	Paragraph	Table
General	3.5.1	
Group A electrical test	3.5.2	IX
Group B testing	3.5.3	X
Group C testing	3.5.4	XI
Group D testing	3.5.5	XII

Table 11.9: Group A Electrical Test

Subgroup	Parameters	LTPD
1	Static test at 25°C	2
2	Static tests at maximum rated operating temperature	3
3	Static tests at minimum rated operating temperature	5
4	Dynamic test at 25°C	2
5	Dynamic tests at maximum rated operating temperature	3
6	Dynamic tests at minimum rated operating temperature	5
7	Functional tests at 25°C	2
8	Functional tests at maximum and minimum rated operating temperatures	5
9	Switching tests at 25°C	2
10	Switching tests at maximum rated operating temperature	3
11	Switching tests at minimum rated operating temperature	5

Table 11.10: Group B Testing

Sub-group	Class S	Class B	Test	MIL-STD-883 Method	Condition	Quantity (accept no.) or LTPD
1	x	x	Physical dimensions	2016		2 (0)
2	x		Particle impact noise detection test	2020	A or B	15 (0)
3	x	x	Resistance to solvents	2015		4 (0)
4	x	x	Internal visual and mechanical	2014		1 (0)
5	x	x	Bond strength	2011		2 (0)
			1. Thermocompression		C or D	
			2. Ultrasonic or wedge		C or D	
			3. Flip-chip		F	
			4. Beam lead		H	
6	x	x	Die shear strength	2019		2 (0)
7	x	x	Solderability	2003	Solder temperature 245°±5°C	1 (0)
8		x	Seal	1014		15
			a. Fine			
			b. Gross			
9	x	x	a. Electrical parameters		Group A-1	
			b. Electrostatic discharge	3015		15 (0)
			c. Electrical parameters		Group A-1	

Table 11.11: Group C Testing

Sub-group	Class S	Class B	Test	MIL-STD-883 Method	Condition	Quantity (accept no.) or LTPD
1	x	x	External visual	2009		15
	x		Temperature cycling	1010	C, 20 cycles	
		x	Temperature cycling or	1010	C, minimum	
		x	Thermal shock	1011	A, minimum	
	x	x	Mechanical shock or	2002	B, Y_1 axis	
	x	x	Constant acceleration	2001	A, Y_1 axis	
	x	x	Seal (fine and gross)	1014		
	x		Radiographic	2012	Y axis	
	x	x	Visual examination			
	x	x	End point electrical			
2	x	x	Steady state life test	1005	1,000 hr at 125°C	10
	x	x	End point electricals			
3	x	x	Internal water vapor content	1018		3 (0) or 5 (1)

Table 11.12: Group D Package Related Tests

Test	MIL-STD-883 Method	MIL-STD-883 Condition	Quantity (accept no.) or LTPD
Thermal shock	1011	C	5 (0)
Stabilization bake	1008	C, 1 hr	5 (0)
Lead integrity	2004	B2 (lead fatigue) D (leadless chip carrier)	1 (0)
Seal	1014	A or B, and C or D	5 (0)

MIL-STD-1772 Certification Requirements for Hybrid Microcircuit Facilities and Lines

MIL-STD-1772, released May 15, 1984, consists of a "checklist" of items that a hybrid manufacturer must meet in order to be certified. The purpose of the specification is to provide a uniform method for evaluating materials and processes for custom hybrid microcircuits and to assure that the process and facility being certified continues to produce satisfactory products during the active certification period.

MIL-STD-1772 was generated by Rome Air Development Center(RADC) together with the Committee on Hybrid Technology (JC-13), so that hybrids achieve recognition by DoD on an equal basis with monolithic devices. There was a need to establish a uniform and standard procedure for conducting a certification audit on hybrids because different agencies had different requirements and each performed its own surveys. It was therefore desirable to standardize on one audit procedure and have one agency responsible for certifying each hybrid manufacturer.

The Defense Electronics Supply Center(DESC) (of the Defense Logistics Agency) in Dayton, Ohio, is responsible for performing the audit. This agency will audit hybrid manufacturers to assure that: all processes and procedures are documented; all processes are being controlled; there is strict conformance to MIL-M-38510, Appendix A and G; there is strict conformance to MIL-STD-883 Method 5008; supplier control is maintained; and there is standardization on documentation, testing, and control of hybrid microcircuits.

MIL-STD-1772 is divided into two sections: Section A, Audit Plan for Manufacturers and Line Certification, and Section B, Qualification of Fabrication Processes.

Section A. This section is comprised of 18 checklists that DESC will use during the audit. The titles of the sections are listed below:

A-1   Product Assurance Program

A-2   Design Documentation

A-3   Process Documentation

A-4   Workmanship

A-5   Cleanliness and Atmospheric Control

A-6   Incoming Material Control

A-7   Substrate Fabrication

A-8   Polymeric Materials

A-9   Circuit Element Attachment to Substrate

A-10   Internal Visual

A-11   Wire Bond

A-12   Cleaning

A-13   Package Seal

A-14   Screening

A-15   Acceptance for Shipment

A-16   Handling and Storage

A-17   Failure Analysis

A-18   Training

The audit based on the above items is required on an annual basis. Table 11.13 is an example of one of the checklists. A hybrid manufacturer must be certified to Section A in order to perform the qual testing of Section B.

Table 11.13: MIL-STD-1772, Section A-14, Screening

Requirement: Screening documentation and operations shall be evaluated to ascertain adequate process control.
References: Method 5008 of MIL-STD-883
Details: Verify the conformance to the following where applicable.

Approval N/A Comments

A. Preseal burn-in, TM 1030
     1. Environmental control
     2. Device protection
     3. Proper burn-in circuit
     4. Proper voltages and signals
     5. Heat sink or other temperature
        control
     6. Burn-in log
B. Nondestructive bond pull, TM 2023
     1. Frequency of calibration
C. Internal visual, precap, TM 2017
D. Stabilization bake, TM 1008
E. Temperature cycling, TM 1010
F. Constant acceleration, TM 2001, or
     Mechanical shock, TM 2002

(continued)

Table 13: (continued)

G. Particle impact noise detection
   (PIND), TM 2020   _____
   1. System calibration   _____
   2. Mounting methods   _____
H. Pre-burn-in electrical test:
   1. Test procedure: appl. rev.   _____
   2. Calibrated electrical instr.   _____
   3. Calibrated temp. probes   _____
   4. Verification of test procedure by
      means of testing a correlation sample   _____
   5. Data recording   _____
I. Burn-in, TM 1015   _____
   1. Proper burn-in circuit   _____
   2. Proper voltages and signals   _____
   3. Heat sink or other temp. contr.   _____
   4. Burn-in log   _____
J. Electrical test:
   1. Test procedure: appl. rev.   _____
   2. Calibrated electrical instr.   _____
   3. Calibrated temperature probes   _____
   4. Verification of test procedure by
      means of testing a correlation unit   _____
   5. Data recording   _____
K. Seal, TM 1014:
   1. Fine: _____
   2. Gross: _____
L. Radiographic, TM 2012   _____
M. External visual, TM 2009   _____

Performed by: _____
Date: _____
Comments: _____
_____

Section B. This section deals with the fabrication processes of hybrid microcircuits. The section is subdivided as follows:

B-1   Thick- and Thin-Film Fabrication

B-2   Element and Substrate Attachment

B-3   Bonding, Internal

B-4   Sealing, Delidding, and Resealing

To be certified to Section B the manufacturer must build samples and test them to the requirements of each section. Figure 11.2 is a flow chart of the testing required for Section B-2. Section B will be performed one time only except where major process changes occur.

MIL-STD-1772 Audit Program. The audit program consists of five phases:

1. Assembling a list of the documentation to be used in the certification process, the Product Assurance Plan, sample travelers, procedures for burn-in, temperature cycle, seal, and internal visual.

2. An informal pre-audit, by DESC, to exchange information and to demonstrate actual audit conditions. DESC will select several Section A checklists and perform an audit.

3. Formal audit to Section A.

4. Follow up audits to insure that nonacceptable conditions have been corrected.

5. Annual reaudits to insure compliance.

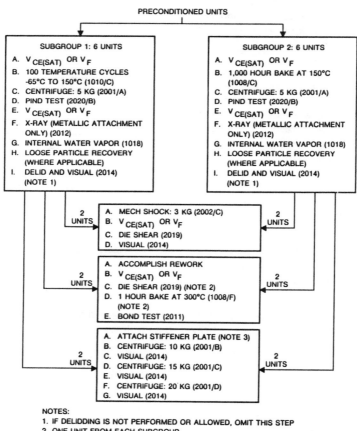

Figure 11.2: Flow chart of tests for evaluation of element and substrate attachment.

12

Failure Analysis

TYPES AND CAUSES OF HYBRID FAILURES

To produce reliable hybrid microcircuits with a high yield, any failures that occur must be analyzed, their causes determined, and corrective action taken to prevent the failure from recurring.

Failures in hybrid microcircuits may be attributed to one or more of six categories:

Devices.

Wire bonds.

Die attachment.

Substrates.

Packages.

Contamination.

Figure 12.1 depicts the percentage of failures for various failure modes. These data, collected by Rome Air Development Center, show that faulty active devices, marginal wire bonds, and contamination are major contributors to failures.[1]

Figure 12.2 identifies potential sites for failure in a cross section of a hybrid, which includes a device mounted on a thick-film substrate, in turn mounted in a hermetically sealed metal package.

Device Failures

Devices are the major contributors to hybrid circuit failures. One estimate is that devices account for over 31 percent of all hybrid failures (Figure 12.1). Devices progress through three phases of failure according to a Weibel or "bathtub" curve (Figure 12.3): (1) Period of infant mortality—where marginal devices fail within a few days of operation, (2) Life span—the period that a properly designed and produced device will function, and

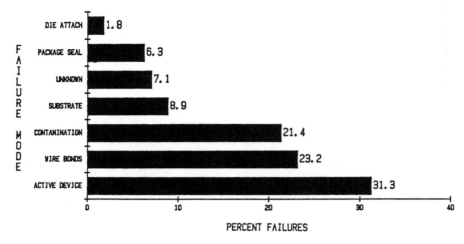

Figure 12.1: Percent failures by failure mode.

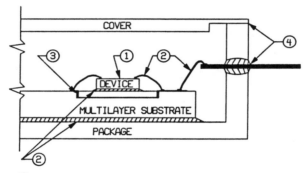

① DEVICE

② INTERCONNECTIONS (WIRE BONDS, DIE AND SUBSTRATE ATTACHMENT)

③ SUBSTRATE (METALLIZATION AND VIAS)

④ PACKAGE (SEALS)

Figure 12.2: Main sites for potential hybrid failures.

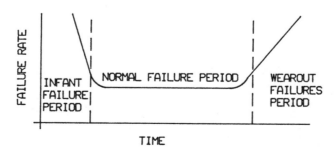

Figure 12.3: Typical Weibel (bathtub) failure curve.

(3) Wear-out period—when most of the devices are expected to fail due to aging. The life expectancy of devices that survive the infant mortality period may be in excess of ten years.

Infant mortality, as the curve depicts, has a decreasing failure rate. Principal causes of failures associated with this stage include:

(1) Weak/marginal components.

(2) Inspection escapes.

(3) Contamination.

(4) Weak wire bonds.

(5) Weak die attachment.

Most infant mortality failures occur during mechanical screen testing and can be repaired. The failure of weak components is further accelerated during burn-in. The major failure mechanism of a component stems from electrical aging,[2] which can be accelereated by temperature and predicted by applying the Arrhenius equation:

$$F = A(e^{-Ea/kT})$$

Where F = Failure rate
      A = A constant
      E_a = Activation energy (ev)
      k = Boltzmann's constant (8.6×10^{-5} ev/K)
      T = Temperature (K)

Interconnection Failures

Interconnection failures, the most prevalent of which occur in wire bonds, account for the next largest category (23 percent) of hybrid failures. Wire bond failures can be minimized by conducting both material and substrate verification.

Material verification. Samples of each lot of devices (ICs, transistors, diodes, chip resistors, substrates, and packages) should be wire bonded using standard production bonders and schedules. After bonding, the wires are destructively pull-tested according to Method 2012 of MIL-STD-883. Bi-metallic bonds should be aged for 1 hour at 250°C to accelerate any intermetallic formation, then destructively pull-tested. If the wire bond pull strengths are greater than a specified value, the lot can be released to production. These tests insure that the device and substrate metallizations are capable of accepting a wire bond and that the bonds will not degrade over a period of time. Material verification is essential in identifying bonding problems prior to committing the parts to production.

Substrate verification. In addition to the wire bond pull test, an adhesion test for the substrate metallization should be performed. A strip of cellophane tape is pressed onto the metallized substrate with a rubber roller, then pulled off in a single, swift motion, at a right angle to the substrate. If the tape removes any metal from the substrate, the adhesion is marginal and the lot should be rejected.

Besides wire bond failures, other interconnection failures may be due to a marginal device or substrate attachment. If insufficient or non-uniform attachment material (epoxy or alloy) is used, the substrate, or large capacitors, can loosen and separate during mechanical screens. This failure mode can be avoided by filleting the edges of the large capacitors with non-conductive epoxy.

Substrate Failures

Substrates contribute about nine percent to the total hybrid failures. A substrate failure may be mechanical or electrical in nature. Examples of mechanical failures include cracked, chipped, or broken substrates, most often resulting from rough handling or high mechanical stresses. Electrical failures are invariably due to design layout errors which result in electrical opens or shorts in the metallization pattern. With thick-film multilayer substrates, the most common failure mode is electrical shorting between conductor layers through pinholes in the dielectric. Consequently, it is recommended that each dielectric layer be double screened to avoid pinholes. Another failure area in multilayer substrates is the via connection from one level to another. If the via is small and not completely filled with conductive paste, a high resistance path or even an open can occur. This is more likely to occur in multilayer substrates that have been designed with stacked vias, making the connection path longer. By designing larger and staggered vias, this failure mode can be avoided or minimized.

There are two ways that the electrical integrity of a thick-film interconnect substrate can be checked prior to release to assembly:

(1)   Inspect and touch up each layer prior to firing.

(2)   Test for opens and shorts using a continuity tester. However, many companies have found that extensive visual inspection will afford reliable substrates without the need for electrical testing.

Package Failures

Packages, which account for about six percent of all failures, may fail for several reasons.

(1)   The package does not pass incoming visual inspection. The plating may be too thin, non-uniform, or have pinholes that can lead to corrosion.

(2)   The glass-to-metal seals (glass beads) around the leads may be cracked or broken through handling, lead forming, or thermal/mechanical screen testing.

(3)   The lid-to-package seal may become non-hermetic during screen testing.

Contamination

Contamination accounts for about 21 percent of all hybrid failures and has long been recognized as a main cause of failures. In fact, many of the other failure categories may ultimately be due to contamination. For example, wire bonds and devices can fail because of trace amounts of ionic contaminants. Conductive particles from epoxy, solder, wire bonds, and packages can become dislodged and cause shorts between two conductors. Due to the high probability of failures due to particle contamination, industry and government have adopted many safeguards among which are the PIND test, use of particle getters, use of particle immobilizing coatings, and extensive cleaning and inspection.

1. Particle-Impact-Noise-Detection (PIND) Testing is a screen test requirement for all Class S hybrids. According to this test, the hybrid circuits are vibrated and shocked while a transducer registers noice due to any loose particles.

2. Particle getters can be applied to the inside of the hybrid covers, then the hybrid is shocked in an upside-down direction, allowing any loose particles to become trapped on the getter.

3. The most effective, though also most expensive, method is to coat the entire inside of the hybrid with a thin particle-immobilizing coating such as Parylene. This coating entraps any particles, preventing them from moving. Chapter 7 discusses this approach in detail.

4. Cleaning and visual inspection are very important in the removal and control of particles. Many manufacturers rely only on extensive inspections under magnification to detect particles. If a manufacturer employs efficient cleaning and inspection procedures, particle failures (as detected by PIND) can be held to below 3 percent.

A primary source of contamination in both semiconductor integrated circuits and hybrid circuits is human skin.[3] Skin cells, oils, and salts are shed continuously and contain organic, particulate, and ionic contaminants. These skin contaminants can deposit the elements listed in Table 12.1. Eliminating harmful contamination at the source can only be done after the contaminant has been identified. If a contaminant is analyzed and found to contain two or more of the elements of Table 12.1, its source is most likely from human handling.

Circuit failures may occur at three stages:

1. In-process (during fabrication and assembly).
2. In final functional test (acceptance).
3. In use (field failures).

The extent of analysis required depends on which of these categories the particular failure fits.

In-process failures are those that are encountered during hybrid assembly and preseal testing. Table 12.2 summarizes some in-process failures.

Table 12.1: Chemical Elements in the Human Body

ELEMENT	PERCENT
OXYGEN	65
CARBON	18
HYDROGEN	10
NITROGEN	3
SULFUR	0.25
CALCIUM	2.2
PHOSPHOROUS	0.8 - 1.2
POTASSIUM	0.35
CHLORINE	0.15
SODIUM	0.15
MAGNESIUM	0.05
IRON	0.004
IODINE	0.0004
SILICON	
COPPER	
FLUORINE	
MANGANESE	TRACE
COBALT	AMOUNTS
BARIUM	
NICKEL	
LITHIUM	

Table 12.2: Summary of In-Process Failures

INDICATED FAILURE	IDENTIFIED BY	PROBABLE CAUSE
WIRE BOND	VISUAL INSPECTION	METALLIZATION CONTAMINATION
	WIRE BOND PULL	BONDING MACHINE SCHEDULE
	(LOW BOND STRENGTH)	(FORCE, TEMP,TIME,POWER)
	ELECTRICAL TEST	WIRE
		WORKMANSHIP
		INTERMETALLIC FORMATION IN Au/Al BONDS
CONTAMINATION	VISUAL INSPECTION	INADEQUATE CLEANING
	INSTRUMENTAL ANAL.	HUMAN HANDLING
	CHEMICAL ANALYSIS	EPOXY BLEEDOUT
		SOLDER FLUX
		LOOSE PARTICLES OF SILICON,METAL,EPOXY
		CORROSION
SUBSTRATE	VISUAL INSPECTION	PIN-HOLES IN DIELECTRIC
	ELECTRICAL TEST	OPENS IN METALLIZATION
		SHORTED METALLIZATION.
DEVICE	VISUAL INSPECTION	NON-OPERATIVE DEVICE
	ELECTRICAL TEST	DAMAGED DEVICE
		WORKMANSHIP
		OVERSTRESSED DEVICE
		ESD
		PART MOUNTED WRONG
		PART BONDED WRONG
		INADEQUATE CLEANING

Acceptance failures are failures detected during the final sell-off of the hybrid. Table 12.3 summarizes these failures.

Table 12.3: Summary of Acceptance Test/Field Failures

INDICATED FAILURE	IDENTIFIED BY	PROBABLE CAUSE
ELECTRICAL FAILURE	ELECTRICAL TEST	DEVICE FAILURE (LEAKAGE CURRENTS, PARAMETER DRIFT, SHORTED)
	VISUAL INSPECTION	
	PIND TEST	WIRE BOND (OPEN, HIGH RESISTANCE WIRE BOND)
		LOOSE COMPONENT
		PACKAGE FAILURE
		CONTAMINATION (CONDUCTIVE PARTICLES SHORTING CONDUCTORS, IONIC RESIDUES)
		ELECTRICAL OVERSTRESS
		ELECTROSTATIC DISCHARGE

Field failures are those that have occurred in hybrids that have been installed in a system that has been operating.

Failure analysis can range from fairly simple for in-process failures (Figure 12.4), to extensive for system returns (Figure 12.5).

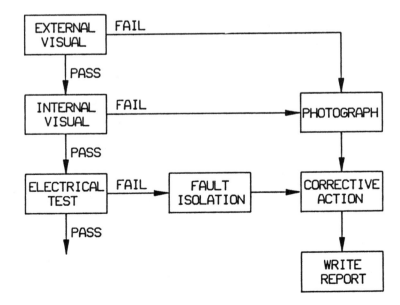

Figure 12.4: Steps in device failure analysis.

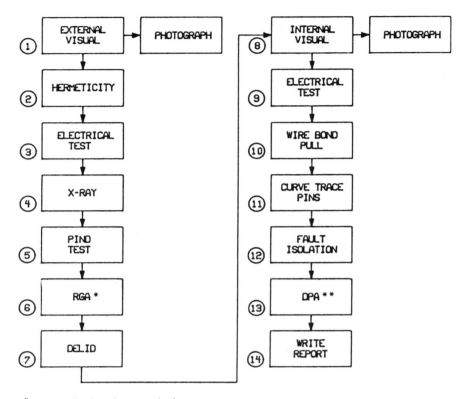

* RGA = Residual gas analysis
**DPA = Destructive physical analysis

Figure 12.5: Complete failure analysis flow chart.

For in-process failures, it is usually only necessary to isolate the failure to a component, replace the component, take corrective action to prevent recurrence, and re-enter the hybrid in the assembly/test flow. In-process failure analysis is aimed at identifying corrective action and preventing the same failure type from recurring.[4]

In analyzing failures that occur after sealing, it is very important to follow a definite sequence so that accurate conclusions can be drawn. Too often, hybrid circuits that have failed after burn-in or after being installed in a system are immediately de-lidded. This destroys data that may be significant in establishing the mode or mechanism of failure. For example, once a hybrid has been de-lidded, a moisture or gas analysis can no longer be performed, nor can PIND or hermeticity tests be run. Furthermore, radiographic inspection is no longer meaningful since metal particles may have entered the package during de-lidding. The failure analysis steps shown in Figure 12.5 provide a logical sequence so that the maximum amount of data can be obtained and analyzed. In Figure 12.5, DPA refers to

Destructive Physical Analysis; more information on this topic may be found in Method 5003 of MIL-STD-883.

Today's failure analysis expert must be part electrical engineer, part chemist, part process engineer, and part super sleuth. Like a detective solving a crime, the specific analytical technique to be used depends on the failure mode.

FAILURE ANALYSIS TECHNIQUES

There are four basic methods that can be used in a failure analysis investigation: Electrical, Chemical, Thermal, and Physical. A summary of these methods is given below but a more comprehensive review may be found in *Failure Analysis Techniques* compiled and edited by E. Doyle of Rome Air Defense Center and B. Morris of General Electric Company.

Electrical Analysis

Electrical analysis is used to isolate an electrical failure to a single component. This is accomplished by applying electrical stimuli to the hybrid and probing until the failed component is located. Electrical analysis should be the first step in any failure analysis. After the failed component has been identified, other failure analysis methods may be used to determine why the component failed.

Chemical Analysis

Wet chemical methods are seldom used in a hybrid failure investigation because of the large amount of sample required and the lengthy time it takes to run the analysis. However, chemical procedures such as colorimetric, gravimetric, and spot testing may be valuable in identifying some ionic residues if they are present in sufficient amounts. More likely, contaminants are present in trace amounts and only the physical/chemical (instrumental) methods are useful in detecting and measuring them.

Thermal Analysis

Thermal analysis by infrared scanning is a valuable tool in locating problem areas in a hybrid. When a thermal map of a failed hybrid is compared with the map of an operating unit, the failure can be identified. A hot spot indicates a short whereas a cool trace identifies an open. As thermal imaging equipment becomes more sophisticated, this will become a more useful tool in hybrid test and analysis.

Physical Analysis

Physical (instrumental) analysis is the most common method used in hybrid failure analysis. It is not possible to discuss here all of the different analysis methods and equipment that can be used in a failure analysis; however, Table 12.4 gives an overview of those most commonly used.

Table 12.4: Summary of Basic Analytical Techniques

Method	Principle	Advantages	Limitations
Infrared spectroscopy	Measures absorption of infrared radiation as a function of wavelength.	Identification of organic functional groups. Virtually no sample limitations. Impurity detection.	Medium sensitivity, no direct information about size of molecule. Does not detect metals or ions. Water interferes with analysis.
Mass spectroscopy	Ionization of molecule by cracking molecule into fragment ions. Separation according to charge to mass ratios.	Precision molecular wt (molecular ion) impurity detection. Excellent for detection of trace amounts of gases and low molecular weight organics.	Does not detect functional groups directly. Slow and destructive.
Gas chromatography	Partitioning and separation in the vapor phase.	General quantitative analysis of volatile organics. Highly efficient technique.	Identifies materials only in special cases. Not applicable to materials of low volatility.
Combined gas chromatography and mass specroscopy	Combines separation efficiency of gas chromatography with sensitivity of mass spectroscopy.	Identification and analysis of trace organic materials. Identification of components in mixtures in low ppm.	Not applicable to materials of low volatility.
Radiographic (x-ray)	X-rays penetrate into cavity. Metal particles are opaque to x-rays and can be detected.	Loose metal particles, distorted wire bonds, and device anomalies may be detected without disturbing the seal. Nondestructive.	Sometimes hard to interpret. Some components may be damaged. Not all views meaningful.
Helium leak test	Helium is forced into sealed package by pressurizing, then detected by mass spectrometry as it escapes from package.	Determines the hermeticity of the package. Eliminates field failures due to moisture and air exposure.	Correlation between leak rate and failure rate not established. Bombing with He can damage large packages.
Scanning electron microscopy (SEM)	Collects secondary electrons ejected when electron beam strikes surface. These electrons modulate a CRT tube scanned in sync with the scanning beam.	High magnification (300,000X) with large depth of focus obtained. Excellent for magnifying abnormalities during failure analysis.	Glassivation must be removed. Vacuum is required. Electron bombardment may destroy surface conditions, causing failure.
Energy dispersive x-ray analysis (EDS or EDAX)	Solid-state crystal detector mounted in SEM chamber. It separates characteristic x-ray radiation according to its energy.	Rapid x-ray spectrum of chemical elements can be analyzed simultaneously.	Detection of chemical elements down to atomic number 6 possible with light element detectors.
Auger electron spectroscopy (AES)	Collects ejected secondary electrons. The electron energy identifies the atom.	Small depth of investigation (4–15 Å)	Cannot detect hydrogen or helium.

Analytical/instrumental techniques have been developed over the years to a degree of sophistication that now, even the chemistry of the first 2-5 nanometers of a surface, can be characterized. All surface analysis methods are based on a similar principle. A sample is bombarded with a probe particle causing either the back-scattering of the probe particle or the emission of secondary particles from the sample.[4] Emitted particles are then analyzed for mass, energy, and wavelength and from these data the elemental composition of the sample can be derived. Each technique gives essentially the same information, differing only in sensitivity. It is largely up to the operator to decide which technique is best suited to a given situation. Figure 12.6 graphically depicts the technique of surface analysis.[5]

Parameters for commonly used surface analysis techniques may be found in Table 12.5.

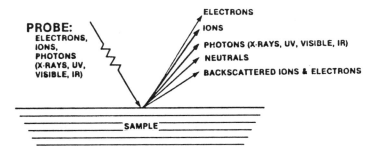

Figure 12.6: Surface analysis probe representation. Under vacuum, probing of a sample's surface with electrons, ions, or photons results in the occurrence of several events, including the emission of ions, electrons, and photons. Analysis of various characteristics of these events provides information concerning the sample.[5]

Table 12.5: Typical Parameters for Common Surface-Analysis Techniques[5]

Analytical Parameter	Technique					
	AES	ESCA	SIMS	SEM/EDX/WDX	RBS	LIMS
Probe particle	Electrons	X-rays	Ions	Electrons	Ions	Photons (laser)
Detected entity	Electrons	Electrons	Ions	Electrons / X-rays	Ions	Ions
Detectable elements	All > helium	All but hydrogen	All	All > beryllium	All > helium	All
Diameter for routine small-area analysis (μm)	Sub μm	150–300*	Sub μm to several μm	1–3 (EDX/WDX); 0.01 (SEM)	~ 1000	~ 1
Surface sensitivity (nm)	~ 1–4	~ 4	~ 0.3–1	> 1000 (EDX/WDX)	Depth resolution is 2.5–20	LD-adsorbed material LI ≈ 100
Routine detection limits† (atom %)	0.3–1	0.1–1	$10^{-7}-10^{-1}$	1 (EDX) 10^{-4} (WDX)	0.01–10	$10^{-4}-10^{-2}$
Routine survey (data-acquisition) analysis time (min)	5	5	5	2 (EDX); 30 (WDX)	15	< seconds

*Many older ESCA systems are limited to areas several millimeters square.

†Detection limits are expressed as a range because different elements exhibit different sensitivities. Minimum figures are difficult to achieve in routine work.

ANALYTICAL TECHNIQUES

AES—Auger Electron Spectroscopy

AES involves the bombardment of the sample with electrons which causes an outer shell electron to transfer its energy to another electron, imparting to that electron enough energy to be ejected from the sample. The ejected, "Auger", electron is analyzed and the atom from which it derived is identified. By scanning across the surface and setting the detection system for a particular element, a video display of the surface distribution is obtained. The major advantage of AES is the small area in which it can operate (1 micrometer or less), and the low levels of elements that it can detect (0.1 percent).

ESCA—Electron Spectroscopy for Chemical Analysis

ESCA, also referred to as XPS (X-ray Photoelectron Spectroscopy), utilizes X-rays (at 2 ev) as the probe causing an electron to be emitted. The energy of the emitted electron is proportional to the difference between the photo energy and the binding energy of the electron. ESCA gives information about the chemical binding state of the elements, which the other techniques cannot provide. ESCA, like AES, can be used to detect low levels of elements (0.1 percent) but with more repeatability. However, it is not useful in detecting hydrogen and has a low sensitivity to helium.

SIMS—Secondary-Ion Mass Spectrometry

SIMS uses a high-energy (1-20 kev) ion beam probe (usually argon, cesium, or oxygen) which ejects neutral atoms, molecules, and other ions. Ions produced by this process are mass analyzed to provide elemental identification and chemical concentrations. SIMS can detect hydrogen and is more sensitive to the detection of elements of low atomic/molecular weight than either AES or ESCA.

SEM—Scanning Electron Microscopy

SEM employs a beam of electrons (500 ev to 40 kev) to eject low-energy secondary electrons. The current from these ejected electrons is used to intensity-modulate the Z-axis of a cathode ray tube, resulting in a video image. The sample must be electrically conductive or coated to provide a conductive surface. The SEM extends the magnification of optical microscopes to 200,000X.

EDX—Energy Dispersive X-Ray Analysis

EDX is often added to SEM to obtain compositional information at the same time as SEM (Figure 12.7).

WDX—Wavelength Dispersive X-Ray Analysis

WDX can also be added to a SEM/EDX system to increase its sensitivity. This technique also extends the elemental range to include boron, carbon, and oxygen. WDX is used to investigate oxides and nitrides in thin-films.

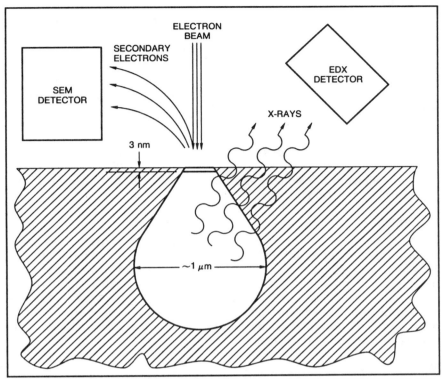

SIMULTANEOUS USE OF SCANNING-ELECTRON MICROSCOPY AND ENERGY-DISPERSIVE X-RAY ANALYSIS. FOCUSED ELECTRON BEAM EXCITES SECONDARY ELECTRONS FOR SEM, AND ALSO PRODUCES X-RAY FLUORESCENCE FOR EDX.

Figure 12.7: EDX detector coupled to SEM.

RBS—Rutherford Backscattering Spectrometry

RBS employs high-energy (1-3 Mev) helium ions to penetrate the sample surface. A small fraction (1×10^{-6} of the ions collide with the sample and are backscattered. The energy loss they experience is characteristic of the sample and the depth at which the collision occurred.

LIMS—Laser Ionization Mass Spectrometry

LIMS sometimes called LMMA (Laser Microprobe Mass Analysis), utilizes a high-energy (10^7 to 10^{12} watts/square centimeter) finely focused (1 micrometer diameter) laser pulse to volatilize and ionize the sample.[4] This technique uses two modes: laser desorption (LD) and laser ionization (LI). A time-of-flight mass spectrometer is then used to analyze the ions ejected from the sample. An advantage of this technique is that it does not charge the surface as do the other analysis methods.

EBIC—Electron Beam Induced Current

EBIC techniques are used to induce a current in a sample which causes electrical paths to be highlighted on a display. Opens or shorts in metallization can be detected by this method.

Infrared

Infrared microscopy and thermography methods of analysis are discussed in Chapter 8.

CASE HISTORIES OF HYBRID CIRCUIT FAILURES

The following are some case histories of failures that have been encountered in hybrid circuits.

Tin Whiskers

Tin whiskers, one-fortieth the diameter of a human hair, have been reported to cause intermittent electrical shorts on an aircraft radar.[6] These whiskers grew spontaneously from tin-plated covers used to seal the hybrids. Whiskers were 0.1 mil in diameter and as long as 0.1 inch. These can detach from the lid and fall onto the circuitry. One hybrid was found to contain 130 free-floating whiskers. Tin whiskers have also been encountered growing on TAB bonded devices; Figure 12.8 depicts four examples of tin-whisker growth on TAB devices.

Metallic Smears

Ceramic substrates or ceramic packages can pick up smears of metals from tweezers, tooling, fixtures, or other metal objects. The danger in these smears is that there is a possibility of an electrical leakage path.[7]

Particles

Particles are a common cause of hybrid failures. A hybrid was returned from the field with a short between power and ground. Visual examination disclosed a metal sliver bridging the power supply and ground. The sliver was analyzed using AES and found to be aluminum with other trace elements, correlating with the composition of aluminum bonding wire.

Flux Residues

Flux residues and tin contaminants from a solder sealing operation are often the cause of hybrid failures. Flux and tin contaminate gold-aluminum wire bond interfaces and cause corrosion or accelerate intermetallic growth that may eventually lead to a bond failure.

Cracked/Broken Die

A hybrid was returned from the field because of failure during a system temperature test. The hybrid was removed from the system and failed ambient electrical test. Upon delidding, a fractured transistor was detected.

The die had a crack that had propagated through the silicon during temperature testing, finally breaking off a piece of silicon. The bare silicon caused a high resistance short when it contacted a wire bond.

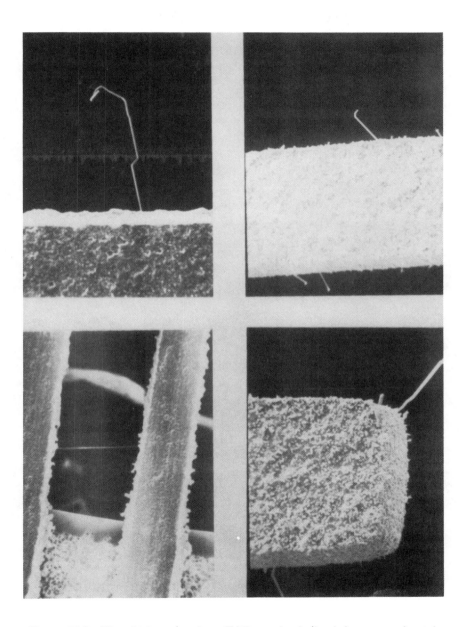

Figure 12.8: Tin whisker growth on TAB tape leads (lead size, approximately 3.0 x 1.4 mils) (Courtesy International MicroIndustries).

Collapsed Wires

Hybrids were returned from the field due to opens in input leads. Upon delidding, the circuits were visually inspected and found to have "smashed" wires. Some wires were nearly touching the substrate (Figure 12.9) while others had touched and been fused open. The wires had no sharp bends to indicate that they had been mechanically disturbed. The only wires affected were 1.5-mil-diameter gold post leads that were greater than 150 mils in length. It was determined through experimentation that the wires collapsed when centrifuged in the Y_2 axis (instead of the Y_1 axis required by MIL-STD-883). These failures resulted from two conditions:

(1) The hybrids were centrifuged at 5,000 g's in the wrong direction.

(2) The wires were longer than the preferred design requirement of 100 mils.

It was also discovered that suspect hybrids (those with possible collapsed wires) could be corrected by re-centrifuging in the Y_1 direction.

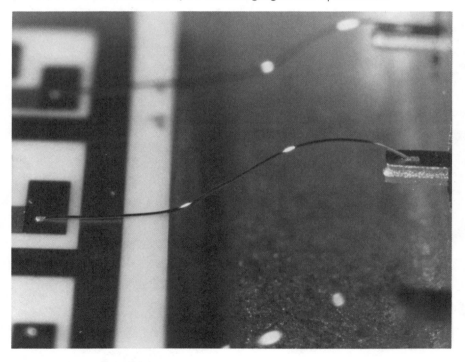

Figure 12.9: Collapsed gold wire (Courtesy Rockwell International).

Package Plating

One hybrid manufacturer experienced a low yield upon using a new lot of gold-plated packages. Gold-plated hybrid packages should have electroless nickel containing phosphorus as a barrier material under the gold. This lot of packages was found to have "electrolytic" nickel containing no phosphorus, which made them extremely difficult to seal.

Package Discoloration

During Quality Assurance inspection of incoming package lots, irregular brown patches of discoloration were observed. Using Auger analysis, the colored stains were found to contain oxygen, iron, and chlorine. The stained packages were cross-sectioned and found to have only 80-90 microinches of nickel plating instead of the normal 100-200 microinches required by specification. Insufficient nickel plating (less than 100 microinches) promotes rapid diffusion of the iron from the base Kovar to the surface of the package. Iron that diffused to the surface oxidized, resulting in the brown stain.

Nickel Ion Contamination

Thin-film hybrid microcircuits containing pnp high-frequency transistor die experienced failure rates greater than 80 percent after burn-in. The hybrid was encased in an electroless nickel-plated Kovar package. An extensive failure analysis consisting of thermal profiling, electrical testing, and chemical analysis showed that nickel ions from the package plating had migrated to the transistor die where they diffused into the silicon dioxide passivation layer causing electrical anomalies. The cases as received from the vendor had not been adequately cleaned and still contained free nickel ions, residues from the plating solutions. The problem was resolved by cleaning the cases, as received, with a 50/50 solution of isopropyl alcohol and deionized water.[8]

The following are case histories of device and hybrid circuit failures that have been selected and edited from the *Microcircuit Manufacturing Control Handbook* 3rd Edition published by Integrated Circuit Engineering Corp., Scottsdale, Arizona and reprinted with their permission. Though some of these failures occurred in single chip devices singly packaged, the failure modes described are such that they can also occur in devices assembled in a hybrid circuit.

Corrosion of Aluminum Wire Bonds (Courtesy of Integrated Circuit Engineering Corp., Scottsdale, AZ.)

Observation. A low resistance open circuit.

Failure Summary. Corrosion of aluminum wire bonds by chlorinated solvents that entered the package through an inadvertent break in the seal of the flat pack.

Description. A 54L00 series TTL device had a low output voltage of 0.7 volt rather than 0.3 volts. This caused a system malfunction. Corrosion of wire bonds at both the die pads and package posts resulted in low resistance opens. Visual inspection showed corrosion products at several wire bond locations. SEM analysis of the corroded parts indicated that the possible cause was chlorinated hydrocarbon solvents and other contaminants. A 400X SEM of the corrosion product on the wire bond is shown in

Figure 12.10. SEM-WDX (wave length detected X-ray) analysis indicated that the corrosion product had extraneous elements such as chlorine, sulfur, lead and carbon. The main corrosion product appeared to be aluminum oxide and aluminum hydroxide.

Figure 12.10: Aluminum wire bond corrosion.

Diagnosis. Increased voltage drop across alumiunum wire bonds resulted because of corrosion that was produced by solvent contamination. Chlorinated solvents entered the package cavity because the lid seal integrity had been destroyed. Other devices from the same lot with hermetic seals showed no corrosion.

Possible Causes. Mechanical stress applied to flat packs during assembly into microwave systems probably destroyed the lid seal integrity. Solvents used to clean the system after assembly introduced contamination to the chip cavity through voids in the lid seal area. The solvents used are alcohols, fluorocarbons, and chlorinated hydrocarbons. Without moisture or sulfur contamination these solvents are chemically nonreactive. With moisture present in chlorinated hydrocarbons an aluminum catalyzed reaction can produce hydrochloric acid. This acid reacts with the aluminum to form aluminum chloride and subsequently oxide/hydroxide corrosion products by hydrolysis. Sulfur can have a synergistic influence on the acid formation and corrosive reaction.

Corrective Action. The main corrective action was to avoid mechanical stress on flat pack lids to assure a hermetic seal so that solvents and other contaminants cannot enter the package.

Process Control Techniques. Monitor torquing or mechanical stress placed on flat packs during system assembly.

Analytical Techniques.

a.   Evaluate increased output low voltage versus sink current.

b.   Test for hermetic seal integrity.

c.   Examine visually and by SEM for corrosion products.

d.   Analyze corrosion products by WDX.

Corrosion of Nichrome Resistors (Courtesy of Integrated Circuit Engineering Corp., Scottsdale, AZ.)

Observation. Discoloration and change in resistance of thin-film Nichrome resistors.

Failure Summary. Nichrome resistors attacked by the phosphorus contained in glass passivation layer.

Description. Dual-input radiation-hardened TTL gates incorporating thin-film Nichrome resistors were failing in burn-in due to very large increases in resistor values. Visual examination of the failed units revealed a non-uniform discoloration of the resistors. The integrated circuits were passivated with a CVD glass except for the bonding pads.

Diagnosis. Electron microprobe of the glass over the discolored resistors showed that is contained 3.7% by weight phosphorus.

Possible Causes. The Nichrome was chemically attacked by the phosphorus contained in the glass; thereby causing discoloration and changes in the Nichrome resistance. Since the concentration of phosphorus in the glass is not high enough for the general formation of P_2O_5, the mechanism of the attack is not clear.

Corrective Action. Remove all phosphorus from the CVD scratch protection glass. **Note:** Pure CVD glass is subject to cracking. A layered structure of pure glass/doped glass/pure glass might be desirable.

Process Control Techniques. Microprobe or chemically analyze a sample from each lot of glassed material for the presence of phosphorus.

Analytical Techniques. Microprobe of failed or suspected units.

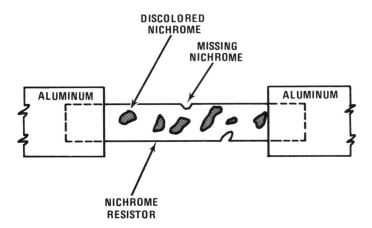

Figure 12.11: Nichrome resistor faults.

Stress Corrosion of Kovar (Courtesy of Integrated Circuit Engineering Corp., Scottsdale, AZ.)

Observation. Rusted and broken Kovar package leads.

Failure Summary: Stress corrosion cracking of Kova.; rust formation.

Description. A number of losses have occurred in equipment life test and field failures have been reported due to breakage of gold-plated Kovar leads. Failures have been reported on both metal flat packages and transistor packages. Failures normally occur after the device or equipment has been subjected to high humidity. This failure has increased recently because of the industry's trend toward the use of thinner gold plating.

It has also been reported that there is a potential failure mode related to the Kovar lid of metal packages. Stress corrosion cracking can occur completely through the lid, destroying the hermetic seal. This problem is now particularly important because of the present practice of plating lids with as little as 30 microinches of gold, instead of the typical 100 microinches.

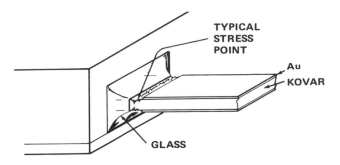

Figure 12.12: Stress of Kovar lead.

Diagnosis. SEM and optical micro-photographs of the fractured or rusted areas show visible evidence of corrosion, usually where the gold plating has been destroyed or was not sufficiently thick. A particularly vulnerable area is where the pins enter the package. Stress corrosion can occur there under high humidity and the presence of flux or chloride residues.

Possible Causes.

a. The presence of a tensile force on a lead accelerates the cracking and thus the breakage of the lead.

b. Insufficient nickel plating on the Kovar and resultant lack of protective gold exacerbates the problem.

c. The presence of contaminants containing free chloride ions can accelerate the action.

d. The practice of bending package leads after gold plating provides an opening for the chemical corrosion action to take place.

Corrective Action.

a. Although thicker gold plating will not provide the complete answer, adequate nickel and gold plating will minimize some of the reported problems.

b. In subsystems, the application of a conformal coating is recommended to minimize the direct effect of humidity and moisture on the leads.

c. It is recommended that devices be mounted utilizing approved guidelines for stress relieving the leads.

d. Thorough cleaning of the devices after PC board assembly to remove solder fluxes is required.

e. The use of chemicals or cleaning agents that may generate chloride ions should be avoided.

Process Control Techniques.

a. Incoming inspection is recommeded on a sampling of all packages. Sample tests should be made on the thickness and uniformity of gold plating.

b. Specifications should be written to ensure minimum lead bending and handling of the package to minimize cracking of leads near the glass seals.

c. Written specifications should include adequate cleaning procedures for removal of solder fluxes after PC board assembly.

Analytical Technique. The defect is normally detected by use of an optical microscope. Inspection with a low power SEM is sometimes desirable to confirm the exact nature of the corrosion or cracking. Additional checking with X-ray fluorescence or Auger spectroscopy may be used.
Reports.

L.J. Weirick, "A Metallurgical Analysis of Stress-Corrosion Cracking of Kovar Package Leads," *Solid State Technology*, March 1975.

B. Reich, "Stress Corrosion Cracking of Gold-Plated Kovar Transistor Leads," *Solid State Technology*, April 1969.

S.C. Kolesar, "Principles of Corrosion," *Proc. of the 12th Annual Symposium on Reliability Physics*, 1974.

Intermetallics in Wire Bonds (Courtesy of Integrated Circuit Engineering Corp., Scottsdale, AZ.)

Observations. Hybrid circuit defective after burn-in.
Failure Summary. High bond resistance caused by intermetallic compounds.

Description. Hybrid circuits constructed with ultrasonically bonded aluminum wire bonds on gold metallization on a ceramic substrate were subjected to a burn-in test of 168 hours at 160°C with current passing through the wire bonds. The hybrid circuits had also been subjected to 10 temperature cycles from 150°C to −65°C, with a maximum of 5 minutes transfer time.

Failure analysis showed that all of the wire bonds had adequate pull strengths (>2 grams). However, upon checking bond resistance with a 4-point probe, it was determined that the bond resistance had risen sharply.

Diagnosis. A detailed investigation of the conditions contributing to a change in bond resistance was undertaken. Twenty tests were conducted to determine the effects of:

a.   elevated temperatures and times,

b.   DC and AC current,

c.   nitrogen and air environment in the package,

d.   temperature cycling, and

e.   Parylene coating on aluminum wire bonds.

The parameters tested were bond strengths, electrical resistance and wire bond break modes. Samples were fabricated from thick-film gold pastes (from two vendors) and from a thin-film gold on ceramic substrates.

Some findings of this extensive evaluation showed that the pull strengths decreased with time and temperatures above 75°C. The pull strengths were still relatively good (6 grams) at 200°C. Little or no effect on the integrity of the bonds was noted due to the type of gas in the package, the temperature cycling pre-conditioning or the current flowing through the wire bonds. The Parylene application had no detrimental effects, rather it enhanced the mechanical integrity of the bond at all temperatures.

It is clear that the most significant information obtained was the sharp increase in bond resistance above 150°C after having remained essentially constant below that temperature. The increase appears to be primarily a temperature and time related phenomenon and is attributed to the formation of intermetallics and the generation of Kirkendall type voids.

Possible Causes. The principal failure mode was found to be an increase in resistance of the aluminum wire bond to the gold metallization. The cause of the increase in resistance appears to be the depletion of the gold under the bond as it is absorbed into intermetallics at the bond interface. The gold depletion causes formation of Kirkendall type voids around the bond effectively isolating the bond from the surrounding metallization.

Note that the intermetallics occurred on a hybrid circuit with aluminum wire ultrasonically bonded to gold metallization. It is similar to the more common case found in monolithic integrated circuits where gold wire bonds are attached to aluminum metallization.

Figure 12.13 shows a curve of the bond resistance versus time for the three different gold films that were used. The films were subjected to 168 hours of storage at the temperatures indicated. The resistances remained flat at temperatures up to 160°C. The resistance increased above this temperature. The constituents of the 99+ gold appear to retard the gold absorption and intermetallic formation.

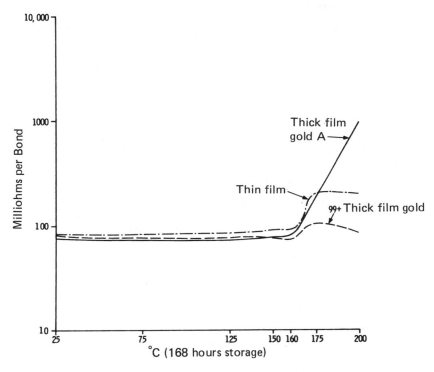

Figure 12.13: Bond resistance as a function of storage temperature (average value per bond of 36–60 bond loops in series).

Corrective Action. For hybrid circuits with the construction described above, exposure and storage temperatures should be kept below 160°C. At this time there is no reason to believe that this caution should not also extend to monolithic circuits using aluminum metallization and gold ball bonds. This caution may conflict with MIL-STD-883 and, in particular, with Method 1008, "High Temperature Storage," in which burn-in at 200°C for 168 hours is permitted. The results of such a burn-in on these hybrid circuits is shown in Figure 12.14.

Higher temperatures for shorter times are also permitted. However it is recommended that the applicability of these tests to a specific product be first verified by making resistance measurements of the wire bonds before and after these tests. Further verification should be made with an SEM microscope.

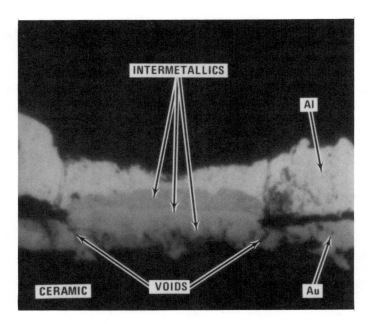

Figure 12.14: Wire bond cross-section of thick-film gold sample after aging 168 hours at 200°C.

Process Control Techniques. Process control techniques consist of first establishing maximum storage temperatures that the circuit may be subjected to without causing detrimental effects and the setting up procedures to prevent the circuit from being subjected to temperatures above this critical value. In some cases this may extend the burn-in time substantially. Wire bond resistance measurements should be made periodically on circuits that have undergone high temperature storage.

Analytical Techniques. The most useful analytical technique was measurement of the bond resistance with a 4-point probe. Current was applied beteween the outermost probes and the voltage drop measured between the inner two probes. The contact resistance of the probes was thus eliminated. The initial resistance measured on most of the bonds was approximately 80 milliohms.

A scanning electron-beam probe was also used on selected bonds to obtain a highly magnified view of a cross-section of the bond.

Die Bond Surface Oxidation (Courtesy of Integrated Circuit Engineering Corp., Scottsdale, AZ.)

Observation. Open Circuits

Failure Summary. Poor wetting of semiconductor die to metallized bonding pad due to oxidation of eutectic surface.

Description. Approximately 25% of a lot of hybrid thin-film integrated circuits hermetically sealed in a 4-pin TO type package failed a die bond shear test. The die sheared off with forces of 60 to 250 grams. Shear strengths were as low as 0.1 gram per square mil of die area, with a mean strength of 0.75 gram per square mil. The semiconductor die was processed by evaporating gold on the back of the wafer and forming a gold-silicon eutectic. The die were bonded to a sputtered gold metallized pad approximately 100-μm thick.

Diagnosis. Shear tests indicated that most of the die would lift off intact and visual examination of both the back of the die and the bond area indicated little or no wetting (see Figure 12.15). The gold-silicon eutectic on the back of the die appeared to be undisturbed along with most of the pad area. Eutectic around the die had a "scum-like" appearance indicative of a surface film.

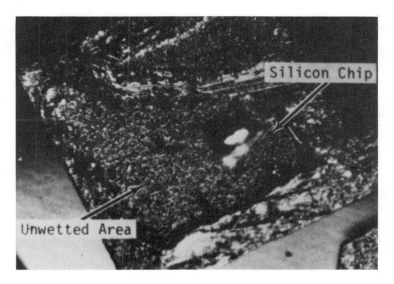

Figure 12.15: Poor wetting of silicon die to header.

Possible Causes. There are several possible causes for this type of die bond failure.

a. The die or pad could be contaminated with an organic material preventing proper wetting.

b. The die bonding was performed in room atmosphere, thus allowing oxidation of the eutectic area.

c. The temperature at the interface may have been less than 400°C. Although this temperature is above the eutectic point, 425°C is recommended for easier wetting.

Corrective Action. Assuming no contamination of the surface or of the die, the basic corrective action would be to perform the die bonding operation in an inert gas ambient. A dry nitrogen gas is recommended. The adjustment of the gas flow rate over the heat column is critical. A very slow laminar flow is recommended to provide an adequate blanket. High flow rates will result in excessive cooling of the die and turbulence which can draw in oxygen and defeat the purpose of the inert gas. A temperature of 425°C is recommneded at the floor of the package. Note: in some applications this die bond temperature is sufficiently high to damage sensitive devices such as high-frequency "washed" emitter transistors and some MOS circuits. Thus, in some cases these devices must be attached with epoxy.

Process Control Techniques. The effective temperature of the heat column should be monitored twice daily using a special package and thermocouple which will measure the temperature at the exact location during normal gas flow conditions. The gas flow rates should also be recorded daily.

Analytical Techniques. The shear strength of the die was measured by using a "gram gauge" in a special holder and a micro-manipulator. The surfaces were inspected through a vertical illumination microscope for wetting. A microprobe analysis of the non-wetted areas was performed to detect any organic contamination.

Reports.

J.E. Johnson, "Die Bond Failure Modes," Westinghouse Electric Corp., 12th Annual Proc. Reliability of Physics, 1974, IEEE Catalog No. 74CHO-389-1PHY.

Loose Particle Short (Courtesy of Ingetrated Circuit Engineering Corp., Scottsdale, AZ.)

Observation. Pin shorted to substrate.

Failure Summary. Conductive particle lodged between bonding wire and scribing grid.

Description. Loose conducting particles (silicon chips, gold slag, weld fragments, aluminum metal, etc.) sometimes are inadvertently sealed into hermetic packages. If the wire bonding in the packages is also done in reverse, e.g., bonded to the package frame first, then the chip pad, the angle of the wire to the pad may be small. The small conductive particles are likely to wedge between the wire and scribe grid, shorting the wire to the substrate. See Figure 12.16.

Diagnosis. The small angle of the wire of "reverse bonded" circuits allows small conductive particles to short the bonding wire to the substrate.

Possible Causes.

a. "Reverse bonding" causes the bonding wire to come off the pad at a small angle.

b. Small conductive particles are sometimes left in a hermetic package.

c. The power leads attract the conductive particles.

d. The conductive particles become wedged between the bonding wire and the scribing pad.

Figure 12.16: SEM photograph of conductive particle under lead after lead has been raised up.

Corrective Action. Carefully screen parts for conductive particles at precap visual.

Process Control Techniques. Reduce occurrence of conductive particles in packages by careful inspection and elimination of the source of particles. Increase the angle of the wire bonds by bonding to the pad first and maintaining the proper loop in the wire during bonding.

Analytical Techniques. Electrical measurement between each pin and the substrate, visual inspection after delidding, and SEM analysis were the principal techniques used.

Wire Bond Short, Case 1 (Courtesy of Integrated Circuit Engineering Corp., Scottsdale, AZ.)

Observation. Pin shorted to substrate.

Failure Summary. Short between wire bond and bare silicon at the edge of a laser-scribed die.

Description. A short which resulted in a blown circuit occurred between the power lead and substrate of an operational amplifier. See Figure 12.17.

Diagnosis. The wire bond came off the bonding pad at a small angle to the surface of the die. Mounds of silicon residue formed during the laser scribe operation were observed along the edge of the die. After the circuit was sealed, a short developed and blow-out occurred when power was applied.

Possible Causes. The short may have been caused by a loose particle, vibration or shock, or electrical attractions, but the failure was the result of two contributing factors:

    a.   Low angle of the wire bond probably resulted from "reverse bonding" (bonding first to the lead frame and second to the pad on the chip).

    b.   Piles of silicon residue along the edge of the die caused by too much power applied during the laser-scribe operation.

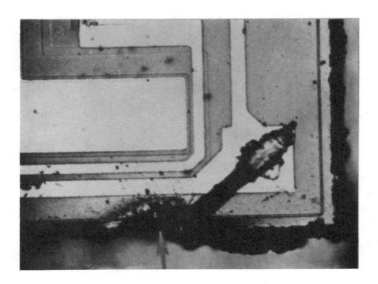

Figure 12.17: Wire bond short to bare silicon.

Corrective Action. The recommeded correction action is two-fold.

    a.   The angle of the wire bond should be increased to allow more space between bonding wire and scribing grid. A space of at least 1 mil is recommended.

    b.   The power of the laser used for scribing should be reduced to prevent build-up of silicon residue along the edge of the die.

Process Control Techniques. Increase the angle of the wire bonds by bonding to the pad first and maintaining the proper loop in the wire during bonding.

Adjust the power of the laser scriber to eliminate the silicon residue build-up along the edge of the die.

Inspect at least once each shift to ensure that the machines remain properly adjusted.

Analytical Techniques. Electrical measurement between each pin and the substrate, visual inspection after delidding and SEM photographs were the principal analytical techniques used.

Wire Bond Short, Case 2 (Courtesy of Integrated Circuit Engineering Corp., Scottsdale, AZ.)

Observation. Electrical shorting from package pins to ground.

Failure Summary. 1-mil diameter wire shorted to edge of die.

Description. Problem with wire sagging onto active area at edge of semiconductor die, and causing electrical short circuit. Problem caused by necessity to "reverse bond" from package post to die because of inadequate space in package. "Reverse bonding" causes the best loop to be where it is not needed—at the package post, while the flatter portion of the wire loop is at the die.

"Forward bonding" is the technique whereby the first bond is made to the semiconductor die pad, and the second to the package post. This places the best loop over the semiconductor die and minimizes the possibility of wire sagging and shorting to the active portion of the die. See Figures 12.18 and 12.19.

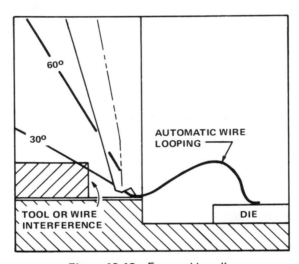

Figure 12.18: Forward bonding.

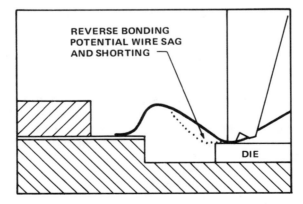

Figure 12.19: Reverse bonding.

The low angle wire of the "reverse" bond makes the circuits susceptible to shorts caused by loose particles wedged between the bonding wire and substrate.

Diagnosis. It is normal practice in wire bonding to make the first bond on the die pad. When using gold ball bonding, this places the gold ball bond on the pad. When using ultrasonic or thermocompression bonding this procedure takes advantage of the natural tendency of the wire to form a high loop after the first bond.

When ultrasonically bonding into a deeply recessed package, it becomes difficult to make the bonds in this sequence because of the interference between the bonding tool, the wires, or the wire clamp, and the side of the package. The common solution to the problem is to reach across the die with the bonding tool and bond first to the package lead frame and second to the die pad. This procedure results in a high loop on the package, where the wire may possibly short to the lid, and a low loop on the die where it is likely to cause some of the above described problems.

Possible Causes. The principle cause of the problem was the practice of "reverse bonding".

Corrective Action. "Forward bonding" is required to ensure consistent, reliable wire bonds with a high loop at the die. If the package does not permit the use of a standard bonding tool with a 30°C angle wire hold, either the package or the tool should be changed. Bonding tools with 45° and 60° wire holes are now available.

Process Control Techniques. Operator instructions should specify that without specific engineering approval only forward bonding be permitted. Quality Assurance visual inspection should be used to assure that adequate looping is occurring.

Analytical Techniques. Optical inspection for proper looping and recording of inspection results.

Weak Wire Bonds, Case 1 (Courtesy of Integrated Circuit Engineering Corp., Scottsdale, AZ.)

Observation. Weak wire bonds

Failure Summary. Crystallographic damage of silicon during wire bonding.

Description. During wire pull testing wire bonds pulled loose from the chip at the $Si-SiO_2$ interface, leaving a small hole in the silicon substrate (see Figure 12.20). This phenomenon has been observed on gold thermocompression ball bonds, as well as aluminum ultrasonic bonds. The strength of the bonds may meet MIL-STD-883 (2 grams for 1-mil Al, 3 grams for 1-mil Au wire), but will be below that of properly made bonds. Components such as diodes placed beneath these pads will show degraded characteristics.

Diagnosis. The crystalline structure of the silicon beneath the pad is damaged during the wire bonding step because of poor wire bonding techniques and/or materials. Severely damaged silicon will lift off the wire bond during a destructive wire pull test. Less severe damage produces no visible effects but can degrade the electrical characteristics of the semiconductor junction beneath the bond.

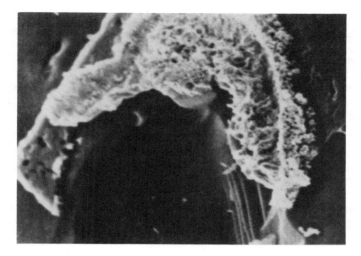

Figure 12.20: Silicon pulled loose during wire bond pull test.

Possible Causes. Cratering in the thermocompression bonding may result from:

   a.   Too high a bonding force.

   b.   Too great a tool-to-substrate impact velocity.

   c.   Contact of the metallization by the bonding tool.

Cratering in ultrasonic bonding may result from:

   a.   Too hard a wire.

   b.   Excessive ultrasonic power.

   c.   Too large a bonding pressure (bond deformation should be 1.5 diameters without any ultrasonic power applied).

   d.   Excessive tool-to-substrate impact velocity.

Corrective Action. Adjustment of wire bond machine at the beginning of each shift. Good quality control of bonding wire.
Process Control Technique.

   a.   Check wire bonder adjustments at beginning of each shift.

   b.   Make a destructive pull test of a small sample of wire bonds during each shift.

   c.   Visually inspect wire bonds for appearance (deformation, angle, looping, etc.)

Analytical Technique. Severe cratering may be observed by pulling the wire bonds to destruction.

Threshold damage: Threshold level damage may be determined by removing the bond and metallization with HCl or aqua regia and then lightly etching the substrate with buffered HF (if the pad is over an oxide) or with a weak HF-HNO$_3$ solution or a special preferential etch (if the pad is over silicon). Damaged areas etch at a faster rate than undamaged ones. Observation of threshold damage revealed through etching is generally easier with an optical microscope using vertical light than with a scanning-electron microscope.

Reports.

G. Harmon, "Metallurgical Failure Modes of Wire Bonds," Proceedings of the 12th Annual Symposium on Reliability Physics, 1974.

Weak Wire Bonds, Case 2 (Courtesy of Integrated Circuit Engineering Corp., Scottsdale, AZ.)

Observation. Gold wire bonds lift during 125°C burn-in.

Failure Summary. Wire bonds show high resistance and lift off.

Description. A "silox" glassivation layer was used over an integrated circuit having aluminum metallization. A buffered HF solution was used to etch windows in the silox for the wire bonding pads. Gold wires were ultrasonically bonded to the aluminum pads. After a 480 hour 125°C burn-in, some bonds developed high electrical resistance. SEM and microprobe analysis showed erratic diffusion between the gold and aluminum to the extent that Kirkendall voids were evident in some areas. In adjoining areas, no reaction had taken place. See Figure 12.21.

A similar problem has been experienced by another manufacturer except that the device was a small-signal transistor with a nitride layer. The wire bond system was also gold thermocompression. compression.

Diagnosis. Some reaction at seemingly random locations caused intermetallics to form and bonds to weaken at relatively low temperatures. The reaction was associated with the glassivation coating and occurred only in the presence of silox and nitride coatings. Bonds made to chips that had no glassivation coating survived several thousand hours at 125°C. Temperatures higher than 150°C are usually required to form appreciable amounts of intermetallics.

Possible Causes. Analysis found that moisture in the buffered HF etchant attacked the aluminum in the bonding pad areas after the glassivation layer had been etched. The resultant aluminum oxide and residues left from the glass etch caused the reliability problems of the wire bonds.

Corrective Action. Addition of a few drops of glycol to the buffered HF etchant is an effective solution to the problem.

Process Control Techniques. High temperature burn-in of a sampling of wire bonded devices discloses if there is a problem with the lot.

As a normal process control procedure, bonds are periodically tested on a nondestructive pull tester during each shift.

Analytical Technique. An increase in resistance was the first symptom of the defective wire bonds. The problem was then confirmed by a wire pull

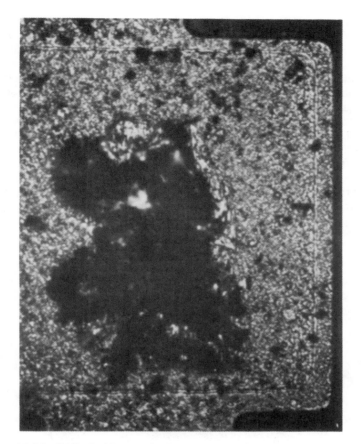

Figure 12.21: Diffusion between wire bond and aluminum pad caused high resistance/weak bonds.

test and by inspection with a metallurgical microscope, SEM and microprobe analysis. Auger analysis was used to identify the aluminum oxide interface layer.

Reports.

J.E. Mann, W.E. Anderson, T.J. Raab, and J.S. Rollins, "Reliability of Deposited Glass," Final Report for Period 1 March 1974 to 1 April 1975, RADC, TR-75. Rockwell International Corp., Anaheim, California.

Open Wire Bonds (Courtesy of Integrated Circuit Engineering Corp., Scottsdale, AZ.)

Observation. Open circuits after ultrasonic cleaning.

Failure Summary. Metallurgical fatigue of bonding wires due to ultrasonic energy.

Description. Circuits bonded with gold thermocompression bonds in flat packs mounted on circuit board subassemblies failed after immersion in an ultrasonic cleaning bath. In one case damage was caused in less than one minute.

Diagnosis. The ultrasonic bath caused the relatively heavy gold wires to resonate and otherwise vibrate with sufficient excitation to cause metal fatigue (see Figure 12.22A and B). Some bonds were broken while others remained undamaged, apparently depending on the geometrical resonant frequency and plane of vibration of each individual bond. Calculations show that typical gold wire bonds may resonate with excitation in the 3 to 5 kHz range while aluminum bonds require excitation frequencies greater than 10 kHz. Centrifugal forces greater than 30,000 g's for gold wire and greater than 100,000 g's for aluminum are required to produce significant stress on typical wire bonds.

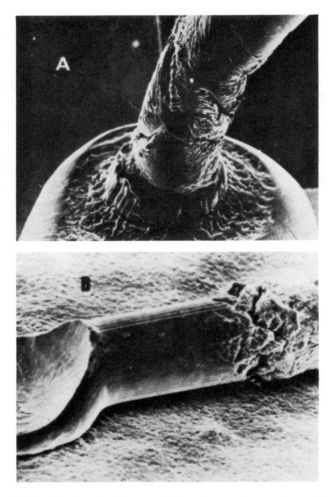

Figure 12.22: Metallurgical fatigue of wire bonds due to ultrasonic energy.

Possible Causes. Ultrasonic cleaning to remove flux residues from PC boards may have sufficient energy at the resonant frequency of the bonding wires to cause damage. The high resonant frequency precludes most other vibrational sources encountered in the field from causing damage to the IC.

Corrective Action. Avoid ultrasonic cleaning processes unless careful study has verified that it causes no wire bond damage.

Process Control Technique. None required.

Analytical Techniques. Electrical testing before PC board assembly and after cleaning. Delidding showed some bonds broken. SEM photographs showed evidence of metal fatigue.

Reports.

J. Beall, "Problems Experienced with Bonds to Gold Thick-Film Conductors," Invited paper presented at ASTM F-1, Interconnection Bonding Section, Palo Alto, California, Sept. 5, 1973.

H.A. Schafft, "Testing and Fabrication of Wire-Bond Electrical Connections, A Comprehensive Survey," Natl. Bureau of Standards, Tech. Note 726, 1972.

T.H. Ramsey, "Metallurgical Behavior of Gold Wire in Thermal Compression Bonding," Solid State Technology 16, 1973.

G. Harmon, "Metallurgical Failure Modes of Wire Bonds," 12th Annual Proceedings IEEE, Reliability Physics, 1974.

REFERENCES

1. Blanton, J., Defense Electronic Supply Center, *How to Prepare for a 1772 Audit*, November 1985.
2. Jones, R.D., *Hybrid Circuit Design and Manufacture*, Marcel Dekker, Inc., 1982.
3. Lange, J., "Sources of Semiconductor Wafer Contamination", *Semiconductor International*, April 1983.
4. Singer, P., "Surface Analysis Technology", *Semiconductor International*, July 1986.
5. Linder, R., Bryaon III, C, Bakale, D., "Surface Analysis for the Semiconductor Industry", *Microelectronic Manufacturing and Testing*, Feb. 1985.
6. Nordwall, B.D., "Air Force Links Radar Problems to Growth of Tin Whiskers", *Aviation Week & Space Technology*, June 30, 1986.
7. Government—Industry Data Exchange Program, E4-F-86-1, June 1986.
8. McAffee, G.D. and Raab, T.J., "Nickel Contamination as a Potential Problem Area in Hybrid Microcircuits", *Proceedings of the 1985 International Symposium on Microelectronics*, 1985.

Index

Active devices - 159–163
  Diodes - 160, 162
  ESD-sensitive - 276, 280–287
  Evaluation and inspection -
    377–378
  Integrated circuits - 160, 162,
    163
  Metallization - 159
  Passivation - 159
  Placement - 328–330
  Transistors - 160, 161
Adhesion
  of Copper conductors - 121
  of Thick-film conductors -
    99–102, 121
Adhesives
  Bond strength - 180
  Electrical conductivity - 181–
    183
  Epoxy attachment - 176–186
  Film/tape forms - 178
  Ionic contaminants - 185, 186
  Outgassing - 184
  Polyimide adhesives - 179–180
  Selection factors - 177–178
  Silver-glass - 188–190
  Thermally conductive - 176
  Weight loss - 183, 184
Air-gap microbridge - 76, 77

Alumina substrates - 29–35
  Co-fired tape - 35, 89–95
  Electrical properties - 32, 33
  Grades - 29, 30
  Thermal properties - 31, 34, 35
Aluminum nitride - 39, 40
Artwork - 299, 340–343
Assembly processes - 174–246
  Cleaning - 213–225
  Die and substrate attachment -
    175–190
  Epoxy attachment - 176–186
  Metallurgical attachment - 186–
    189
  Sealing - 235–246
  Solder attachment - 177
  Vacuum baking - 232–234
  Wire bonding - 191–200
Automated assembly (see Wire
  bonding)
Automated testing - 252–254

Beryllia substrates - 35–39
  Co-fired tape - 90
  Electrical properties - 37–39
  Mechanical properties - 39
  Thermal conductivity - 36

Capacitance - 303–308

Capacitors
  Chip - 164–168
  Thick-film - 114–116
Ceramics (see also Substrates,
    Alumina, Beryllia) - 4
Cermet resistors - 62, 63
Cleaning
  Solvents - 213–216
  Sources of contaminants -
    213, 214
Cleaning processes
  Batch cleaning - 218–222
  Manual cleaning - 218
  Plasma cleaning - 222–226
Clean rooms - 274, 276–279
Coatings
  Epoxy - 5
  Fluorocarbon - 5
  Particle-immobilizing - 225–
    231
  Parylene - 225–231
  Silicones - 5, 231
  Solvent-soluble - 231
Co-fired ceramic substrates - 35
  High-temperature process -
    90–94
  Low-temperature process -
    94–96
Computer-aided design - 298,
    299
Computer programs
  Capacitance calculations -
    307–308
  Thermal analysis - 321–326
Conductivity (see also Resistivity) -
    1, 98, 335
Conductors
  Copper, thick-film - 116–124
  Design guidelines - 344–347
  Gold, thick and thin film - 296
  Non-noble metals - 116–124
  Polymer thick films - 126–128
  Solderability - 107
  Thick films - 97–107, 335
  Uses in electronics - 1–3
Copper
  Adhesion of thick films - 121
  Electrical conductivity - 119–120

Thick-film conductors - 116–124,
    335
Wire-bond reliability - 119

Design
  Artwork - 299
  Computer-aided - 298, 299
  Effect of systems requirements -
    292
  Guidelines, for thick film - 340–
    360
  Guidelines, for thin film - 360–
    365
  Layout - 297–299, 326–360
  Materials and processes - 294
  Partitioning - 293
  Power dissipation - 294
  Review - 300–302
  Thermal considerations - 315–
    326
  Transmittal documents - 288–
    292
  Verification - 302, 303
Destructive Physical Analysis (DPA) -
    269
Dielectrics, thick-film - 111–114
  Capacitors - 114–116
  Electrical properties - 111, 112
  Nitrogen-fired - 124, 125
  Polymer thick films - 129
  Porosity - 112, 113
  Via resolution - 114
Documentation - 367–388
  Design flow diagram - 297
  Design review checklist - 301,
    302
  Design transmittal - 288–290
  Parts list - 296, 328
  Product checklist - 291, 292
  Specifications - 369–388

Electrostatic discharge handling -
    276, 280–287
Enameled metal substrates - 41, 42
Epoxy adhesives - 176–186
  Bond strengths - 180
  Compliance with specifications -
    177

Epoxy adhesives (continued)
  Electrically conductive - 181–183
  Film or tape forms - 178
  Ionic contaminants - 185-186
  Lid sealing - 244
  Outgassing - 184, 185
  Selection factors - 177, 178, 189
  Thermally conductive - 176
  Weight loss - 183, 184
Etching
  Dry etching (plasma) - 74–76
  Wet chemical - 73-75
Eutectic alloys - 181, 186–189

Failure analysis
  Analytical techniques - 398–402
  Case histories - 402–423
  Methodology - 396–397
Failures
  Contamination - 217, 392–395, 404-407, 414, 415
  Devices - 389
  ESD - 284
  Packages - 392
  Substrates - 392
  Types and causes - 389
  Wirebonds - 217, 391, 395, 404, 405, 409–412, 415–423

Gold conductors
  Thick film - 296, 335
  Thin film - 296

Handling
  Active and passive devices - 274-276
  ESD sensitive devices - 276, 280–287
Hermeticity - 244–246
Hybrid microcircuit applications
  Commercial - 17, 18
  Military/space - 16, 18–20
  Power - 20-24
Hybrid microcircuits
  Applications - 15–24

Assembly - 10, 174–246
  Comparison with ICs - 14, 15
  Comparison with printed wiring boards - 13, 14
  Design guidelines - 288-365
  Failure analysis - 389–423
  Inspection - 256
  Market - 15-24
  Sizes - 11, 12
  Specifications - 367–388
  Testing - 251–256
  Thick films - 9, 10, 79-131
  Thin films - 9, 10, 44-63
  Types and characteristics - 9–13
  Yields - 250, 343, 345

Inductive parasitics - 309–314
Inductors - 170, 172, 173
Infra-red imaging - 265, 266
Inspection - 256
Insulation resistance
  Humidity effects - 5
  of Polymer coatings - 5
  of Substrates - 25
Insulators - 3-6
Ionic contaminants (in adhesives) - 185, 186

Laser trimming - 133–141
Laser welding - 242-243

Masks - 67, 80
Materials
  Classification - 1-8
  Conductors - 1–3
  Insulators - 3–6
  Semiconductors - 6–8
Metal migration - 102-107, 114
Metals - 1-3, 98
Microbridge crossover circuits - 76, 77
Military applications - 16, 18–20
Moisture
  Analysis - 270-272
  Penetration into packages - 245
Multilayer processes
  Co-fired ceramic tape - 89–95
  Thick films - 85-89

Nichrome resistors - 53–60
   Etching - 74

Outgassing (of adhesives) - 184

Packages
   Cavity types - 151, 153, 154
   Ceramic - 157
   Epoxy-sealed - 151, 157
   Flatpaks - 150, 152, 156, 294
   Integral lead - 151, 153
   Plastic encapsulated - 151, 157
   Plug-in types - 150, 151, 154,
      155, 294
   testing - 158
Packaging - 294
Particle getters - 231, 232, 393
Particle-immobilizing coatings -
   225–231
Particle Impact Noise Detection
   (PIND) - 225, 263–265,
   393
Partitioning - 293
Parts selection - 149–173
   Devices, active - 159–163
   Devices, passive - 164–173
   Packages, testing - 158
   Packages, types - 150–158
Parylene - 225–231, 393
Photolithography - 64–73, 77
Photoresists - 63–73
   Negative types - 64–69
   Positive types - 69–71
   Processing - 71–73
Plasma etching (see Etching)
Polyimides
   Adhesives - 179, 180
   Tape for TAB - 203–205
Polymer coatings - 5, 225–231,
   393
Porcelainized steel substrates -
   41
Power applications - 20–24
Power dissipation - 294, 314–326,
   359, 362, 364, 365
Processes
   Assembly processes - 174–246
   Classification of - 8, 9

Quality assurance - 296

Reliability
   of Wire bonds - 122, 123
Resistivities
   of Cermet resistors - 63
   as Function of temperature - 3
   of Insulators - 4
   of Metals - 2
   of Semiconductors - 8
Resistors
   Cermet - 62, 63
   Chip resistors - 166, 168–171
   Nichrome - 53–60
   Nitrogen-fired - 125
   Overglaze - 337
   Polymer thick film - 128, 129
   Power rating - 359–360, 364,
      365
   Precision - 14, 57, 58
   Probe cards - 142
   Probing - 142–144
   Tantalum nitride - 59, 61, 62
   TCR tracking - 358
   Thick film - 107–111, 335, 336
   Thick film, design - 352–358
   Thin film - 52–63, 361–365
   Trim cuts - 145–147
   Trimming - 14, 132–148, 362–
      365
Rework - 250, 251

Screen tests
   Burn-in - 258, 260–263
   Non-destructive - 257, 259
   PIND Testing - 225, 263–265,
      393
Sealing - 235–246
   Epoxy sealing - 244
   Glass sealing - 243
   Solder sealing - 235–238
   Welding (laser) - 242–243
   Welding (seam) - 238–242
Seam sealing - 238–242
Semiconductors - 6–8
Silicones - 231
Silver-filled epoxies - 181–184
Silver-glass adhesives - 188–190

Silver migration - 102–107, 128, 178
Soldering - 339, 350
Space applications - 18, 19
Specifications
   for Electrostatic control - 283
   General types - 367–368
   Military/Government - 369–388
Sputter deposition - 46–52
   DC sputtering - 46–48
   Reactive sputtering - 49, 51
   RF sputtering - 49, 50
Substrates
   Alumina - 29–35, 333, 361
   Aluminum nitride - 39, 40
   Beryllia - 35–39
   Camber - 28
   Co-fired ceramic - 35, 90–96
   Dimensions - 334, 335
   Drilling - 333, 334, 348
   Enameled metal - 41, 42
   Functions of - 25, 26
   Grain size - 29
   Manufacture of - 40
   Quality assurance - 42, 43
   Silicon carbide - 40
   Surface roughness - 26, 27
   Test methods - 42, 43
   for Thick-film circuits - 30, 31, 35, 333, 334
   for Thin-film circuits - 31–35, 361

Tantalum nitride resistors - 58, 59, 61, 62
Tape automated bonding (TAB) - 203–208
Testing
   Acoustic microscopy - 268, 269
   Automated - 252
   Destructive - 268–272
   Electrical - 249–256
   Infra-red imaging - 265
   Mechanical screen tests - 257
   PIND - 263–265
   Thermography - 266–268
Thermal conductivity
   of Alumina - 34

of Berylia - 36
Computer analysis - 321–326
Conduction mechanisms - 314–315
of Electronic materials - 36, 317
Thermocompression bonding (see Wire bonding)
Thermosonic bonding (see Wire bonding)
Thick-film pastes
   Composition - 97
   Conductor pastes - 97–107, 335
   Dielectric pastes - 111–114, 340
   Fritless and fritted - 83, 99, 100
   Polymer thick films - 126–129
   Resistor pastes - 107–111, 335–337
Thick films - 79–131
   Adhesion, mechanisms for conductors - 99–102
   Artwork - 340–343
   Capacitors - 114
   Comparison with thin films - 295
   Conductors - 296, 97–107, 116–124
   Copper - 116–124
   Design guidelines - 340–360
   Dielectrics - 111–114
   Drying - 83
   Firing - 83–85
   Layout guidelines - 326–365
   Multilayer - 343
   Multilayer process - 85–96
   Non-noble metal - 116–124
   Polymer thick films - 126–129
   Resistors - 86, 107–111
   Resistors, design - 352–360
   Screen-printing - 79–82, 86
   Silver migration - 102–107
Thin films
   Comparison with thick films - 295
   Conductors - 296
   Design - 361–365
   Flash evaporation - 46
   Layout guidelines - 326–340
   Resistors - 52–63, 361–365

Thin films (continued)
    Sputter deposition - 46–52
    Vapor deposition - 44–46
Trimming, resistors - 132–148
    Abrasive trimming - 142
    Guidelines - 353–354
    Laser trimming - 133–141, 364

Ultrasonic bonding (see Wire bonding)

VHSIC (Very High Speed Integrated Circuit) - 14
Vias - 347–348
VLSI (Very Large Scale Integration) - 14

Welding
    Electron beam - 242
    Laser welding - 242–343
    Seam welding - 238–242
Wire bonding
    Automated - 203–208

    Beam-lead - 201
    Guidelines for - 330–331, 348–349
    Microgap - 199–200
    Pressure-sensitive devices - 202
    Thermocompression - 191–195, 197
    Thermosonic - 197–199
    Ultrasonic - 195–199
    Wedge - 193
Wire bonds
    Beam leads - 201
    Bonding pads, design - 348–351
    Failure causes - 404, 405, 409–412, 415–423
    Microgap - 199–200
    Quality and Reliability - 208–213
    Reliability to copper conductors - 122
    Resistance change with aging - 123
    Thermocompression bonds - 191–195, 197
    Thermosonic bond - 197–199
    Ultrasonic bonds - 195–199